A COURSE IN FINITE GROUP REPRESENTATION THEORY

This graduate-level text provides a thorough grounding in the representation theory of finite groups over fields and rings. The book provides a balanced and comprehensive account of the subject, detailing the methods needed to analyze representations that arise in many areas of mathematics. Key topics include the construction and use of character tables, the role of induction and restriction, projective and simple modules for group algebras, indecomposable representations, Brauer characters, and block theory.

This classroom-tested text provides motivation through a large number of worked examples, with exercises at the end of each chapter that test the reader's knowledge, provide further examples and practice, and include results not proven in the text. Prerequisites include a graduate course in abstract algebra and familiarity with the properties of groups, rings, field extensions, and linear algebra.

Peter Webb is a Professor of Mathematics at the University of Minnesota. His research interests focus on the interactions between group theory and other areas of algebra, combinatorics, and topology.

CAMBRIDGE STUDIES IN ADVANCED MATHEMATICS

All the titles listed below can be obtained from good booksellers or from Cambridge University Press. For a complete series listing, visit: www.cambridge.org/mathematics.

Already published

124 K. Lux & H. Pahlings *Representations of groups*
125 K. S. Kedlaya *p-adic differential equations*
126 R. Beals & R. Wong *Special functions*
127 E. de Faria & W. de Melo *Mathematical aspects of quantum field theory*
128 A. Terras *Zeta functions of graphs*
129 D. Goldfeld & J. Hundley *Automorphic representations and L-functions for the general linear group, I*
130 D. Goldfeld & J. Hundley *Automorphic representations and L-functions for the general linear group, II*
131 D. A. Craven *The theory of fusion systems*
132 J. Väänänen *Models and games*
133 G. Malle & D. Testerman *Linear algebraic groups and finite groups of Lie type*
134 P. Li *Geometric analysis*
135 F. Maggi *Sets of finite perimeter and geometric variational problems*
136 M. Brodmann & R. Y. Sharp *Local cohomology (2nd Edition)*
137 C. Muscalu & W. Schlag *Classical and multilinear harmonic analysis, I*
138 C. Muscalu & W. Schlag *Classical and multilinear harmonic analysis, II*
139 B. Helffer *Spectral theory and its applications*
140 R. Pemantle & M. C. Wilson *Analytic combinatorics in several variables*
141 B. Branner & N. Fagella *Quasiconformal surgery in holomorphic dynamics*
142 R. M. Dudley *Uniform central limit theorems (2nd Edition)*
143 T. Leinster *Basic category theory*
144 I. Arzhantsev, U. Derenthal, J. Hausen & A. Laface *Cox rings*
145 M. Viana *Lectures on Lyapunov exponents*
146 J.-H. Evertse & K. Győry *Unit equations in Diophantine number theory*
147 A. Prasad *Representation theory*
148 S. R. Garcia, J. Mashreghi & W. T. Ross *Introduction to model spaces and their operators*
149 C. Godsil & K. Meagher *Erdős–Ko–Rado theorems: Algebraic approaches*
150 P. Mattila *Fourier analysis and Hausdorff dimension*
151 M. Viana & K. Oliveira *Foundations of ergodic theory*
152 V. I. Paulsen & M. Raghupathi *An introduction to the theory of reproducing kernel Hilbert spaces*
153 R. Beals & R. Wong *Special functions and orthogonal polynomials*
154 V. Jurdjevic *Optimal control and geometry: Integrable systems*
155 G. Pisier *Martingales in Banach spaces*
156 C. T. C. Wall *Differential topology*
157 J. C. Robinson, J. L. Rodrigo & W. Sadowski *The three-dimensional Navier–Stokes equations*
158 D. Huybrechts *Lectures on K3 surfaces*
159 H. Matsumoto & S. Taniguchi *Stochastic analysis*
160 A. Borodin & G. Olshanski *Representations of the infinite symmetric group*
161 P. Webb *A course in finite group representation theory*
162 C. J. Bishop & Y. Peres *Fractals in probability and analysis*
163 A. Bovier *Gaussian processes on trees*

A Course in Finite Group Representation Theory

PETER WEBB
University of Minnesota

Shaftesbury Road, Cambridge CB2 8EA, United Kingdom

One Liberty Plaza, 20th Floor, New York, NY 10006, USA

477 Williamstown Road, Port Melbourne, VIC 3207, Australia

314–321, 3rd Floor, Plot 3, Splendor Forum, Jasola District Centre, New Delhi – 110025, India

103 Penang Road, #05–06/07, Visioncrest Commercial, Singapore 238467

Cambridge University Press is part of Cambridge University Press & Assessment, a department of the University of Cambridge.

We share the University's mission to contribute to society through the pursuit of education, learning and research at the highest international levels of excellence.

www.cambridge.org
Information on this title: www.cambridge.org/9781107162396

First published 2016

A catalogue record for this publication is available from the British Library

Library of Congress Cataloging-in-Publication data
Names: Webb, Peter, 1954–
Title: A course in finite group representation theory / Peter Webb, University of Minnesota.
Description: Cambridge : Cambridge University Press, [2016] | Series: Cambridge studies in advanced mathematics | Includes bibliographical references and index.
Identifiers: LCCN 2016013913 | ISBN 9781107162396 (hardback : alk. paper)
Subjects: LCSH: Finite groups – Textbooks. | Group theory – Textbooks. | Modules (Algebra) – Textbooks. | Representations of groups – Textbooks. | Rings (Algebra) – Textbooks.
Classification: LCC QA177 .W43 2016 | DDC 512/.23–dc23
LC record available at http://lccn.loc.gov/2016013913

ISBN 978-1-107-16239-6 Hardback

Contents

Preface

The representation theory of finite groups has a long history, going back to the nineteenth century and earlier. A milestone in the subject was the definition of characters of finite groups by Frobenius in 1896. Prior to this there was some use of the ideas that we can now identify as representation theory (characters of cyclic groups as used by number theorists, the work of Schönflies, Fedorov, and others on crystallographic groups, invariant theory, for instance), and during the twentieth century, there was continuously active development of the subject. Nevertheless, the theory of complex characters of finite groups, with its theorem of semisimplicity and the orthogonality relations, is a stunning achievement that remains a cornerstone of the subject. It is probably what many people think of first when they think of finite group representation theory.

This book is about character theory, and it is also about other things: the character theory of Frobenius occupies less than one-third of the text. The rest of the book comes about because we allow representations over rings other than fields of characteristic zero. The theory becomes more complicated, and also extremely interesting, when we consider representations over fields of characteristic dividing the group order. It becomes still more complicated over rings of higher Krull-dimension, such as rings of integers. An important case is the theory over a discrete valuation ring, because this provides the connection between representations in characteristic zero and in positive characteristic. We describe these things in this text.

Why should we want to know about representations over rings that are not fields of characteristic zero? It is because they arise in many parts of mathematics. Group representations appear any time we have a group of symmetries where there is some linear structure present, over some commutative ring. That ring need not be a field of characteristic zero. Here are some examples:

- In number theory, groups arise as Galois groups of field extensions, giving rise not only to representations over the ground field but also to integral representations over rings of integers (in case the fields are number fields). It is natural to reduce these representations modulo a prime ideal, at which point we have modular representations.
- In the theory of error-correcting codes, many important codes have a nontrivial symmetry group and are vector spaces over a finite field, thereby providing a representation of the group over that field.
- In combinatorics, an active topic is to obtain "q-analogs" of enumerative results, exemplified by replacing binomial coefficients (which count subsets of a set) by q-binomial coefficients (which count subspaces of vector spaces over $\mathbb{F}_q$). Structures permuted by a symmetric group are replaced by linear structures acted on by a general linear group, thereby giving representations in positive characteristic.
- In topology, a group may act as a group of self-equivalences of a topological space. Thereby, giving representations of the group on the homology groups of the space. If there is torsion in the homology, these representations require something other than ordinary character theory to be understood.

This book is written for students who are studying finite group representation theory beyond the level of a first course in abstract algebra. It has arisen out of notes for courses given at the second-year graduate level at the University of Minnesota. My aim has been to write the book for the course. It means that the level of exposition is appropriate for such students, with explanations that are intended to be full but not overly lengthy.

Most students who attend an advanced course in group representation theory do not go on to be specialists in the subject, for otherwise the class would be much smaller. Their main interests may be in other areas of mathematics, such as combinatorics, topology, number theory, or commutative algebra. These students need a solid, comprehensive grounding in representation theory that enables them to apply the theory to their own situations as the occasion demands. They need to be able to work with complex characters, and they also need to be able to say something about representations over other fields and rings. While they need the theory to be able to do this, they do not need to be presented with overly deep material whose main function is to serve the internal workings of the subject.

With these goals in mind, I have made a choice of material covered. My main criterion has been to ask whether a topic is useful outside the strict confines of representation theory and, if it is, to include it. At the same time, if there is a theorem that fails the test, I have left it out or put it in the exercises. I have sometimes omitted standard results where they appear not to have sufficiently compelling applications. For example, the theorem of Frobenius on Frobenius groups does not appear, because I do not consider that we need this theorem to understand these groups at the level of this text. I have also omitted Brauer's characterization of characters, leading to the determination of a minimal splitting field for a group and its subgroups. That result is stated without proof, and we do prove what is needed, namely that there exists a finite degree field extension that is a splitting field. For the students who go on to be specialists in representation theory there is no shortage of more advanced monographs. They can find these results there—but they may also find it helpful to start with this book! One of my aims has been to make it possible to read this book from the beginning without having to wade through chapters full of preliminary technicalities, and omitting some results aids in this.

I have included many exercises at the ends of the chapters, and they form an important part of this book. The benefit of learning actively by having to apply the theory to calculate with examples and solve problems cannot be overestimated. Some of these exercises are easy, some more challenging. In a number of instances, I use the exercises as a place to present extensions of results that appear in the text or as an indication of what can be done further.

I have assumed that the reader is familiar with the first properties of groups, rings, and field extensions and with linear algebra. More specifically the reader should know about Sylow subgroups, solvable, and nilpotent groups, as well as the examples that are introduced in a first group theory course, such as the dihedral, symmetric, alternating, and quaternion groups. The reader should also be familiar with tensor products, Noetherian properties of commutative rings, the structure of modules over a principal ideal domain, and the first properties of ideals as well as with Jordan and rational canonical forms for matrices. These topics are covered in a standard graduate-level algebra course. I develop the properties of algebraic integers, valuation theory, and completions within the text since they usually fall outside such a course.

Many people have read sections of this book, worked through the exercises, and been very generous with the comments they have made. I wish to thank them all. They include Cihan Bahran, Dave Benson, Daniel Hess, John Palmieri, Sverre Smalø, and many others.

1

Representations, Maschke's Theorem, and Semisimplicity

In this chapter, we present the basic definitions and examples to do with group representations. We then prove Maschke's theorem, which states that in many circumstances representations are completely reducible. We conclude by describing the properties of semisimple modules.

1.1 Definitions and Examples

Informally, a representation of a group is a collection of invertible linear transformations of a vector space (or, more generally, of a module for a ring) that multiply together in the same way as the group elements. The collection of linear transformations thus establishes a pattern of symmetry of the vector space, which copies the symmetry encoded by the group. Because symmetry is observed and understood so widely, and is even one of the fundamental notions of mathematics, there are applications of representation theory across the whole of mathematics as well as in other disciplines.

For many applications, especially those having to do with the natural world, it is appropriate to consider representations over fields of characteristic zero such as $\mathbb{C}$, $\mathbb{R}$, or $\mathbb{Q}$ (the fields of complex numbers, real numbers, or rational numbers). In other situations that might arise in topology or combinatorics or number theory, for instance, we find ourselves considering representations over fields of positive characteristic, such as the field with p elements $\mathbb{F}_p$, or over rings that are not fields, such as the ring of integers $\mathbb{Z}$. Many aspects of representation theory do change as the ring varies, but there are also parts of the theory that are similar regardless of the field characteristic or even if the ring is not a field. We develop the theory independently of the choice of ring where possible so as to be able to apply it in all situations and to establish a natural context for the results.

Let G denote a finite group, and let R be a commutative ring with a 1. If V is an R-module, we denote it by $GL(V)$ the group of all invertible R-module homomorphisms $V \to V$. In case, $V \cong R^n$ is a free module of rank n, this group is isomorphic to the group of all nonsingular $n \times n$ matrices over R, and we denote it by $GL(n, R)$ or $GL_n(R)$, or in case $R = \mathbb{F}_q$ is the finite field with q elements by $GL(n, q)$ or $GL_n(q)$. We point out also that unless otherwise stated, modules will be left modules and morphisms will be composed reading from right to left so that matrices in $GL(n, R)$ are thought of as acting from the left on column vectors.

A (*linear*) *representation* of G (over R) is a group homomorphism

$$\rho : G \to GL(V).$$

In a situation where V is free as an R-module, on taking a basis for V, we may write each element of $GL(V)$ as a matrix with entries in R, and we obtain for each $g \in G$ a matrix $\rho(g)$. These matrices multiply together in the manner of the group, and we have a *matrix representation* of G. In this situation, the rank of the free R-module V is called the *degree* of the representation. Sometimes, by abuse of terminology, the module V is also called the representation, but it is more properly called the *representation module* or *representation space* (if R is a field).

To illustrate some of the possibilities that may arise, we consider some examples.

Example 1.1.1. For any group G and commutative ring R, we can take $V = R$ and $\rho(g) = 1$ for all $g \in G$, where 1 denotes the identify map $R \to R$. This representation is called the *trivial representation*, and it is often denoted simply by its representation module R. Although this representation turns out to be extremely important in the theory, it does not at this point give much insight into the nature of a representation.

Example 1.1.2. A representation on a space $V = R$ of rank 1 is in general determined by specifying a homomorphism $G \to R^\times$. Here $R^\times$ is the group of units of R, and it is isomorphic to $GL(V)$. For example, if $G = \langle g \rangle$ is cyclic of order n and $k = \mathbb{C}$ is the field of complex numbers, there are n possible such homomorphisms, determined by $g \mapsto e^{\frac{2r\pi i}{n}}$ where $0 \le r \le n-1$. Another important example of a degree 1 representation is the *sign representation* of the symmetric group S_n on n symbols, given by the group homomorphism that assigns to each permutation its sign, regarded as an element of the arbitrary ring R.

Example 1.1.3. Let $R = \mathbb{R}$, $V = \mathbb{R}^2$, and $G = S_3$. This group G is isomorphic to the group of symmetries of an equilateral triangle. The symmetries are the three reflections in the lines that bisect the equilateral triangle, together with three rotations:

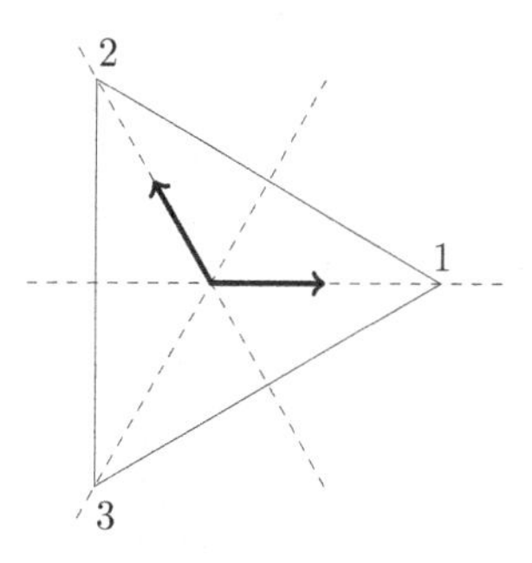

Positioning the center of the triangle at the origin of V and labeling the three vertices of the triangle as 1, 2, and 3, we get a representation

$$() \mapsto \begin{bmatrix} 1 & 0 \\ 0 & 1 \end{bmatrix},$$

$$(1,2) \mapsto \begin{bmatrix} 0 & 1 \\ 1 & 0 \end{bmatrix},$$

$$(1,3) \mapsto \begin{bmatrix} -1 & 0 \\ -1 & 1 \end{bmatrix},$$

$$(2,3) \mapsto \begin{bmatrix} 1 & -1 \\ 0 & -1 \end{bmatrix},$$

$$(1,2,3) \mapsto \begin{bmatrix} 0 & -1 \\ 1 & -1 \end{bmatrix},$$

$$(1,3,2) \mapsto \begin{bmatrix} -1 & 1 \\ -1 & 0 \end{bmatrix},$$

where we have taken basis vectors in the directions of vertices 1 and 2, making an angle of $\frac{2\pi}{3}$ to each other. In fact these matrices define a representation of degree 2 over any ring R, because although the representation was initially constructed over $\mathbb{R}$ the matrices have integer entries, and these may be interpreted in every ring. No matter what the ring is, the matrices always multiply together to give a copy of S_3.

At this point, we have constructed three representations of S_3: the trivial representation, the sign representation, and one of dimension 2.

Example 1.1.4. Let $R = \mathbb{F}_p$, $V = R^2$, and let $G = C_p = \langle g \rangle$ be cyclic of order p generated by an element g. We see that the assignment

$$\rho(g^r) = \begin{bmatrix} 1 & 0 \\ r & 1 \end{bmatrix}$$

is a representation. In this case, the fact that we have a representation is very much dependent on the choice of R as the field $\mathbb{F}_p$: in any other characteristic it would not work, because the matrix shown would no longer have order p.

We can think of representations in various ways. One of them is that a representation is the specification of an action of a group on an R-module, as we now explain. Given a representation $\rho : G \to GL(V)$, an element $v \in V$, and a group element $g \in G$, we get another module element $\rho(g)(v)$. Sometimes we write just $g \cdot v$ or gv for this element. This rule for multiplication satisfies

$$\begin{aligned} g \cdot (\lambda v + \mu w) &= \lambda g \cdot v + \mu g \cdot w, \\ (gh) \cdot v &= g \cdot (h \cdot v), \\ 1 \cdot v &= v \end{aligned}$$

for all $g \in G$, $v, w \in V$, and $\lambda, \mu \in R$. A rule for multiplication $G \times V \to V$ satisfying these conditions is called a *linear action* of G on V. To specify a linear action of G on V is the same thing as specifying a representation of G on V, since given a representation, we obtain a linear action as indicated earlier, and evidently, given a linear action, we may recover the representation.

Another way to define a representation of a group is in terms of the group algebra. We define the *group algebra* RG (or $R[G]$) of G over R to be the free R-module with the elements of G as an R-basis and with multiplication given on the basis elements by group multiplication. The elements of RG are the (formal) R-linear combinations of group elements, and the multiplication of the basis elements is extended to arbitrary elements using bilinearity of the operation. What this means is that a typical element of RG is an expression $\sum_{g \in G} a_g g$ where $a_g \in R$, and the multiplication of these elements is given symbolically by

$$\left(\sum_{g \in G} a_g g \right) \left(\sum_{h \in G} b_h h \right) = \sum_{k \in G} \left(\sum_{gh = k} a_g b_h \right) k.$$

More concretely, we exemplify this definition by listing some elements of the group algebra $\mathbb{Q}S_3$. We write elements of S_3 in cycle notation, such as $(1, 2)$. This group element gives rise to a basis element of the group algebra which we write either as $1 \cdot (1, 2)$ or simply as $(1, 2)$ again. The group identity element $()$ also serves as the identity element of $\mathbb{Q}S_3$. In general, elements of $\mathbb{Q}S_3$ may look like $(1,2)-(2,3)$ or $\frac{1}{5}(1,2,3)+6(1,2)-\frac{1}{7}(2,3)$. Here is a computation:

$$\begin{aligned}(3(1,2,3)+(1,2))(()-2(2,3)) &= 3(1,2,3)+(1,2)-6(1,2)-2(1,2,3)\\ &= (1,2,3)-5(1,2).\end{aligned}$$

An (associative) *R-algebra* is defined to be a (not necessarily commutative) ring A with a 1, equipped with a (unital) ring homomorphism $R \to A$ whose image lies in the center of A. The group algebra RG is indeed an example of an R-algebra.

Having defined the group algebra, we may now define a representation of G over R to be a unital RG-module. The fact that this definition coincides with the previous ones is the content of the next proposition. Throughout this text, we may refer to group representations as modules (for the group algebra).

Proposition 1.1.5. *A representation of G over R has the structure of a unital RG-module. Conversely, every unital RG-module provides a representation of G over R.*

Proof. Given a representation $\rho : G \to GL(V)$, we define a module action of RG on V by $(\sum a_g g)v - \sum a_g \rho(g)(v)$.

Given an RG-module V, the linear map $\rho(g) : v \mapsto gv$ is an automorphism of V and $\rho(g_1)\rho(g_2) = \rho(g_1 g_2)$ so $\rho : G \to GL(V)$ is a representation. □

The group algebra gives another example of a representation, called the *regular representation*. In fact, for any ring A, we may regard A itself as a left A-module with the action of A on itself given by multiplication of the elements. We denote this left A-module by ${}_AA$ when we wish to emphasize the module structure, and this is the (left) regular representation of A. When $A = RG$, we may describe the action on ${}_{RG}RG$ by observing that each element $g \in G$ acts on ${}_{RG}RG$ by permuting the basis elements in the fashion $g \cdot h = gh$. Thus, each g acts by a *permutation matrix*, namely a matrix in which, in every row and column, there is precisely one nonzero entry, and that nonzero entry is 1. The regular representation is an example of a *permutation representation*, namely one in which every group element acts by a permutation matrix.

Regarding representations of G as RG-modules has the advantage that many definitions, we wish to make may be borrowed from module theory. Thus, we may study RG-submodules of an RG-module V, and if we wish, we may call them *subrepresentations* of the representation afforded by V. To specify an RG-submodule of V, it is necessary to specify an R-submodule W of V that is closed under the action of RG. This is equivalent to requiring that $\rho(g)w \in W$ for all $g \in G$ and $w \in W$. We say that a submodule W satisfying this condition is *stable* under G or that it is an *invariant submodule* or *invariant subspace* (if R happens to be a field). Such an invariant submodule W gives rise to a homomorphism $\rho_W : G \to GL(W)$ that is the subrepresentation afforded by W.

Example 1.1.6. 1. Let $C_2 = \{1, -1\}$ be cyclic of order 2 and consider the representation

$$\rho : C_2 \to GL(\mathbb{R}^2),$$
$$1 \mapsto \begin{bmatrix} 1 & 0 \\ 0 & 1 \end{bmatrix},$$
$$-1 \mapsto \begin{bmatrix} 1 & 0 \\ 0 & -1 \end{bmatrix}.$$

There are just four invariant subspaces, namely $\{0\}$, $\langle \binom{1}{0} \rangle$, $\langle \binom{0}{1} \rangle$, $\mathbb{R}^2$, and no others. The representation space $\mathbb{R}^2 = \langle \binom{1}{0} \rangle \oplus \langle \binom{0}{1} \rangle$ is the direct sum of two invariant subspaces.

Example 1.1.7. In Example 1.1.4, an elementary calculation shows that $\langle \binom{0}{1} \rangle$ is the only 1-dimensional invariant subspace, and so it is not possible to write the representation space V as the direct sum of two nonzero invariant subspaces.

We make use of the notions of a *homomorphism* and an *isomorphism* of RG-modules. Since RG has as a basis the elements of G, to check that an R-linear homomorphism $f : V \to W$ is in fact a homomorphism of RG-modules, it suffices to check that $f(gv) = gf(v)$ for all $g \in G$—we do not need to check for every $x \in RG$. By means of the identification of RG-modules with representations of G (in the first definition given here) we may refer to homomorphisms and isomorphisms of group representations. In many books the algebraic condition on the representations that these notions entail is written out explicitly, and two representations that are isomorphic are also said to be *equivalent*.

If V and W are RG-modules then we may form their (external) *direct sum* $V \oplus W$, which is the same as the direct sum of V and W as R-modules together with an action of G given by $g(v, w) = (gv, gw)$. We also have the notion of the internal direct sum of RG-modules and write $U = V \oplus W$ to mean that U has RG-submodules V and W satisfying $U = V + W$ and $V \cap W = 0$. In this situation, we also say that V and W are *direct summands* of U. We just met this property in Example 1.1.6, which gives a representation that is a direct sum of two nonzero subspaces; by contrast, Example 1.1.7 provides an example of a subrepresentation that is not a direct summand.

1.2 Semisimple Representations

We come now to our first nontrivial result, one that is fundamental to the study of representations over fields of characteristic zero, or characteristic not dividing the group order. This surprising result says that in this situation representations always break apart as direct sums of smaller representations. We do now require the ring R to be a field, and in this situation, we will often use the symbols F or k instead of R.

Theorem 1.2.1 (Maschke). *Let V be a representation of the finite group G over a field F in which $|G|$ is invertible. Let W be an invariant subspace of V. Then there exists an invariant subspace W_1 of V such that $V = W \oplus W_1$ as representations.*

Proof. Let $\pi : V \to W$ be any projection of V onto W as vector spaces, that is, a linear transformation such that $\pi(w) = w$ for all $w \in W$. Since F is a field, we may always find such a projection by finding a vector space complement to W in V and projecting off the complementary factor. Then $V = W \oplus \operatorname{Ker}(\pi)$ as vector spaces, but $\operatorname{Ker}(\pi)$ is not necessarily invariant under G. Consider the map

$$\pi' = \frac{1}{|G|}\sum_{g\in G} g\pi g^{-1} : V \to V.$$

Then π' is linear, and if $w \in W$ then

$$\begin{aligned}\pi'(w) &= \frac{1}{|G|}\sum_{g\in G} g\pi(g^{-1}w)\\ &= \frac{1}{|G|}\sum_{g\in G} gg^{-1}w\\ &= \frac{1}{|G|}|G|w\\ &= w.\end{aligned}$$

Since furthermore $\pi'(v) \in W$ for all $v \in V$, π' is a projection onto W and so $V = W \oplus \operatorname{Ker}(\pi')$. We show finally that $\operatorname{Ker}(\pi')$ is an invariant subspace by verifying that π' is an FG-module homomorphism: if $h \in G$ and $v \in V$ then

$$\begin{aligned}\pi'(hv) &= \frac{1}{|G|}\sum_{g\in G} g\pi(g^{-1}hv)\\ &= \frac{1}{|G|}\sum_{g\in G} h(h^{-1}g)\pi((h^{-1}g)^{-1}v)\\ &= h\pi'(v)\end{aligned}$$

because as g ranges over the elements of G, so does $h^{-1}g$. Now if $v \in \operatorname{Ker}(\pi')$ then $hv \in \operatorname{Ker}(\pi')$ also (since $\pi'(hv) = h\pi'(v) = 0$) and so $\operatorname{Ker}(\pi')$ is an invariant subspace. □

Because the next results apply more generally than to group representations, we let A be a ring with a 1 and consider its modules. A nonzero A-module V is said to be *simple* or *irreducible* if V has no A-submodules other than 0 and V.

Example 1.2.2. When A is an algebra over a field, every module of dimension 1 is simple. In Example 1.1.3, we have constructed three representations of $\mathbb{R}S_3$, and they are all simple. The trivial and sign representations are simple because they have dimension 1, and the 2-dimensional representation is simple because, visibly, no 1-dimensional subspace is invariant under the group action. We will see in Example 2.1.6 that this is a complete list of the simple representations of S_3 over $\mathbb{R}$.

We see immediately that a nonzero module is simple if and only if it is generated by each of its nonzero elements. Furthermore, the simple A-modules are exactly those of the form A/I for some maximal left ideal I of A: every such module is simple, and given a simple module S with a nonzero element $x \in S$ the A-module homomorphism $A \to S$ specified by $a \mapsto ax$ is surjective with kernel a maximal ideal I, so that $S \cong A/I$. Because all simple modules appear inside A in this way, we may deduce that if A is a finite-dimensional algebra over a field there are only finitely many isomorphism types of simple modules, these appearing among the composition factors of A when regarded as a module. As a consequence, the simple A-modules are all finite-dimensional.

A module that is the direct sum of simple submodules is said to be *semisimple* or *completely reducible*. We saw in Examples 1.1.6 and 1.1.7 two examples of modules, one of which was semisimple and the other of which was not. Every module of finite composition length is somehow built up out of its composition factors, which are simple modules, and we know from the Jordan–Hölder theorem that these composition factors are determined up to isomorphism, although there may be many composition series. The most rudimentary way these composition factors may be fitted together is as a direct sum, giving a semisimple module. In this case, the simple summands are the composition factors of the module, and their isomorphism types and multiplicities are uniquely determined. There may, however, be many ways to find simple submodules of a semisimple module so that the module is their direct sum.

We will now relate the property of semisimplicity to the property that appears in Maschke's theorem, namely that every submodule of a module is a direct summand. Our immediate application of this will be an interpretation of Maschke's theorem, but the results have application in greater generality in situations where R is not a field, or when $|G|$ is not invertible in R. To simplify the exposition, we have imposed a finiteness condition in the statement of each result, thereby avoiding arguments that use Zorn's lemma. These finiteness conditions can be removed, and we leave the details to Exercise 14 at the end of this chapter.

In the special case when the ring A is a field and A-modules are vector spaces, the next result is familiar from linear algebra.

Lemma 1.2.3. *Let A be a ring with a 1, and suppose that $U = S_1 + \cdots + S_n$ is an A-module that can be written as the sum of finitely many simple modules $S_1, \ldots, S_n$. If V is any submodule of U, there is a subset $I = \{i_1, \ldots, i_r\}$ of $\{1, \ldots, n\}$ such that $U = V \oplus S_{i_1} \oplus \cdots S_{i_r}$. In particular,*

(1) V is a direct summand of U, and
(2) (taking $V = 0$), U is the direct sum of some subset of the S_i and hence is necessarily semisimple.

Proof. Choose a subset I of $\{1, \ldots, n\}$ maximal subject to the condition that the sum $W = V \oplus (\bigoplus_{i \in I} S_i)$ is a direct sum. Note that $I = \emptyset$ has this property, so we are indeed taking a maximal element of a nonempty collection of subsets. We show that $W = U$. If $W \neq U$ then $S_j \not\subseteq W$ for some j. Now $S_j \cap W = 0$, being a proper submodule of S_j, so $S_j + W = S_j \oplus W$, and we obtain a contradiction to the maximality of I. Therefore, $W = U$. The consequences (1) and (2) are immediate. □

Proposition 1.2.4. *Let A be a ring with a 1 and let U be an A-module. The following are equivalent:*

(1) U can be expressed as a direct sum of finitely many simple A-submodules.
(2) U can be expressed as a sum of finitely many simple A-submodules.
(3) U has finite composition length and has the property that every submodule of U is a direct summand of U.

When these three conditions hold, every submodule of U and every factor module of U may also be expressed as the direct sum of finitely many simple modules.

Proof. The implication (1) $\Rightarrow$ (2) is immediate and the implications (2) $\Rightarrow$ (1) and (2) $\Rightarrow$ (3) follow from Lemma 1.2.3. To show that (3) $\Rightarrow$ (1), we argue by induction on the composition length of U and first observe that hypothesis (3) passes to submodules of U. For if V is a submodule of U and W is a submodule of V then $U = W \oplus X$ for some submodule X, and now $V = W \oplus (X \cap V)$ by the modular law (Exercise 2 at the end of this chapter). Proceeding with the induction argument, when U has length 1 it is a simple module, and so the induction starts. If U has length greater than 1, it has a submodule V and by condition (3), $U = V \oplus W$ for some submodule W. Now both V and W inherit condition (3) and are of shorter length, so by induction they are direct sums of simple modules and hence so is U.

We have already observed that every submodule of U inherits condition (3) and so satisfies condition (1) also. Every factor module of U has the form U/V for some submodule V of U. If condition (3) holds then $U = V \oplus W$ for some submodule W that we have just observed satisfies condition (1), and hence so does U/V, because $U/V \cong W$. □

We now present a different version of Maschke's theorem. The assertion remains correct if the words "finite-dimensional" are removed from it, but we leave the proof of this to the exercises.

Corollary 1.2.5. *Let F be a field in which $|G|$ is invertible. Then every finite-dimensional FG-module is semisimple.*

Proof. This combines Theorem 1.2.1 with the equivalence of the statements of Proposition 1.2.4. □

This result puts us in very good shape if we want to know about the representations of a finite group over a field in which $|G|$ is invertible—for example any field of characteristic zero. To obtain a description of all possible finite-dimensional representations, we need only describe the simple ones, and then arbitrary ones are direct sums of these.

The following corollaries to Lemma 1.2.3 will be used on many occasions when we are considering modules that are not semisimple.

Corollary 1.2.6. *Let A be a ring with a 1, and let U be an A-module of finite composition length.*

(1) The sum of all the simple submodules of U is a semisimple module, that is the unique largest semisimple submodule of U.
(2) The sum of all submodules of U isomorphic to some given simple module S is a submodule isomorphic to a direct sum of copies of S. It is the unique largest submodule of U with this property.

Proof. The submodules described can be expressed as the sum of finitely many submodules by the finiteness condition on U. They are the unique largest submodules with their respective properties since they contain all simple submodules (in case (1)), and all submodules isomorphic to S (in case (2)). □

The largest semisimple submodule of a module U is called the *socle* of U, and is denoted $\mathrm{Soc}(U)$. There is a dual construction called the radical of U, denoted $\mathrm{Rad}\, U$, that we will study in Chapter 6. It is defined to be the intersection of all the maximal submodules of U, and has the property that it is the smallest submodule of U with semisimple quotient.

Corollary 1.2.7. *Let $U = S_1^{a_1} \oplus \cdots \oplus S_r^{a_r}$ be a semisimple module over a ring A with a 1, where the S_i are nonisomorphic simple A-modules and the a_i are their multiplicities as summands of U. Then each submodule $S_i^{a_i}$ is uniquely determined and is characterized as the unique largest submodule of U expressible as a direct sum of copies of S_i.*

Proof. It suffices to show that $S_i^{a_i}$ contains every submodule of U isomorphic to S_i. If T is any nonzero submodule of U not contained in $S_i^{a_i}$ then for some $j \neq i$, its projection to a summand S_j must be nonzero. If we assume that T is simple, this projection will be an isomorphism $T \cong S_j$. Thus, all simple submodules isomorphic to S_i are contained in the summand $S_i^{a_i}$. □

1.3 Summary of Chapter 1

- Representations of G over R are the same thing as RG-modules.
- Semisimple modules may be characterized in several different ways. They are modules that are the direct sum of simple modules, or equivalently the sum of simple modules, or equivalently modules for which every submodule is a direct summand.
- If F is a field in which G is invertible, FG-modules are semisimple.
- The sum of all simple submodules of a module is the unique largest semisimple submodule of that module: the socle.

1.4 Exercises for Chapter 1

1. In Example 1.1.6, prove that there are no invariant subspaces other than the ones listed.

2. (The modular law.) Let A be a ring and $U = V \oplus W$ an A-module that is the direct sum of A-modules V and W. Show by example that if X is any submodule of U then it need not be the case that $X = (V \cap X) \oplus (W \cap X)$. Show that if we make the assumption that $V \subseteq X$ then it is true that $X = (V \cap X) \oplus (W \cap X)$.

3. Suppose that ρ is a finite-dimensional representation of a finite group G over $\mathbb{C}$. Show that for each $g \in G$ the matrix $\rho(g)$ is diagonalizable.

4. Let $\phi : U \to V$ be a homomorphism of A-modules. Show that $\phi(\operatorname{Soc} U) \subseteq \operatorname{Soc} V$, and that if ϕ is an isomorphism then ϕ restricts to an isomorphism $\operatorname{Soc} U \to \operatorname{Soc} V$.

5. Let $U = S_1 \oplus \cdots \oplus S_r$ be an A-module that is the direct sum of finitely many simple modules $S_1, \ldots, S_r$. Show that if T is any simple submodule of U then $T \cong S_i$ for some i.

6. Let V be an A-module for some ring A and suppose that V is a sum $V = V_1 + \cdots + V_n$ of simple submodules. Assume further that the V_i are pairwise nonisomorphic. Show that the V_i are the only simple submodules of V and that $V = V_1 \oplus \cdots \oplus V_n$ is their direct sum.

7. Let $G = \langle x, y \mid x^2 = y^2 = 1 = [x, y] \rangle$ be the Klein four-group, $R = \mathbb{F}_2$, and consider the two representations ρ_1 and ρ_2 specified on the generators of G by

$$\rho_1(x) = \begin{bmatrix} 1 & 1 & 0 \\ 0 & 1 & 0 \\ 0 & 0 & 1 \end{bmatrix}, \quad \rho_1(y) = \begin{bmatrix} 1 & 0 & 1 \\ 0 & 1 & 0 \\ 0 & 0 & 1 \end{bmatrix}$$

and

$$\rho_2(x) = \begin{bmatrix} 1 & 0 & 0 \\ 0 & 1 & 1 \\ 0 & 0 & 1 \end{bmatrix}, \quad \rho_2(y) = \begin{bmatrix} 1 & 0 & 1 \\ 0 & 1 & 0 \\ 0 & 0 & 1 \end{bmatrix}.$$

Calculate the socles of these two representations. Show that neither representation is semisimple.

8. Let $G = C_p = \langle x \rangle$ and $R = \mathbb{F}_p$ for some prime $p \geq 3$. Consider the two representations ρ_1 and ρ_2 specified by

$$\rho_1(x) = \begin{bmatrix} 1 & 1 & 0 \\ 0 & 1 & 1 \\ 0 & 0 & 1 \end{bmatrix} \quad \text{and} \quad \rho_2(x) = \begin{bmatrix} 1 & 1 & 1 \\ 0 & 1 & 0 \\ 0 & 0 & 1 \end{bmatrix}.$$

Calculate the socles of these two representations and show that neither representation is semisimple. Show that the second representation is nevertheless the direct sum of two nonzero subrepresentations.

9. Let k be an infinite field of characteristic 2, and $G = \langle x, y \rangle \cong C_2 \times C_2$ be the noncyclic group of order 4. For each $\lambda \in k$, let $\rho_\lambda(x), \rho_\lambda(y)$ be the matrices

$$\rho_\lambda(x) = \begin{bmatrix} 1 & 0 \\ 1 & 1 \end{bmatrix}, \quad \rho_\lambda(y) = \begin{bmatrix} 1 & 0 \\ \lambda & 1 \end{bmatrix},$$

regarded as linear maps $U_\lambda \to U_\lambda$ where U_λ is a k-vector space of dimension 2 with basis $\{e_1, e_2\}$.

(a) Show that ρ_λ defines a representations of G with representation space U_λ.
(b) Find a basis for $\operatorname{Soc} U_\lambda$.
(c) By considering the effect on $\operatorname{Soc} U_\lambda$, show that any kG-module homomorphism $\alpha : U_\lambda \to U_\mu$ has a triangular matrix $\alpha = \begin{bmatrix} a & 0 \\ b & c \end{bmatrix}$ with respect to the given bases.
(d) Show that if $U_\lambda \cong U_\mu$ as kG-modules then $\lambda = \mu$. Deduce that kG has infinitely many nonisomorphic 2-dimensional representations.

10. Let

$$\rho_1 : G \to GL(V)$$
$$\rho_2 : G \to GL(V)$$

be two representations of G on the same R-module V that are injective as homomorphisms. (We say that such a representation is *faithful*.) Consider the three properties that

(1) the RG-modules given by ρ_1 and ρ_2 are isomorphic,
(2) the subgroups $\rho_1(G)$ and $\rho_2(G)$ are conjugate in $GL(V)$,
(3) for some automorphism $\alpha \in \operatorname{Aut}(G)$, the representations ρ_1 and $\rho_2\alpha$ are isomorphic.

Show that (1) $\Rightarrow$ (2) and that (2) $\Rightarrow$ (3). Show also that if $\alpha \in \operatorname{Aut}(G)$ is an inner automorphism (i.e., one of the form "conjugation by g" for some $g \in G$) then ρ_1 and $\rho_1\alpha$ are isomorphic.

11. One form of the Jordan–Zassenhaus Theorem asserts that for each n, $GL(n, \mathbb{Z})$ (i.e., $\operatorname{Aut}(\mathbb{Z}^n)$) has only finitely many conjugacy classes of subgroups of finite order. Assuming this, show that for each finite group G and each integer n there are only finitely many isomorphism classes of representations of G on $\mathbb{Z}^n$.

12. (a) Write out a proof of Maschke's Theorem in the case of representations over $\mathbb{C}$ along the following lines.

Given a representation $\rho : G \to GL(V)$ where V is a vector space over $\mathbb{C}$, let $(\ ,\ )$ be any positive definite Hermitian form on V. Define a new form $(\ ,\ )_1$ on V by

$$(v, w)_1 = \frac{1}{|G|} \sum_{g \in G} (gv, gw).$$

Show that $(\, , \,)_1$ is a positive definite Hermitian form, preserved under the action of G; that is, $(v, w)_1 = (gv, gw)_1$ always.
If W is a subrepresentation of V, show that $V = W \oplus W^\perp$ as representations.

(b) Show that any finite subgroup of $GL(n, \mathbb{C})$ is conjugate to a subgroup of $U(n, \mathbb{C})$ (the unitary group, consisting of $n \times n$ complex matrices A satisfying $A\bar{A}^T = I$). Show that any finite subgroup of $GL(n, \mathbb{R})$ is conjugate to a subgroup of $O(n, \mathbb{R})$ (the orthogonal group consisting of $n \times n$ real matrices A satisfying $AA^T = I$).

13. (a) Using Proposition 1.2.4, show that if A is a ring for which the regular representation ${}_AA$ is semisimple, then every finitely generated A-module is semisimple.

(b) Extend the result of part (a), using Zorn's Lemm, to show that if A is a ring for which the regular representation ${}_AA$ is semisimple, then every A-module is semisimple.

14. Let U be a module for a ring A with a 1. Show that the following three statements are equivalent:

(1) U is a direct sum of simple A-submodules.
(2) U is a sum of simple A-submodules.
(3) Every submodule of U is a direct summand of U.

[Use Zorn's Lemma to prove a version of Lemma 1.2.3 that has no finiteness hypothesis and then copy Proposition 1.2.4. This deals with all implications except (3) $\Rightarrow$ (2). For that, use the fact that A has a 1 and hence every (left) ideal is contained in a maximal (left) ideal, combined with condition (3), to show that every submodule of U has a simple submodule. Consider the sum of all simple submodules of U and show that it equals U.]

15. Let RG be the group algebra of a finite group G over a commutative ring R with 1. Let S be a simple RG-module and let I be the anihilator in R of S, that is

$$I = \{r \in R \mid rx = 0 \text{ for all } x \in S\}.$$

Show that I is a maximal ideal in R.
[This question requires some familiarity with standard commutative algebra. We conclude from this result that when considering simple RG modules we may reasonably assume that R is a field, since S may naturally be regarded as an $(R/I)G$-module and R/I is a field.]

2

The Structure of Algebras for Which Every Module Is Semisimple

In this chapter, we present the Artin–Wedderburn structure theorem for semisimple algebras and its immediate consequences. This theorem is the ring-theoretic manifestation of the module theoretic hypothesis of semisimplicity that was introduced in Chapter 1, and it shows that the kind of algebras that can arise when all modules are semisimple is very restricted. The theorem applies to group algebras over a field in which the group order is invertible (as a consequence of Maschke's Theorem), but since the result holds in greater generality we will assume we are working with a finite-dimensional algebra A over a field k.

2.1 Schur's Lemma and Wedderburn's Theorem

Possibly the most important single technique in representation theory is to consider endomorphism rings. It is the main technique of this chapter, and we will see it in use throughout this book. The first result is basic and will be used time and time again.

Theorem 2.1.1 (Schur's Lemma). *Let A be a ring with a 1 and let S_1 and S_2 be simple A-modules. Then* $\mathrm{Hom}_A(S_1, S_2) = 0$ *unless* $S_1 \cong S_2$, *in which case the endomorphism ring* $\mathrm{End}_A(S_1)$ *is a division ring. If A is a finite-dimensional algebra over an algebraically closed field k, then every A-module endomorphism of S_1 is multiplication by some scalar. Thus,* $\mathrm{End}_A(S_1) \cong k$ *in this case.*

Proof. Suppose $\theta : S_1 \to S_2$ is a nonzero homomorphism. Then $0 \neq \theta(S_1) \subseteq S_2$, so $\theta(S_1) = S_2$ by simplicity of S_2, and we see that θ is surjective. Thus, $\mathrm{Ker}\,\theta \neq S_1$, so $\mathrm{Ker}\,\theta = 0$ by simplicity of S_1, and θ is injective. Therefore, θ is invertible, $S_1 \cong S_2$, and $\mathrm{End}_A(S_1)$ is a division ring.

If A is a finite-dimensional k-algebra and k is algebraically closed then S_1 is a finite-dimensional vector space. Let θ be an A-module endomorphism of S_1 and let λ be an eigenvalue of θ. Now $(\theta - \lambda I) : S_1 \to S_1$ is a singular endomorphism of A-modules, so $\theta - \lambda I = 0$ and $\theta = \lambda I$. □

We have just seen that requiring k to be algebraically closed guarantees that the division rings $\mathrm{End}_A(S)$ are no larger than k, and this is often a significant simplifying condition. In what follows we sometimes make this requirement, also indicating how the results go more generally. At other times requiring k to be algebraically closed is too strong, but we still want k to have the property that $\mathrm{End}_A(S) = k$ for all simple A-modules S. In this case, we call k a *splitting field* for the k-algebra A. The theory of splitting fields will be developed in Chapter 9; for the moment it suffices know that algebraically closed fields are always splitting fields.

The next result is the main tool in recovering the structure of an algebra from its representations. We use the notation A^{op} to denote the *opposite* ring of A, namely, the ring that has the same set and the same addition as A, but with a new multiplication given by $a \cdot b = ba$.

Lemma 2.1.2. *For any ring A with a* 1, $\mathrm{End}_A({}_AA) \cong A^{\mathrm{op}}$.

Proof. We prove the result by writing down homomorphisms in both directions that are inverse to each other. The inverse isomorphisms are

$$\phi \mapsto \phi(1)$$
$$(a \mapsto ax) \leftarrow x.$$

There are several things here that need to be checked: that the second assignment does take values in $\mathrm{End}_A({}_AA)$, that the morphisms are ring homomorphisms, and that they are mutually inverse. We leave most of this to the reader, observing only that under the first homomorphism a composite $\theta\phi$ is sent to $(\theta\phi)(1) = \theta(\phi(1)) = \theta(\phi(1)1) = \phi(1)\theta(1)$, so that it is indeed a homomorphism to A^{op}. □

Observe that the proof of Lemma 2.1.2 establishes that every endomorphism of the regular representation is of the form "right multiplication by some element."

A ring A with 1 all of whose modules are semisimple is itself called *semisimple*. By Exercise 13 of Chapter 1, it is equivalent to suppose that the regular representation ${}_AA$ is semisimple. It is also equivalent, if A is a finite-dimensional algebra over a field, to suppose that the Jacobson radical of the ring is zero, but the Jacobson radical has not yet been defined, and we will not deal with this point of view until Chapter 6.

Theorem 2.1.3 (Artin–Wedderburn). *Let A be a finite-dimensional algebra over a field k with the property that every finite-dimensional module is semisimple. Then A is a direct sum of matrix algebras over division rings. Specifically, if*

$$_AA \cong S_1^{n_1} \oplus \cdots \oplus S_r^{n_r},$$

where the $S_1, \dots, S_r$ are nonisomorphic simple modules occuring with multiplicities $n_1, \dots, n_r$ in the regular representation, then

$$A \cong M_{n_1}(D_1) \oplus \cdots \oplus M_{n_r}(D_r),$$

where $D_i = \operatorname{End}_A(S_i)^{\mathrm{op}}$. Furthermore, if k is algebraically closed then $D_i = k$ for all i.

More is true: every such direct sum of matrix algebras is a semisimple algebra. Each matrix algebra over a division ring is a *simple* algebra (namely, one that has no 2-sided ideals apart from the zero ideal and the whole ring), and it has up to isomorphism a unique simple module (see the exercises). Furthermore, the matrix algebra summands are uniquely determined as subsets of A (although the module decomposition of $_AA$ is usually only determined up to isomorphism). The uniqueness of the summands will be established in Proposition 3.6.1.

Proof. We first observe that if we have a direct sum decomposition

$$U = U_1 \oplus \cdots \oplus U_r$$

of a module U then $\operatorname{End}_A(U)$ is isomorphic to the algebra of $r \times r$ matrices in which the i, j entries lie in $\operatorname{Hom}_A(U_j, U_i)$. This is because any endomorphism $\phi : U \to U$ may be writen as a matrix of components $\phi = (\phi_{ij})$ where $\phi_{ij} : U_j \to U_i$, and when viewed in this way endomorphisms compose in the manner of matrix multiplication. Since $\operatorname{Hom}_A(S_j^{n_j}, S_i^{n_i}) = 0$ if $i \neq j$ by Schur's Lemma, the decomposition of $_AA$ shows that

$$\operatorname{End}_A(_AA) \cong \operatorname{End}_A(S_1^{n_1}) \oplus \cdots \oplus \operatorname{End}_A(S_r^{n_r})$$

and furthermore $\operatorname{End}_A(S_i^{n_i}) \cong M_{n_i}(D_i^{\mathrm{op}})$. Evidently, $M_{n_i}(D_i^{\mathrm{op}})^{\mathrm{op}} \cong M_{n_i}(D_i)$ and by Lemma 2.1.2, we identify $\operatorname{End}_A(_AA)$ as A^{op}. Putting these pieces together gives the matrix algebra decomposition. Finally, if k is algebraically closed, it is part of Schur's Lemma that $D_i = k$ for all i. □

Corollary 2.1.4. *Let A be a finite-dimensional semisimple algebra over a field k. In any decomposition,*

$$_AA = S_1^{n_1} \oplus \cdots \oplus S_r^{n_r}$$

where the S_i are pairwise nonisomorphic simple modules we have that $S_1, \ldots, S_r$ is a complete set of representatives of the isomorphism classes of simple A-modules. When k is algebraically closed $n_i = \dim_k S_i$ and $\dim_k A = n_1^2 + \cdots + n_r^2$.

Proof. All isomorphism types of simple modules must appear in the decomposition because every simple module can be expressed as a homomorphic image of ${}_AA$ (as observed at the start of this chapter), and so must be a homomorphic image of one of the modules S_i. When k is algebraically closed all the division rings D_i coincide with k by Schur's Lemma, and $\mathrm{End}_A(S_i^{n_i}) \cong M_{n_i}(k)$. The ring decomposition $A = M_{n_1}(k) \oplus \cdots \oplus M_{n_r}(k)$ of Theorem 2.1.3 immediately gives $\dim_k A = n_1^2 + \cdots + n_r^2$.

We obtained this decomposition by identifying A with $\mathrm{End}({}_AA)^{\mathrm{op}}$ in such a way that an element $a \in A$ is identified with the endomorphism "right multiplication by a," by Lemma 2.1.2. From this, we see that right multiplication of an element of $S_j^{n_j}$ by an element of $M_{n_i}(k)$ is 0 if $i \neq j$, and hence $S_i^{n_i}$ is the unique summand of A (in the initial decomposition of A) containing elements on which $M_{n_i}(k)$ acts in a nonzero fashion from the right. We deduce that $M_{n_i}(k) \cong S_i^{n_i}$ as left A-modules, since the term on the left is isomorphic to the quotient of A by the left submodule consisting of elements that the summand $M_{n_i}(k)$ annihilates by right multiplication, the term on the right is an image of this quotient, and to have $\dim_k A = \sum_i \dim_k S_i^{n_i}$ they must be isomorphic. Hence

$$\dim_k M_{n_i}(k) = n_i^2 = \dim_k S_i^{n_i} = n_i \dim S_i,$$

and so $\dim S_i = n_i$. □

Let us now restate what we have proved specifically in the context of group representations.

Corollary 2.1.5. *Let G be a finite group and k a field in which $|G|$ is invertible.*

(1) As a ring, kG is a direct sum of matrix algebras over division rings.
(2) Suppose in addition that k is algebraically closed. Let $S_1, \ldots, S_r$ be pairwise non-isomorphic simple kG-modules and let $d_i = \dim_k S_i$ be the degree of S_i. Then d_i equals the multiplicity with which S_i is a summand of the regular representation of G, and $|G| \geq d_1^2 + \cdots + d_r^2$ with equality if and only if $S_1, \ldots, S_r$ is a complete set of representatives of the simple kG-modules.

Proof. This follows from Maschke's Theorem 1.2.1, the Artin–Wedderburn Theorem 2.1.3, and Corollary 2.1.4. □

Part (2) of this result provides a numerical criterion that enables us to say when we have constructed all the simple modules of a group over an algebraically closed field k in which $|G|$ is invertible: we check that $\sum d_i^2 = |G|$. While this is an easy condition to verify, it will be superseded later on by the even more straightforward criterion that the number of simple kG-modules (with the same hypotheses on k) equals the number of conjugacy classes of elements of G. Once we have proved this, the formula $\sum d_i^2 = |G|$ allows the degree of the last simple representation to be determined once the others are known.

Example 2.1.6. In Example 1.2.2, we have seen three representations of S_3 over $\mathbb{R}$ that are simple: the trivial representation, the sign representation, and a 2-dimensional representation. Since $1^2 + 1^2 + 2^2 = |S_3|$, we have constructed all the simple representations of S_3 over $\mathbb{R}$.

At this point, we make a deduction about representations of finite abelian groups. Looking ahead to later results, we will obtain a partial converse of the next result in Theorem 4.1.5, and in Theorem 5.3.3, we will obtain an extension to fields in which $|G|$ is not invertible. A more detailed description of representations of abelian groups when $|G|$ is not invertible follows from Example 8.2.1.

Corollary 2.1.7. *Let G be a finite abelian group. Over an algebraically closed field k in which $|G|$ is invertible, every simple representation of G has degree 1 and the number of nonisomorphic simple representations equals $|G|$. In particular, we may deduce that every invertible matrix of finite order, with order relatively prime to the characteristic of k, is diagonalizable.*

Proof. We know that kG is semisimple, and because kG is a commutative ring the matrix summands that appear in Theorem 2.1.3 must all have size 1, and the division rings that appear must be commutative. In fact, since we have supposed that k is algebraically closed, the division rings must all be k. This means that the degrees of the irreducible representations are all 1, and so the number of them must be $|G|$ since this is the sum of the degrees.

A matrix of finite order gives a representation of the cyclic group it generates on the space of vectors on which the matrix acts and, by invertibility of the order, the representation is semisimple. It is a direct sum of 1-dimensional spaces by what we have just shown. On choosing basis vectors to lie in these 1-dimensional spaces, the matrix is diagonal. □

We do not have to exploit the theory we have developed to show that a matrix of finite, invertible order is diagonalizable over an algebraically closed field. A different approach is to consider its Jordan canonical form and observe that all Jordan blocks have size 1, because blocks of size 2 or more have order that is either infinite (in characteristic zero) or a multiple of the field characteristic.

2.2 Summary of Chapter 2

- Endomorphism algebras of simple modules are division rings.
- Semisimple algebras are direct sums of matrix algebras over division rings.
- For a semisimple algebra over an algebraically closed field, the sum of the squares of the degrees of the simple modules equals the dimension of the algebra.

2.3 Exercises for Chapter 2

1. Let A be a finite-dimensional semisimple algebra. Show that A has only finitely many isomorphism types of modules in each dimension. [This is not in general true for algebras that are not semisimple: we saw in Chapter 1 Exercise 9 that $k[C_2 \times C_2]$ has infinitely many nonisomorphic 2-dimensional representations when k is an infinite field of characteristic 2.]

2. Let D be a division ring and n a natural number.

(a) Show that the natural $M_n(D)$-module, consisting of column vectors of length n with entries in D, is a simple module.

(b) Show that $M_n(D)$ is semisimple and has up to isomorphism only one simple module.

(c) Show that every algebra of the form

$$M_{n_1}(D_1) \oplus \cdots \oplus M_{n_r}(D_r)$$

is semisimple.

(d) Show that $M_n(D)$ is a simple ring, namely, one in which the only 2-sided ideals are the zero ideal and the whole ring.

3. Show that for any field k, we have $M_n(k) \cong M_n(k)^{\mathrm{op}}$, and in general for any division ring D that given any positive integer n, $M_n(D) \cong M_n(D)^{\mathrm{op}}$ if and only if $D \cong D^{\mathrm{op}}$.

4. Let U be a module for a semisimple finite-dimensional algebra A. Show that if $\mathrm{End}_A(U)$ is a division ring then U is simple.

5. Prove the following extension of Corollary 2.1.4:

Theorem. *Let A be a finite-dimensional semisimple algebra, S a simple A-module and $D = \mathrm{End}_A(S)$. Then S may be regarded as a module over D and the multiplicity of S as a summand of ${}_AA$ equals* $\dim_D S$.

6. Let k be a field of characteristic 0 and suppose the simple kG-modules are $S_1, \ldots, S_r$ with degrees $d_i = \dim_k S_i$. Show that $\sum_{i=1}^{r} d_i^2 \geq |G|$ with equality if and only if $\mathrm{End}_{kG}(S_i) = k$ for all i.

7. Using Exercise 10 from Chapter 1, Exercise 1 from this chapter, and Maschke's Theorem, show that if k is any field of characteristic 0 then for each natural number m, $GL_n(k)$ has only finitely many conjugacy classes of subgroups of order m. [In view of the comment to Exercise 1 the same is not true when $k = \overline{\mathbb{F}}_2$.]

8. Using the fact that $M_n(k)$ has a unique simple module up to isomorphism, prove the Noether–Skolem Theorem: every algebra automorphism of $M_n(k)$ is inner, that is, of the form conjugation by some invertible matrix.

9. Let A be a ring with a 1, and let V be an A-module. An element e in any ring is called *idempotent* if and only if $e^2 = e$.

(a) Show that an endomorphism $e : V \to V$ is a projection onto a subspace W if and only if e is idempotent as an element of $\mathrm{End}_A(V)$. (The term *projection* was defined at the start of the proof of Theorem 1.2.1. It is a linear mapping onto a subspace that is the identity on restriction to that subspace.)

(b) Show that direct sum decompositions $V = W_1 \oplus W_2$ as A-modules are in bijection with expressions $1 = e + f$ in $\mathrm{End}_A(V)$, where e and f are idempotent elements with $ef = fe = 0$. (In case $ef = fe = 0$, e and f are called *orthogonal*.)

(c) A nonzero idempotent element e is called *primitive* if it cannot be expressed as a sum of orthogonal idempotent elements in a nontrivial way. Show that $e \in \mathrm{End}_A(V)$ is primitive if and only if $e(V)$ has no (nontrivial) direct sum decomposition. (In this case, $e(V)$ is said to be *indecomposable*.)

(d) Suppose that V is semisimple with finitely many simple summands and let $e_1, e_2 \in \mathrm{End}_A(V)$ be idempotent elements. Show that $e_1(V) \cong e_2(V)$ as A-modules if and only if e_1 and e_2 are conjugate by an invertible element of $\mathrm{End}_A(V)$ (i.e., there exists an invertible A-endomorphism $\alpha : V \to V$ such that $e_2 = \alpha e_1 \alpha^{-1}$).

(e) Let k be a field. Show that all primitive idempotent elements in $M_n(k)$ are conjugate under the action of the unit group $GL_n(k)$. Write down explicitly any primitive idempotent element in $M_3(k)$. (It may help to use Exercise 2.)

10. Prove the following theorem of Burnside: let G be a finite group, k an algebraically closed field in which $|G|$ is invertible, and $\rho : G \to GL(V)$ be a representation over k. By taking a basis of V write each endomorphism $\rho(g)$ as a matrix. Let $\dim V = n$. Show that the representation is simple if and only if there exist n^2 elements $g_1, \dots, g_{n^2}$ of G so that the matrices $\rho(g_1), \dots, \rho(g_{n^2})$ are linearly independent, and that this happens if and only if the algebra homomorphism $kG \to \mathrm{End}_k(V)$ is surjective. (Note that ρ itself is generally not surjective.)

11. (We exploit results from a basic algebra course in our suggested approach to this question.) Let G be a cyclic group of order n and k a field.

(a) By considering a homomorphism $k[X] \to kG$ or otherwise, where $k[X]$ is a polynomial ring, show that $kG \cong k[X]/(X^n - 1)$ as rings.

(b) Suppose that the characteristic of k does not divide n. Use the Chinese Remainder Theorem and separability of $X^n - 1$ to show that when kG is expressed as a direct sum of irreducible representations, no two of the summands are isomorphic, and that their degrees are the same as the degrees of the irreducible factors of $X^n - 1$ in $k[X]$. Deduce, as a special case of Corollary 2.1.7, that when k is algebraically closed all irreducible representations of G have degree 1.

(c) When n is prime and $k = \mathbb{Q}$, use irreducibility of $X^{n-1} + X^{n-2} + \cdots X + 1$ to show that G has a simple module S of degree $n - 1$, and that $\mathrm{End}_{kG}(S) \cong \mathbb{Q}(e^{2\pi i/n})$.

(d) When $k = \mathbb{R}$ and n is odd show that G has $\frac{n-1}{2}$ simple representations of degree 2 as well as the trivial representation of degree 1. When $k = \mathbb{R}$ and n is even show that G has $\frac{n-2}{2}$ simple representations of degree 2 as well as two simple representations of degree 1. If S is one of the simple representations of degree 2, show that $\mathrm{End}_{kG}(S) = \mathbb{C}$.

12. Let $\mathbb{H}$ be the algebra of quaternions, that has a basis over $\mathbb{R}$ consisting of elements $1, i, j, k$ and multiplication determined by the relations

$$i^2 = j^2 = k^2 = -1,\ ij = k,\ jk = i,\ ki = j,\ ji = -k,\ kj = -i,\ ik = -j.$$

You may assume that $\mathbb{H}$ is a division ring. The elements $\{\pm 1, \pm i, \pm j, \pm k\}$ under multiplication form the quaternion group Q_8 of order 8, and it acts on $\mathbb{H}$ by left multiplication, so that $\mathbb{H}$ is a 4-dimensional representation of Q_8 over $\mathbb{R}$.

(a) Show that $\mathrm{End}_{\mathbb{R}Q_8}(\mathbb{H}) \cong \mathbb{H}$, and that $\mathbb{H}$ is simple as a representation of Q_8 over $\mathbb{R}$. [Consider the image of $1 \in \mathbb{H}$ under an endomorphism.]

(b) In the decomposition $\mathbb{R}Q_8 = \bigoplus_{i=1}^{t} M_{n_i}(D_i)$ predicted by Corollary 2.1.5, compute the number of summands t, the numbers n_i, and the divisions rings D_i. Show that $\mathbb{R}Q_8$ has no simple representation of dimension 2. [Observe that there are four homomorphisms $Q_8 \to \{\pm 1\} \subset \mathbb{R}$ that give four 1-dimensional representations. Show that, together with the representation of dimension 4, we have a complete set of simple representations.]

(c) The span over $\mathbb{R}$ of the elements $1, i \in \mathbb{H}$ is a copy of the field of complex numbers $\mathbb{C}$, so that $\mathbb{H}$ contains $\mathbb{C}$ as a subfield. We may regard $\mathbb{H}$ as a vector space over $\mathbb{C}$ by letting elements of $\mathbb{C}$ act as scalars on $\mathbb{H}$ by multiplication from the *right*. Show that with the action of Q_8 from the left and of $\mathbb{C}$ from the right, $\mathbb{H}$ becomes a left $\mathbb{C}Q_8$-module. With respect to the basis $\{1, j\}$ for $\mathbb{H}$ over $\mathbb{C}$, write down matrices for the action of the elements $i, j \in Q_8$ on $\mathbb{H}$. Show that this 2-dimensional $\mathbb{C}Q_8$-module is simple, and compute its endomorphism ring $\mathrm{End}_{\mathbb{C}Q_8}(\mathbb{H})$.

(d) Show that $\mathbb{C} \otimes_{\mathbb{R}} \mathbb{H} \cong M_2(\mathbb{C})$.

3
Characters

Characters are an extremely important tool for handling the simple representations of a group. In this chapter, we will see them in the form that applies to representations over a field of characteristic zero, and these are called *ordinary characters*. Since representations of finite groups in characteristic zero are semisimple, knowing about the simple representations in some sense tells us about all representations. Later, in Chapter 10, we will study characters associated to representations in positive characteristic, the so-called *Brauer characters*.

Characters are very useful when we have some specific representation and wish to compute its decomposition as a direct sum of simple representations. The information we need to do this is contained in the character table of the group, which we introduce in this chapter. We also establish many important theoretical properties of characters that enable us to calculate them more easily and to check that our calculations are correct. The most spectacular of these properties is the orthogonality relations, which may serve to convince the reader that something extraordinary and fundamental is being studied. We establish numerical properties of the character degrees, and a description of the center of the group algebra that aids in decomposition the group algebra as a sum of matrix algebras. This would be of little significance unless we could use characters to prove something outside their own area. Aside from their use as a computational tool, we use them to prove Burnside's $p^a q^b$ theorem: every group whose order is divisible by only two primes is solvable.

3.1 The Character Table

Assume that $\rho : G \to GL(V)$ is a finite-dimensional representation of G over the field of complex numbers $\mathbb{C}$ or one of its subfields. We define the *character*

χ of ρ to be the function $\chi : G \to \mathbb{C}$ given by

$$\chi(g) = \mathrm{tr}(\rho(g)),$$

the trace of the linear map $\rho(g)$. The *degree* of the character is $\dim V$, which equals $\chi(1)$. For example, the 2-dimensional representation of S_3 we considered in Chapter 1 has character given on the group elements by forming the representing matrices and taking the trace as follows:

$$\begin{array}{ccccc}
() & \mapsto & \begin{bmatrix} 1 & 0 \\ 0 & 1 \end{bmatrix} & \mapsto & 2, \\
(1,2) & \mapsto & \begin{bmatrix} 0 & 1 \\ 1 & 0 \end{bmatrix} & \mapsto & 0, \\
(1,3) & \mapsto & \begin{bmatrix} -1 & 0 \\ -1 & 1 \end{bmatrix} & \mapsto & 0, \\
(2,3) & \mapsto & \begin{bmatrix} 1 & -1 \\ 0 & -1 \end{bmatrix} & \mapsto & 0, \\
(1,2,3) & \mapsto & \begin{bmatrix} 0 & -1 \\ 1 & -1 \end{bmatrix} & \mapsto & -1, \\
(1,3,2) & \mapsto & \begin{bmatrix} -1 & 1 \\ -1 & 0 \end{bmatrix} & \mapsto & -1.
\end{array}$$

We say that the representation ρ and the representation space V *afford* the character χ, and we may write χ_ρ or χ_V when we wish to specify this character more precisely. The restriction to subfields of $\mathbb{C}$ is not significant: we will see in Chapter 9 that representations of a finite group over a field characteristic 0 may always be realized over $\mathbb{C}$.

In the next result, we list some immediate properties of characters. The converse of part (7) will be established later on in Corollary 3.3.3. As it is, part (7) provides a useful way to show that representations are not isomorphic, by showing that their characters are different.

Proposition 3.1.1. *Let χ be the character of a representation ρ of G over $\mathbb{C}$ and let $g, h \in G$. Then*

(1) $\chi(1)$ is the degree of ρ, namely, the dimension of the representation space of ρ;
(2) if g has order n then $\chi(g)$ is a sum of nth roots of 1 (including roots whose order divides n);
(3) $|\chi(g)| \le \chi(1)$, with equality if and only if $\rho(g)$ is scalar multiplication;
(4) $\chi(g) = \chi(1)$ if and only if $\rho(g) = 1$, that is, g lies in the kernel of ρ;

(5) $\chi(g^{-1}) = \overline{\chi(g)}$, *the complex conjugate;*
(6) $\chi(hgh^{-1}) = \chi(g)$;
(7) *if* V *and* W *are isomorphic* $\mathbb{C}G$*-modules then* $\chi_V = \chi_W$ *as functions on* G.

Proof. (1) is immediate because the identity of the group must act as the identity matrix, and its trace is the degree of ρ, since this is the dimension of the representation space of ρ.

(2) Recall from Corollary 2.1.7 or the comment after it that $\rho(g)$ is diagonalizable, so that $\chi(g)$ is the sum of its eigenvalues $\lambda_1, \ldots, \lambda_d$, where d is the degree of χ. (In fact, it is sufficient for now to let $\lambda_1, \ldots, \lambda_d$ be the diagonal entries in the Jordan canonical form of $\rho(g)$, without knowing that this form is diagonal.) These eigenvalues are roots of unity since g has finite order, and the roots of unity have orders dividing n.

(3) Each root of unity has absolute value 1, so adding d of them and applying the triangle inequality repeatedly we get $|\chi(g)| \leq |\lambda_1| + \cdots + |\lambda_d| = d$. The only way we can have equality is if $\lambda_1 = \cdots = \lambda_d$ so that $\rho(g)$ is scalar (since it is diagonalizable).

(4) If $\rho(g) = 1$ then $\chi(g) = d$. Conversely, if $\chi(g) = d$ then by part (3), $\rho(g)$ is multiplication by a scalar λ. Now $\chi(g) = d\lambda = d$, so $\lambda = 1$.

(5) If $\rho(g)$ has eigenvalues $\lambda_1, \ldots, \lambda_d$ then $\rho(g^{-1})$ has eigenvalues $\lambda_1^{-1}, \ldots, \lambda_d^{-1}$, and $\lambda_i^{-1} = \overline{\lambda}_i$ for each i since these are roots of unity. Thus,

$$\chi(g^{-1}) = \overline{\lambda}_1 + \cdots + \overline{\lambda}_d = \overline{\chi(g)}.$$

(6) This results from the fact that $\mathrm{tr}(ab) = \mathrm{tr}(ba)$ for endomorphisms a and b, so that $\chi(hgh^{-1}) = \mathrm{tr}\rho(hgh^{-1}) = \mathrm{tr}(\rho(h)\rho(g)\rho(h^{-1})) = \mathrm{tr}\rho(g) = \chi(g)$.

(7) Suppose that ρ_V and ρ_W are the representations of G on V and W, and that we have an isomorphism of $\mathbb{C}G$-modules $\alpha : V \to W$. Then $\alpha\rho_V(g) = \rho_W(g)\alpha$ for all $g \in G$, so that

$$\chi_W(g) = \mathrm{tr}\rho_W(g) = \mathrm{tr}(\alpha\rho_V(g)\alpha^{-1}) = \mathrm{tr}\rho_V(g) = \chi_V(g). \qquad \square$$

As an application of Proposition 3.1.1 part (4), we see that certain normal subgroups, namely, the kernels of representations, are determined by knowing the characters of the representations. We will see in Exercise 7 that all normal subgroups of a finite group may be found from knowledge of the characters of representations, as intersections of the kernels of simple characters. This means that whether or not a group is simple may be easily read from this information on characters.

We see in Proposition 3.1.1 part (6) that characters are constant on conjugacy classes, so that in listing values of characters on group elements we only need take one element from each conjugacy class. The table of complex numbers whose rows are indexed by the isomorphism types of simple representations of G, whose columns are indexed by the conjugacy classes of G and whose entries are the values of the characters of the simple representations on representatives of the conjugacy classes is called the *character table* of G. It is usual to index the first column of a character table by the (conjugacy class of the) identity, and to put the character of the trivial representation as the top row. With this convention the top row of every character table will be a row of 1s, and the first column will list the degrees of the simple representations. Above the table, it is usual to list two rows, the first of which is a list of representatives of the conjugacy classes of elements of G, in some notation. The row underneath lists the value of $|C_G(g)|$ for each element g in the top row.

Example 3.1.2. We present the character table of S_3. We saw at the end of Chapter 2 that we already have a complete list of the simple modules for S_3, and the values of their characters on representatives of the conjugacy classes of S_3 are computed from the matrices that give these representations.

S_3
Ordinary characters

g $\lvert C_G(g)\rvert$	$()$ 6	(12) 2	(123) 3
χ_1	1	1	1
χ_{sign}	1	-1	1
χ_2	2	0	-1

We will see that the character table has remarkable properties, among which are that it is always square, and its rows (and also its columns) satisfy certain orthogonality relations. Our next main goal is to state and prove these results. To do this, we first introduce three ways to construct new representations of a group from existing ones. These constructions have validity no matter what ring R we work over, although in the application to the character table we will suppose that $R = \mathbb{C}$.

Suppose V and W are representations of G over R. The R-module $V \otimes_R W$ acquires an action of G by means of the formula $g \cdot (v \otimes w) = gv \otimes gw$, thereby making the tensor product into a representation. This is what is called the *tensor product* of the representations V and W, but it is not the only occurrence of tensor products in representation theory, and as the other ones are different this one is sometimes also called the *Kronecker product*. The action of G on the Kronecker product is called the *diagonal* action. To do things properly, we should check that the formula for the diagonal action does indeed define a representation of G. This is immediate, but the fact that we can make the definition at all is special for finite groups and group algebras: it does not work for algebras in general.

For the second construction, we form the R-module $\text{Hom}_R(V, W)$. This acquires an action of G by means of the formula $(g \cdot f)(v) = gf(g^{-1}v)$ for each R-linear map $f : V \to W$ and $g \in G$. It is worth checking that the negative exponent in the formula is really necessary so that we have a left action of G. Thus, $\text{Hom}_R(V, W)$ becomes an RG-module.

The third construction is the particular case of the second in which we take W to be the trivial module R. We write $V^* = \text{Hom}_R(V, R)$ and the action is $(g \cdot f)(v) = f(g^{-1}v)$ for each $f : V \to R$ and $g \in G$. This representation is called the *dual* or *contragredient* representation of V. It is usually only considered when V is free (or at least projective) of finite rank as an R-module, in which case we have $V \cong V^{**}$ as RG-modules. The exponent -1 in the action of G is necessary to make V^* a left RG-module. Without this exponent we would get a right RG-module.

If R happens to be a field and we have bases $v_1, \ldots, v_m$ for V and $w_1, \ldots, w_n$ for W then $V \otimes W$ has a basis $\{v_i \otimes w_j \mid 1 \leq i \leq m, 1 \leq j \leq n\}$ and V^* has a dual basis $\hat{v}_1, \ldots, \hat{v}_m$. With respect to these bases an element $g \in G$ acts on $V \otimes W$ with the matrix that is the tensor product of the two matrices giving its action on V and W, and on V^* it acts with the transpose of the inverse of the matrix of its action on V. The tensor product of two matrices is not seen so often these days. If (a_{pq}), (b_{rs}) are an $m \times m$ matrix and an $n \times n$ matrix, their tensor product is the $mn \times mn$ matrix (c_{ij}) where if $i = (p-1)n + r$ and $j = (q-1)n + s$ with $1 \leq p, q \leq m$ and $1 \leq r, s \leq n$ then $c_{ij} = a_{pq}b_{rs}$. For example,

$$\begin{bmatrix} a & b \\ c & d \end{bmatrix} \otimes \begin{bmatrix} e & f \\ g & h \end{bmatrix} = \begin{bmatrix} ae & af & be & bf \\ ag & ah & bg & bh \\ ce & cf & de & df \\ cg & ch & dg & dh \end{bmatrix}.$$

If $\alpha : V \to V$ and $\beta : W \to W$ are endomorphisms, then the matrix of

$$\alpha \otimes \beta : V \otimes W \to V \otimes W$$

is the tensor product of the matrices that represent α and β (provided the basis elements $v_i \otimes w_j$ are taken in an appropriate order). We see from this that $\text{tr}(\alpha \otimes \beta) = \text{tr}(\alpha)\text{tr}(\beta)$.

In the following result, we consider the sum and product of characters, which are defined in the usual manner for functions by the formulas

$$(\chi_V + \chi_W)(g) = \chi_V(g) + \chi_W(g)$$
$$(\chi_V \cdot \chi_W)(g) = \chi_V(g) \cdot \chi_W(g).$$

Proposition 3.1.3. *Let V and W be finite-dimensional representations of G over a field k of characteristic zero.*

(1) $V \oplus W$ has character $\chi_V + \chi_W$.
(2) $V \otimes W$ has character $\chi_V \cdot \chi_W$.
(3) V^* has character $\chi_{V^*}(g) = \chi_V(g^{-1}) = \overline{\chi_V(g)}$, the complex conjugate.
(4) Let M and N be representations of G over any ground ring R. Suppose that, as an R-module, at least one of M and N is free of finite rank, then $\operatorname{Hom}_R(M, N) \cong M^* \otimes_R N$ as RG-modules. When $R = k$ is a field of characteristic zero this representation has character equal to $\chi_{M^*} \cdot \chi_N$.

Proof. (1), (2), and (3) are immediate from Proposition 3.1.1 and the subsequent remarks.

As for (4), we define an R-module homomorphism

$$\begin{aligned} \alpha : M^* \otimes_R N &\to \operatorname{Hom}_R(M, N) \\ \phi \otimes v &\mapsto (u \mapsto \phi(u) \cdot v), \end{aligned}$$

this being the specification on basic tensors. We show that if either M or N is free as an R-module of finite rank then α is an isomorphism.

Suppose that M is R-free, with basis $u_1, \dots, u_m$. Then M^* is R-free with basis $\hat{u}_1, \dots, \hat{u}_m$ where $\hat{u}_i(u_j) = \delta_{i,j}$. Any morphism $f : M \to N$ is determined by its values $f(u_i)$ on the basis elements by means of the formula $f(u) = \sum_{i=1}^m \hat{u}_i(u) f(u_i)$ and it follows that $f = \alpha(\sum_{i=1}^m \hat{u}_i \otimes f(u_i))$ so that α is surjective. If $\alpha(\sum_{i=1}^m \hat{u}_i \otimes v_i) = 0$ for certain elements $v_i \in N$ then applying this map to u_j gives v_j, so we deduce $v_j = 0$ for all j, and hence $\sum_{i=1}^m \hat{u}_i \otimes v_i = 0$. Thus, α is injective.

If N is free as an R-module with basis $v_1, \dots, v_n$, let $p_i : N \to R$ be projection onto coordinate i. Any morphism $f : M \to N$ equals $\alpha(\sum_{i=1}^n p_i f \otimes v_i)$, so α is surjective. If $\alpha(\sum_{i=1}^n \phi_i \otimes v_i) = 0$ for certain $\phi_i \in M^*$ then for all $u \in M$, $\sum_{i=1}^n \phi_i(u) v_i = 0$, which implies that $\phi_i(u) = 0$ for all i and u. This means that ϕ_i is the zero map for all i, so that $\sum_{i=1}^n \phi_i \otimes v_i = 0$ and α is injective.

We now observe that α is a map of RG-modules, since for $g \in G$,

$$\begin{aligned} \alpha(g(\phi \otimes w)) &= \alpha(g\phi \otimes gw) \\ &= (v \mapsto (g\phi)(v) \cdot gw) \\ &= (v \mapsto g(\phi(g^{-1}v)w)) \\ &= g(v \mapsto \phi(v)w) \\ &= g\alpha(\phi \otimes w). \end{aligned}$$

Finally, the last formula for characters follows from parts (2) and (3). □

Of course, if R is taken to be a field in part (4) of Proposition 3.1.3 then the argument can be simplified. Since M and N are always free as R-modules in this situation, only one of the two arguments given is needed. After showing that α is either surjective or injective, the other follows by observing that the dimensions on the two sides are equal.

3.2 Orthogonality Relations and Bilinear Forms

We start with some preliminary constructions that will be used in the proof of the orthogonality relations for characters. A fundamental notion in dealing with group actions is that of fixed points. If V is an RG-module, we define the *fixed points*

$$V^G = \{v \in V \mid gv = v \text{ for all } g \in G\}.$$

This is the largest RG-submodule of V on which G has trivial action.

Lemma 3.2.1. *Over any ring R, $\mathrm{Hom}_R(V, W)^G = \mathrm{Hom}_{RG}(V, W)$.*

Proof. An R-linear map $f : V \to W$ is a morphism of RG-modules precisely if it commutes with the action of G, which is to say that $f(gv) = gf(v)$ for all $g \in G$ and $v \in V$, or in other words $gf(g^{-1}v) = f(v)$ always. This is exactly the condition that f is fixed under the action of G. □

The next result is an abstraction of the idea that was used in proving Maschke's Theorem, where the application was to the RG-module $\mathrm{Hom}_R(V, V)$. By a *projection*, we mean a linear transformation $\pi : V \to V$ which is the identity on its image so that, equivalently, $\pi^2 = \pi$. The projection operator about to be described appears throughout representation theory.

Lemma 3.2.2. *Let V be an RG-module where R is a ring in which $|G|$ is invertible. Then*

$$\frac{1}{|G|}\sum_{g\in G} g : V \to V^G$$

is a map of RG-modules which is projection onto the fixed points of V. In particular, V^G is a direct summand of V as an RG-module. When R is a field of characteristic zero, we have

$$\mathrm{tr}\left(\frac{1}{|G|}\sum_{g\in G} g\right) = \dim V^G,$$

where tr *denotes the trace.*

Proof. Let $\pi : V \to V$ denote the map "multiplication by $\frac{1}{|G|}\sum_{g\in G} g$." Clearly π is a linear map, and it commutes with the action of G: if $h \in G$ and $v \in V$, we have

$$\begin{aligned}\pi(hv) &= \left(\frac{1}{|G|}\sum_{g\in G} gh\right)v\\ &= \pi(v)\\ &= \left(\frac{1}{|G|}\sum_{g\in G} hg\right)v\\ &= h\pi(v),\end{aligned}$$

since as g ranges through the elements of G so do gh and hg. The same equations show that every vector of the form $\pi(v)$ is fixed by G. Furthermore, if $v \in V^G$ then

$$\pi(v) = \frac{1}{|G|}\sum_{g\in G} gv = \frac{1}{|G|}\sum_{g\in G} v = v,$$

so π is indeed projection onto V^G. □

There is one more ingredient we describe before stating the orthogonality relations for characters: an inner product on characters. It does not make sense without some further explanation, because an inner product must be defined on a vector space and characters do not form a vector space. They are, however, elements in a vector space, namely, the vector space of class functions on G.

A *class function* on G is a function $G \to \mathbb{C}$ that is constant on each conjugacy class of G. Such functions are in bijection with the functions from the set of conjugacy classes of G to $\mathbb{C}$, a set of functions that we denote $\mathbb{C}^{\mathrm{cc}(G)}$, where $\mathrm{cc}(G)$ is the set of conjugacy classes of G. These functions become an algebra when we define addition, multiplication, and scalar multiplication pointwise on the values of the function. In other words, $(\chi \cdot \psi)(g) = \chi(g)\psi(g)$, $(\chi + \psi)(g) = \chi(g) + \psi(g)$, and $(\lambda\chi)(g) = \lambda\chi(g)$ where χ, ψ are class functions and $\lambda \in \mathbb{C}$. If G has n conjugacy classes, this algebra is isomorphic to $\mathbb{C}^n$, the direct sum of n copies of $\mathbb{C}$, and is semisimple. We have seen in Proposition 3.1.1 that characters of representations of G are examples of class functions on G.

We define a Hermitian form on the complex vector space of class functions on G by means of the formula

$$\langle \chi, \psi \rangle = \frac{1}{|G|} \sum_{g \in G} \overline{\chi(g)} \psi(g).$$

Since the functions χ and ψ are constant on conjugacy classes, this can also be written

$$\begin{aligned}\langle \chi, \psi \rangle &= \frac{1}{|G|} \sum_{g \in [\mathrm{cc}(G)]} \frac{|G|}{|C_G(g)|} \overline{\chi(g)} \psi(g) \\ &= \sum_{g \in [\mathrm{cc}(G)]} \frac{1}{|C_G(g)|} \overline{\chi(g)} \psi(g),\end{aligned}$$

where $[\mathrm{cc}(G)]$ denotes a set of representatives of the conjugacy classes of G, since the number of conjugates of g is $|G : C_G(g)|$. As well as the usual identities that express bilinearity and the fact that the form is Hermitian, it satisfies

$$\langle \chi\phi, \psi \rangle = \langle \chi, \phi^* \psi \rangle,$$

where $\phi^*(g) = \overline{\phi(g)}$ is the class function obtained by complex conjugation. If χ and ψ happen to be characters of a representation, we have $\overline{\chi(g)} = \chi(g^{-1})$ and that ψ^* is the character of the contragredient representation, and in this case, we obtain further expressions for the bilinear form:

$$\begin{aligned}\langle \chi, \psi \rangle &= \frac{1}{|G|} \sum_{g \in G} \chi(g^{-1}) \psi(g) \\ &= \frac{1}{|G|} \sum_{g \in G} \chi(g) \psi(g^{-1}) \\ &= \frac{1}{|G|} \sum_{g \in G} \chi(g) \overline{\psi(g)} \\ &= \langle \psi, \chi \rangle,\end{aligned}$$

where the second equality is obtained by observing that as g ranges over the elements of G, so does g^{-1}. We emphasize that we have assumed that χ and ψ are actually characters of representations to obtain these equalities. With this assumption, $\langle \chi, \psi \rangle = \langle \psi, \chi \rangle = \overline{\langle \psi, \chi \rangle}$ must be real, and we will see in the coming results that, for characters, this number must be a nonnegative integer.

With all this preparation, we now present the orthogonality relations for the rows of the character table. The picture will be completed once we have shown that the character table is square and deduced the orthogonality relations for columns in Theorem 3.4.3 and Corollary 3.4.4.

Theorem 3.2.3 (Row orthogonality relations). *Let G be a finite group. If V, W are simple complex representations of G with characters χ_V, χ_W then*

$$\langle \chi_V, \chi_W \rangle = \begin{cases} 1 & \text{if } V \cong W \\ 0 & \text{otherwise.} \end{cases}$$

Proof. By Proposition 3.1.3 the character of $\mathrm{Hom}_{\mathbb{C}}(V, W)$ is $\overline{\chi_V} \cdot \chi_W$. By Lemma 3.2.1 and Lemma 3.2.2

$$\begin{aligned} \dim \mathrm{Hom}_{\mathbb{C}G}(V, W) &= \mathrm{tr}\left(\frac{1}{|G|}\sum_{g\in G} g\right) \text{ in its action on } \mathrm{Hom}_{\mathbb{C}}(V, W) \\ &= \frac{1}{|G|}\sum_{g\in G} \overline{\chi_V(g)}\chi_W(g) \\ &= \langle \chi_V, \chi_W \rangle. \end{aligned}$$

Schur's Lemma 2.1.1 asserts that this number is 1 if $V \cong W$, and 0 if $V \ncong W$. □

3.3 Consequences of the Orthogonality Relations

We will describe many consequences of the orthogonality relations, and the first is that they provide a way of determining the decomposition of a given representation as a direct sum of simple representations. This procedure is similar to the way of finding the coefficients in the Fourier expansion of a function using orthogonality of the functions $\sin(mx)$ and $\cos(nx)$.

Corollary 3.3.1. *Let V be a $\mathbb{C}G$-module. In any expression,*

$$V = S_1^{n_1} \oplus \cdots \oplus S_r^{n_r}$$

in which $S_1, \ldots, S_r$ are nonisomorphic simple modules, we have

$$n_i = \langle \chi_V, \chi_{S_i} \rangle,$$

where χ_V, χ_{S_i} are the characters of V and S_i. In particular, n_i is determined by V independently of the choice of decomposition.

Example 3.3.2. Let $G = S_3$ and denote by $\mathbb{C}$ the trivial representation, ϵ the sign representation, and V the 2-dimensional simple representation over $\mathbb{C}$. We decompose the 4-dimensional representation $V \otimes V$ as a direct sum of simple representations. Since the values of the character χ_V give the row of the character table

χ_V	2	0	-1

we see that $V \otimes V$ has character values

$\chi_{V\otimes V}$	4	0	1

Thus,

$$\langle \chi_{V\otimes V}, \chi_{\mathbb{C}} \rangle = \frac{1}{6}(4 \cdot 1 + 0 + 2 \cdot 1 \cdot 1) = 1,$$

$$\langle \chi_{V\otimes V}, \chi_{\epsilon} \rangle = \frac{1}{6}(4 \cdot 1 + 0 + 2 \cdot 1 \cdot 1) = 1,$$

$$\langle \chi_{V\otimes V}, \chi_{V} \rangle = \frac{1}{6}(4 \cdot 2 + 0 - 2 \cdot 1 \cdot 1) = 1,$$

and we deduce that

$$V \otimes V \cong \mathbb{C} \oplus \epsilon \oplus V.$$

Corollary 3.3.3. *For finite-dimensional complex representations V and W, we have $V \cong W$ if and only if $\chi_V = \chi_W$.*

Proof. We saw in Proposition 3.1.1 that if V and W are isomorphic then they have the same character. Conversely, if they have the same character, they both may be decomposed as a direct sum of simple representations by Corollary 1.2.5, and by Corollary 3.3.1, the multiplicities of the simples in these two decompositions must be the same. Hence the representations are isomorphic. □

The next result is a criterion for a representation to be simple. An important step in studying the representation theory of a group is to construct its character table, and one proceeds by compiling a list of the simple characters which at the end of the calculation will be complete. At any stage one has a partial list of simple characters, and considers some (potentially) new character. One finds the multiplicity of each previously obtained simple character as a summand of the new character, and subtracts off the these simple characters to the correct multiplicity. What is left is a character all of whose simple components are new simple characters. This character will itself be simple precisely if the following easily verified criterion is satisfied.

Corollary 3.3.4. *If χ is the character of a complex representation V then $\langle \chi, \chi \rangle$ is a positive integer, and equals 1 if and only if V is simple.*

Proof. We may write $V \cong S_1^{n_1} \oplus \cdots \oplus S_r^{n_r}$ and then $\langle \chi, \chi \rangle = \sum_{i=1}^{r} n_i^2$ is a positive integer, which equals 1 precisely if one n_i is 1 and the others are 0. □

Example 3.3.5. We construct the character table of S_4, since it illustrates some techniques in finding simple characters. It is as follows.

S_4
Ordinary characters

g $\lvert C_G(g)\rvert$	$()$ 24	(12) 4	$(12)(34)$ 8	(1234) 4	(123) 3
χ_1	1	1	1	1	1
χ_{sign}	1	−1	1	−1	1
χ_2	2	0	2	0	−1
χ_{3a}	3	−1	−1	1	0
χ_{3b}	3	1	−1	−1	0

Above the horizontal line, we list representatives of the conjugacy classes of elements of S_4, and below them the orders of their centralizers, obtained using standard facts from group theory. The first row below the line is the character of the trivial representation, and below that is the character of the sign representation. These are always characters of a symmetric group.

We will exploit the isomorphism between S_4 and the group of rotations of $\mathbb{R}^3$ that preserve a cube. We may see that these groups are isomorphic from the fact that the group of such rotations permutes the four diagonals of the cube. This action on the diagonals is faithful, and since every transposition of diagonals may be realized through some rotation, so can every permutation of the diagonals. Hence the full group of such rotations is isomorphic to S_4.

There is a homomorphism $\sigma : S_4 \to S_3$ which sends

$$\begin{aligned} () &\mapsto (), \\ (12) &\mapsto (12), \\ (12)(34) &\mapsto (), \\ (1234) &\mapsto (13), \\ (123) &\mapsto (123), \end{aligned}$$

which has kernel the normal subgroup $\langle (12)(34), (13)(24) \rangle$. One way to obtain this homomorphism from the identification of S_4 with the group of rotations of a cube is to observe that each rotation gives rise to a permutation of the three pairs of opposite faces. Any representation $\rho : S_3 \to GL(V)$ gives rise to a representation $\rho\sigma$ of S_4 obtained by composition with σ, and if we start with a simple representation of S_3 we will obtain a simple representation of S_4 since σ is surjective and the invariant subspaces for ρ and $\rho\sigma$ are the same. Thus, the simple characters of S_3 give a set of simple characters of S_4 obtained by applying σ and evaluating the character of S_3. This procedure, which in general works whenever one group is a homomorphic image of another, gives the trivial, sign and 2-dimensional representations of S_4. At this point, we have computed the top three rows of the character table of S_4.

The isomorphism from S_4 to the group of rotations of the cube also gives an action of S_4 on $\mathbb{R}^3$. The character χ_{3a} of this action is the fourth row of the character table. To compute the traces of the matrices that represent the different elements, we do not actually have to work out what those matrices are, relying instead on the observation that every rotation of $\mathbb{R}^3$ has matrix conjugate to a matrix

$$\begin{bmatrix} \cos\theta & -\sin\theta & 0 \\ \sin\theta & \cos\theta & 0 \\ 0 & 0 & 1 \end{bmatrix},$$

where θ is the angle of rotation, noting that conjugate matrices have the same trace. Thus, for example, (12) and (123) must act as rotations through π and $\frac{2\pi}{3}$, respectively, so act via matrices that are conjugates of

$$\begin{bmatrix} -1 & 0 & 0 \\ 0 & -1 & 0 \\ 0 & 0 & 1 \end{bmatrix} \quad \text{and} \quad \begin{bmatrix} -\frac{1}{2} & -\frac{\sqrt{3}}{2} & 0 \\ \frac{\sqrt{3}}{2} & -\frac{1}{2} & 0 \\ 0 & 0 & 1 \end{bmatrix},$$

which have traces -1 and 0. We may show that this character is simple either geometrically, since the group of rotations of a cube has no nontrivial invariant subspace, or by computing $\langle \chi_{3a}, \chi_{3a} \rangle = 1$.

At this point, we have computed four rows of the character table of S_4, and we will see several ways to complete the table. We start with one. There is an action of S_4 on $\mathbb{C}^4$ given by the permutation action of S_4 on four basis vectors. Since the trace of a permutation matrix equals the number of points fixed by the permutation, this has character

χ	4	2	0	0	1

and we compute

$$\langle \chi, 1 \rangle = \frac{4}{24} + \frac{2}{4} + 0 + 0 + \frac{1}{3} = 1.$$

Thus, $\chi = 1 + \psi$ where ψ is the character of a 3-dimensional representation:

ψ	3	1	−1	−1	0

Again, we have

$$\langle \psi, \psi \rangle = \frac{9}{24} + \frac{1}{4} + \frac{1}{8} + \frac{1}{4} + 0 = 1$$

so $\chi_{3b} := \psi$ is simple by Corollary 3.3.4, and this is the bottom row of the character table.

We now need to know that there are no more simple characters. This will follow from Theorem 3.4.3 which says that the character table is always square, but we do not need this result here: the simple characters are independent functions in $\mathbb{C}^{cc(G)}$, by orthogonality, and since we have five of them in a 5-dimensional space there can be no more.

There are other ways to obtain the final character. Perhaps the easiest is to construct the bottom row as the tensor product $\chi_{3b} = \chi_{3a} \otimes \chi_{\text{sign}}$. Other approaches use orthogonality. Having computed four of the five rows, the fifth is determined by the fact that it is orthogonal to the other four and the fact, to be seen in Corollary 3.3.7, that the sum of the squares of the degrees of the characters equals 24. We may also use the column orthogonality relations, in the manner of Example 3.4.5.

Our next main goal is to prove that the character table of a finite group is square and to deduce the column orthogonality relations, which we do in Theorem 3.4.3 and Corollary 3.4.4. Before doing this, we show in the next two results how part of the column orthogonality relations may be derived in a direct way.

Consider the regular representation of G on $\mathbb{C}G$, and let $\chi_{\mathbb{C}G}$ denote the character of this representation.

Lemma 3.3.6.

$$\chi_{\mathbb{C}G}(g) = \begin{cases} |G| & \textit{if } g = 1 \\ 0 & \textit{otherwise.} \end{cases}$$

Proof. Each $g \in G$ acts by the permutation matrix corresponding to the permutation $h \mapsto gh$. Now $\chi_{\mathbb{C}G}(g)$ equals the number of 1s down the diagonal of this matrix, which equals $|\{h \in G \mid gh = h\}|$. □

We may deduce an alternative proof of Corollary 2.1.5 (in case $k = \mathbb{C}$), and also a way to do the computation of the final row of the character table once the others have been determined.

Corollary 3.3.7. *Let $\chi_1, \dots, \chi_r$ be the simple complex characters of G, with degrees $d_1, \dots, d_r$. Then $\langle \chi_{\mathbb{C}G}, \chi_i \rangle = d_i$, and hence*

(1) $\sum_{i=1}^r d_i^2 = |G|$, *and*
(2) $\sum_{i=1}^r d_i \chi_i(g) = 0$ *if* $g \neq 1$.

Proof. Direct evaluation gives

$$\langle \chi_{\mathbb{C}G}, \chi_i \rangle = \frac{1}{|G|} |G| \chi_i(1) = d_i$$

and hence $\chi_{\mathbb{C}G} = d_1\chi_1 + \cdots + d_r\chi_r$. Evaluating at 1 gives (1), and at $g \neq 1$ gives (2). □

3.4 The Number of Simple Characters

It is an immediate deduction from the fact that the rows of the character table are orthogonal that the number of simple complex characters of a group is at most the number of conjugacy classes of elements in the group. We shall now prove that there is always equality here. The proof follows a surprising approach in which examine the center of the group algebra. For any ring A, we denote by $Z(A)$ the *center* of A.

Lemma 3.4.1. *(1) For any commutative ring R, the center*

$$Z(M_n(R)) = \{\lambda I \mid \lambda \in R\} \cong R.$$

(2) The number of simple complex characters of G equals $\dim Z(\mathbb{C}G)$.

Proof. (1) Let E_{ij} denote the matrix which is 1 in place i, j and 0 elsewhere. If $X = (x_{ij})$ is any matrix then

$E_{ij}X$ = the matrix with row j of X moved to row i, 0 elsewhere,
XE_{ij} = the matrix with column i of X moved to column j, 0 elsewhere.

If $X \in Z(M_n(R))$ these two are equal, and we deduce that $x_{ii} = x_{jj}$ and all other entries in row j and column i are 0. Therefore, $X = x_{11}I$.

(2) In Theorem 2.1.3, we constructed an isomorphism

$$\mathbb{C}G \cong M_{n_1}(\mathbb{C}) \oplus \cdots \oplus M_{n_r}(\mathbb{C}),$$

where the matrix summands are in bijection with the isomorphism classes of simple modules. On taking centers, each matrix summand contributes 1 to $\dim Z(\mathbb{C}G)$. □

Lemma 3.4.2. *Let $x_1, \dots, x_t$ be representatives of the conjugacy classes of elements of G and let R be any ring. For each i, let $\overline{x}_i \in RG$ denote the sum of the elements in the same conjugacy class as x_i. Then $Z(RG)$ is free as an R-module, with basis $\overline{x}_1, \dots, \overline{x}_t$.*

Proof. We first show that $\overline{x}_i \in Z(RG)$. Write $\overline{x}_i = \sum_{y \sim x_i} y$, where $\sim$ denotes conjugacy. Then

$$g\overline{x}_i = \sum_{y \sim x_i} gy = \left(\sum_{y \sim x_i} gyg^{-1} \right) g = \overline{x}_i g$$

since as y runs through the elements of G conjugate to x_i, so does gyg^{-1}, and from this, it follows that $\overline{x}_i$ is central.

Next suppose $\sum_{g \in G} a_g g \in Z(RG)$. We show that if $g_1 \sim g_2$ then $a_{g_1} = a_{g_2}$. Suppose that $g_2 = hg_1h^{-1}$. The coefficient of g_2 in $h(\sum_{g \in G} a_g g)h^{-1}$ is a_{g_1} and in $(\sum_{g \in G} a_g g)$ is a_{g_2}. Since elements of G are independent in RG, these coefficients must be equal. From this, we see that every element of $Z(RG)$ can be expressed as an R-linear combination of the $\overline{x}_i$.

Finally, we observe that the $\overline{x}_i$ is independent over R, since each is a sum of group elements with support disjoint from the supports of the other $\overline{x}_j$. □

Theorem 3.4.3. *Let G be a finite group. The following three numbers are equal:*

- *the number of simple complex characters of G,*
- *the number of isomorphism classes of simple complex representations of G,*
- *the number of conjugacy classes of elements of G.*

The character table of G is square. The characters simple characters form an ortho-normal basis of the space $\mathbb{C}^{cc(G)}$ of class functions on G.

Proof. It follows from the definition of a simple character as the character of a simple representation and Corollary 3.3.3 that the first two numbers are equal.

In Lemma 3.4.1, we showed that the number of simple characters equals the dimension of the center $Z(\mathbb{C}G)$, and in Lemma 3.4.2, we showed that this is equal to the number of conjugacy classes of G. □

From the fact that the character table is square, we get orthogonality relations between its columns.

Corollary 3.4.4 (Column orthogonality relations). *Let X be the character table of G, regarded as a matrix, and let*

$$C = \begin{bmatrix} |C_G(x_1)| & 0 & \cdots & 0 \\ 0 & |C_G(x_2)| & & \\ \vdots & & \ddots & \vdots \\ 0 & & \cdots & |C_G(x_r)| \end{bmatrix}$$

where $x_1, \ldots, x_r$ are representatives of the conjugacy classes of elements of G. Then

$$X^T\overline{X} = C,$$

where the bar denotes complex conjugation.

Proof. The orthogonality relations between the rows may be written $\overline{X}C^{-1}X^T = I$. Since all these matrices are square with independent rows or columns, they are invertible, and in fact, $(\overline{X}C^{-1})^{-1} = X^T = C\overline{X}^{-1}$. Therefore, $X^T\overline{X} = C$. □

Another way to state the column orthogonality relations is

$$\sum_{i=1}^{r} \chi_i(g)\overline{\chi_i(h)} = \begin{cases} |C_G(g)| & \text{if } g \sim h \\ 0 & \text{if } g \not\sim h. \end{cases}$$

This means that the column of the character table indexed by $g \in G$ is orthogonal in the usual sense to all the other columns, except the column indexed by g^{-1} (because $\chi(g^{-1}) = \overline{\chi(g)}$), and the scalar product of those two columns is $|C_G(g)|$. The special case of this in which $g = 1$ has already been seen in Corollary 3.3.7.

Example 3.4.5. The column orthogonality relations are useful in finding the final row of a character table when all except one character have been computed. Suppose we have computed the following rows of a character table:

g $\lvert C_G(g)\rvert$	1 20	a 5	b 4	b^2 4	b^3 4
χ_0	1	1	1	1	1
χ_1	1	1	i	-1	$-i$
χ_2	1	1	-1	1	-1
χ_3	1	1	$-i$	-1	i

We see from this that the group has five conjugacy classes, so that one simple character is missing, and also that the group has order 20. By Corollary 3.3.7, which is also part of Corollary 3.4.4, the sum of the squares of the degrees of the characters is $|G|$:

$$20 = 1^2 + 1^2 + 1^2 + 1^2 + d^2,$$

so that the remaining character has degree $d = 4$. Each of columns 2–5 has scalar product 0 with column 1 in the table:

g	1	a	b	b^2	b^3
$\|C_G(g)\|$	20	5	4	4	4
χ_0	1	1	1	1	1
χ_1	1	1	i	-1	$-i$
χ_2	1	1	-1	1	-1
χ_3	1	1	$-i$	-1	i
χ_4	4	r	s	t	u

from which we immediately deduce $[r, s, t, u] = [-1, 0, 0, 0]$ and the last character is

χ_4	4	-1	0	0	0

That completes the calculation of the character table without knowing anything more about any group whose character table it might be. In fact, the following group has this character table:

$$G = C_5 \rtimes C_4 = \langle a, b \mid a^5 = b^4 = 1,\ bab^{-1} = a^2 \rangle.$$

We sketch the computation of the conjugacy classes and centralizers. Observe first that all four nonidentity powers of a are conjugate, but $C_G(a) \supseteq \langle a \rangle$ has order ≥ 5, so these powers of a are the conjugacy class of a. The subgroup $\langle a \rangle$ is normal in G with quotient group C_4 generated by the image of b. Since the four powers of b are not conjugate in the quotient, they are not conjugate in G either. Each is centralized by $\langle b \rangle$, but is centralized by only the identity from $\langle a \rangle$, so the centralizer of each nonidentity power of b is $\langle b \rangle$ of order 4, giving five conjugates of that element. This accounts for 20 elements of G, so we have a complete list of conjugacy class representatives.

The top four rows of the character table of G are the characters of the quotient group C_4, lifted to become characters of G via the quotient homomorphism. This computes the character table as far as it was presented at the start of this example.

In Exercise 16 at the end of this chapter, you are asked to show that this G is the only group that has this character table.

3.5 Algebraic Integers and Divisibility of Character Degrees

So far we have established that the degrees of the irreducible complex characters of G have the properties that their number equals the number of conjugacy classes of G, and the sum of their squares is $|G|$. We now establish a further important property, which is that the character degrees all divide $|G|$—a big restriction on the possible degrees that may occur. It is proved using properties of algebraic integers, which we now develop. They will be used again in the proof of Burnside's Theorem 3.7.1, but not elsewhere in this text.

Suppose that S is a commutative ring with 1 and R is a subring of S with the same 1. An element $s \in S$ is said to be *integral* over R if $f(s) = 0$ for some *monic* polynomial $f \in R[X]$, that is, a polynomial in which the coefficient of the highest power of X is 1. We say that the ring S is integral over R if every element of S is integral over R. An element of $\mathbb{C}$ integral over $\mathbb{Z}$ is called an *algebraic integer*. We summarize the properties of integrality that we will need.

Theorem 3.5.1. *Let R be a subring of a commutative ring S.*

(1) The following are equivalent for an element $s \in S$:
  (a) s is integral over R,
  (b) $R[s]$ is contained in some R-submodule M of S, which is finitely generated over R and such that $sM \subseteq M$.
(2) The elements of S integral over R form a subring of S.
(3) $\{x \in \mathbb{Q} \mid x \text{ is integral over } \mathbb{Z}\} = \mathbb{Z}$.
(4) Let g be any element of a finite group G and χ any character of G. Then $\chi(g)$ is an algebraic integer.

In this theorem, $R[s]$ denotes the subring of S generated by R and s.

Proof. (1) (a) $\Rightarrow$ (b). Suppose s is an element integral over R, satisfying the equation

$$s^n + a_{n-1}s^{n-1} + \cdots + a_1 s + a_0 = 0,$$

where $a_i \in R$. Then $R[s]$ is generated as an R-module by $1, s, s^2, \ldots, s^{n-1}$. This is because the R-span of these elements is also closed under multiplication by s, using the fact that

$$s \cdot s^{n-1} = -a_{n-1}s^{n-1} - \cdots - a_1 s - a_0,$$

and hence equals $R[s]$. We may take $M = R[s]$.

(b) $\Rightarrow$ (a). Suppose $R[s] \subseteq M = Rx_1 + \cdots Rx_n$ with $sM \subseteq M$. Thus, for each i, we have $sx_i = \sum_{j=1}^{n} \lambda_{ij}x_j$ for certain $\lambda_{ij} \in R$. Consider the $n \times n$ matrix $A = sI - (\lambda_{ij})$ with entries in S, where I is the identity matrix. We have $Ax = 0$

where x is the vector $(x_1, \ldots, x_n)^T$, and so $\mathrm{adj}(A)Ax = 0$ where $\mathrm{adj}(A)$ is the adjugate matrix of A satisfying $\mathrm{adj}(A)A = \det(A) \cdot I$. Hence $\det(A) \cdot x_i = 0$ for all i. Since $1 \in R \subseteq M$ is a linear combination of the x_i we have $\det(A) = 0$, and so s is a root of the monic polynomial $\det(X \cdot I - (\lambda_{ij}))$.

(2) We show that if $a, b \in S$ are integral over R then $a + b$ and ab are also integral over R. These lie in $R[a, b]$, and we show that this is finitely generated as an R-module. We see from the proof of part (1) that each of $R[a]$ and $R[b]$ is finitely generated as an R-module. If $R[a]$ is generated by $x_1, \ldots, x_m$ and $R[b]$ is generated by $y_1, \ldots, y_n$, then $R[a, b]$ is evidently generated as an R-module by all the products $x_i y_j$. Now $R[a, b]$ also satisfies the remaining condition of part (b) of (1), and we deduce that $a + b$ and ab are integral over R.

(3) Suppose that $\frac{a}{b}$ is integral over $\mathbb{Z}$, where a, b are coprime integers. Then

$$\left(\frac{a}{b}\right)^n + c_{n-1}\left(\frac{a}{b}\right)^{n-1} + \cdots + c_1\frac{a}{b} + c_0 = 0$$

for certain integers c_i, and so

$$a^n + c_{n-1}a^{n-1}b + \cdots + c_1 ab^{n-1} + c_0 b^n = 0.$$

Since b divides all terms in this equation except perhaps the term a^n, b must also be a factor of a^n. Since a and b are coprime, this is only possible if $b = \pm 1$, and we deduce that $\frac{a}{b} \in \mathbb{Z}$.

(4) $\chi(g)$ is the sum of the eigenvalues of g in its action on the representation which affords χ. Since $g^n = 1$ for some n these eigenvalues are all roots of $X^n - 1$ and so are integers. □

In the next result, we identify $\mathbb{Z}$ with the subring $\mathbb{Z} \cdot 1$, which is contained in the center $Z(\mathbb{Z}G)$ of $\mathbb{Z}G$.

Proposition 3.5.2. *The center $Z(\mathbb{Z}G)$ is integral over $\mathbb{Z}$. Hence if $x_1, \ldots, x_r$ are representatives of the conjugacy classes of G, $\overline{x}_i \in \mathbb{Z}G$ is the sum of the elements conjugate to x_i, and $\lambda_1, \ldots, \lambda_r \in \mathbb{C}$ are algebraic integers then the element $\sum_{i=1}^r \lambda_i \overline{x}_i \in Z(\mathbb{C}G)$ is integral over $\mathbb{Z}$.*

Proof. It is the case that every commutative subring of $\mathbb{Z}G$ is integral over $\mathbb{Z}$, using condition 1(b) of Theorem 3.5.1, since such a subring is in particular a subgroup of the finitely generated free abelian group $\mathbb{Z}G$, and hence is finitely generated as a $\mathbb{Z}$-module.

We have seen in Lemma 3.4.2 that the elements $\overline{x}_1, \ldots, \overline{x}_r$ lie in $Z(\mathbb{Z}G)$, so they are integral over $\mathbb{Z}$, and by part (2) of Theorem 3.5.1, the linear combination $\sum_{i=1}^r \lambda_i \overline{x}_i$ is integral also. (We note that the $\overline{x}_i$ are in fact a finite set of generators for $Z(\mathbb{Z}G)$ as an abelian group, but we did not need to know this for the proof.) □

Let $\rho_1, \dots, \rho_r$ be the simple representations of G over $\mathbb{C}$ with degrees $d_1, \dots, d_r$ and characters $\chi_1, \dots, \chi_r$. Then each $\rho_i : G \to M_{d_i}(\mathbb{C})$ extends by linearity to a $\mathbb{C}$-algebra homomorphism

$$\rho_i : \mathbb{C}G = \bigoplus_{i=1}^{r} M_{d_i}(\mathbb{C}) \to M_{d_i}(\mathbb{C})$$

projecting onto the ith matrix summand. The fact that the group homomorphism ρ_i extends to an algebra homomorphism in this way comes formally from the construction of the group algebra. The fact that this algebra homomorphism is projection onto the ith summand arises from the way we decomposed $\mathbb{C}G$ as a sum of matrix algebras, in which each matrix summand acts on the corresponding simple module as matrices on the space of column vectors.

Proposition 3.5.3. *Fixing the suffix i, if $x \in Z(\mathbb{C}G)$ then $\rho_i(x) = \lambda I$ for some $\lambda \in \mathbb{C}$. In fact*

$$\rho_i(x) = \frac{1}{d_i} \cdot \mathrm{tr}(\rho_i(x)) \cdot I,$$

where tr *is the trace and I is the identity matrix. Writing $x = \sum_{g \in G} a_g g$, we have*

$$\rho_i(x) = \frac{1}{d_i} \sum_{g \in G} a_g \chi_i(g) \cdot I.$$

Proof. Since x is central the matrix $\rho_i(x)$ commutes with the matrices $\rho_i(g)$ for all $g \in G$. Therefore by Schur's Lemma 2.1.1, since ρ_i is a simple complex representation, $\rho_i(x) = \lambda I$ for some scalar λ that we now compute. Evidently, $\lambda = \frac{1}{d_i}\mathrm{tr}(\lambda I)$. Substituting $\rho_i(x)$ into the right-hand side and multiplying both sides by I gives the first expression for $\rho_i(x)$. Replacing x by the expression in the statement of the proposition, we obtain

$$\begin{aligned} \lambda &= \frac{1}{d_i} \mathrm{tr}\left(\rho_i\left(\sum_{g \in G} a_g g\right)\right) \\ &= \frac{1}{d_i} \sum_{g \in G} a_g \mathrm{tr}(\rho_i(g)) \\ &= \frac{1}{d_i} \sum_{g \in G} a_g \chi_i(g), \end{aligned}$$

which gives the last expression for $\rho_i(x)$. □

Theorem 3.5.4. *The degrees d_i of the simple complex representations of G all divide $|G|$.*

Proof. Let $x = \sum_{g \in G} \chi_i(g^{-1})g$. This element is central in $\mathbb{C}G$ by Lemma 3.4.2, since the coefficients of group elements are constant on conjugacy classes. By Proposition 3.5.3,

$$\begin{aligned}\rho_i(x) &= \frac{1}{d_i} \sum_{g \in G} \chi_i(g^{-1})\chi_i(g) \cdot I \\ &= \frac{|G|}{d_i} \cdot I,\end{aligned}$$

the second equality arising from the fact that $\langle \chi_i, \chi_i \rangle = 1$. Now x is integral over $\mathbb{Z} \cdot 1$ by Proposition 3.5.2 since the coefficients $\chi_i(g^{-1})$ are algebraic integers, so $\rho_i(x)$ is integral over $\rho_i(\mathbb{Z} \cdot 1) = \mathbb{Z} \cdot I$. Thus, $\frac{|G|}{d_i}$ is integral over $\mathbb{Z}$ and hence $\frac{|G|}{d_i} \in \mathbb{Z}$. We deduce that $d_i \bigm| |G|$. □

3.6 The Matrix Summands of the Complex Group Algebra

Given a set of rings with identity $A_1, \ldots, A_r$, we may form their direct sum $A = A_1 \oplus \cdots \oplus A_r$, and this itself becomes a ring with componentwise addition and multiplication. In this situation, each ring A_i may be identified as the subset of A consisting of elements that are zero except in component i, but this subset is not a subring of A because it does not contain the identity element of A. It is, however, a 2-sided ideal. Equally, in any decomposition of a ring A as a direct sum of 2-sided ideals, these ideals have the structure of rings with identity.

Decompositions of a ring as direct sums of other rings are closely related to idempotent elements in the center of the ring. We have seen idempotent elements introduced in Exercise 9 from Chapter 2, and they will retain importance throughout this book. An element e of a ring A is said to be *idempotent* if $e^2 = e$. It is a *central idempotent element* if it lies in the center $Z(A)$. Two idempotent elements e and f are *orthogonal* if $ef = fe = 0$. An idempotent element e is called *primitive* if whenever $e = e_1 + e_2$ where e_1 and e_2 are orthogonal idempotent elements then either $e_1 = 0$ or $e_2 = 0$. We say that e is a *primitive central idempotent element* if it is primitive as an idempotent element in $Z(A)$, that is, e is central and has no proper decomposition as a sum of orthogonal central idempotent elements.

We comment that the term "idempotent element" is very often abbreviated to "idempotent," thereby elevating the adjective to the status of a noun. This does have justification, in that "idempotent element" is unwieldy, and we will usually conform to the shorter usage.

Proposition 3.6.1. *Let A be a ring with identity. Decompositions*

$$A = A_1 \oplus \cdots \oplus A_r$$

as direct sums of 2-sided ideals A_i biject with expressions

$$1 = e_1 + \cdots + e_r$$

as a sum of orthogonal central idempotent elements, in such a way that e_i is the identity element of A_i and $A_i = Ae_i$. The A_i are indecomposable as rings if and only if the e_i are primitive central idempotent elements. If every A_i is indecomposable as a ring then the subsets A_i and also the primitive central idempotents e_i are uniquely determined; furthermore, every central idempotent can be written as a sum of certain of the e_i.

Proof. Given any ring decomposition $A = A_1 \oplus \cdots \oplus A_r$, we may write $1 = e_1 + \cdots + e_r$ where $e_i \in A_i$, and it is clear that the e_i are orthogonal central idempotent elements. Conversely, given an expression $1 = e_1 + \cdots + e_r$ where the e_i are orthogonal central idempotent elements, we obtain a decomposition $A = Ae_1 \oplus \cdots \oplus Ae_r$ as rings.

To say that the ring A_i is indecomposable means that it cannot be expressed as a direct sum of rings, except in the trivial way, and evidently this happens precisely if the corresponding idempotent element cannot be decomposed as a sum of orthogonal central idempotent elements.

We now demonstrate the (perhaps surprising) fact that there is at most one decomposition of A as a sum of indecomposable rings. Suppose we have two such decompositions, and that the corresponding primitive central idempotent elements are labelled e_i and f_j, so that

$$1 = e_1 + \cdots + e_r = f_1 + \cdots + f_s.$$

We have

$$e_i = e_i \cdot 1 = \sum_{j=1}^{s} e_i f_j,$$

and so $e_i = e_i f_j$ for some unique j and $e_i f_k = 0$ if $k \neq j$, by primitivity of e_i. Also

$$f_j = 1 \cdot f_j = \sum_{k=1}^{r} e_k f_j$$

so that $e_k f_j \neq 0$ for some unique k. Since $e_i f_j \neq 0$, we have $k = i$ and $e_i f_j = f_j$. Thus, $e_i = f_j$. We proceed by induction on r, starting at $r = 1$. If $r > 1$, we now work with the ring $A \cdot \sum_{k \neq i} e_k = A \cdot \sum_{k \neq j} f_k$ in which the identity is expressible as sums of primitive central idempotent elements $\sum_{k \neq i} e_k = \sum_{k \neq j} f_k$. The first of these expressions has $r - 1$ terms, so by induction the e_ks are the same as the f_ks after some permutation.

If e is any central idempotent and the e_i are primitive then ee_i is either e_i or 0 since $e = ee_i + e(1 - e_i)$ is a sum of orthogonal central idempotents. Thus,

$$e = e \sum_{k=1}^{r} e_i = \sum_{k=1}^{r} ee_i$$

is a sum of certain of the e_i. □

Notice that it follows from Proposition 3.6.1 that distinct primitive central idempotents must necessarily be orthogonal, a conclusion that is false without the word "central." The primitive central idempotents, and also the indecomposable ring summands to which they correspond according to Proposition 3.6.1, are known as *blocks*. We will study blocks in detail in Chapter 12, and also at the end of Chapter 9. In the case of the complex group algebra $\mathbb{C}G$, that is a direct sum of matrix rings, the block idempotents are the elements that are the identity in one matrix summand and zero in the others. We shall now give a formula for these block idempotents. It is the same (up to a scalar) as the formula used in the proof of Theorem 3.5.4.

Theorem 3.6.2. *Let $\chi_1, \ldots, \chi_r$ be the simple complex characters of G with degrees $d_1, \ldots, d_r$. The primitive central idempotent elements in $\mathbb{C}G$ are the elements*

$$\frac{d_i}{|G|} \sum_{g \in G} \chi_i(g^{-1})g$$

where $1 \leq i \leq r$, the corresponding indecomposable ring summand of $\mathbb{C}G$ having a simple representation that affords the character χ_i.

Proof. Using the notation of Proposition 3.5.3, we have that the representation ρ_i which affords χ_i yields an algebra map $\rho_i : \mathbb{C}G \to M_{d_i}(\mathbb{C})$ that is projection onto the ith matrix summand in a decomposition of $\mathbb{C}G$ as a sum of matrix rings. For any field k, the matrix ring $M_n(k)$ is indecomposable, since we have seen in Lemma 3.4.1 that $Z(M_n(k)) \cong k$ and the only nonzero idempotent element in a field is 1. Thus, the decomposition of $\mathbb{C}G$ as a direct sum of matrix rings is the unique decomposition of $\mathbb{C}G$ as a sum of indecomposable ring summands. The corresponding primitive central idempotent elements are the identity matrices in the various summands, and so they are the elements $e_i \in \mathbb{C}G$ such that $\rho_i(e_i) = I$ and $\rho_j(e_i) = 0$ if $i \neq j$. From the formula of Proposition 3.5.3 and the orthogonality relations, we have (using the Kronecker δ)

$$\rho_j\left(\frac{d_i}{|G|}\sum_{g\in G}\chi_i(g^{-1})g\right) = \frac{d_i}{|G|d_j}\sum_{g\in G}\chi_i(g^{-1})\chi_j(g)\cdot I$$
$$= \frac{d_i}{d_j}\langle\chi_i,\chi_j\rangle\cdot I$$
$$= \frac{d_i}{d_j}\delta_{i,j}\cdot I$$
$$= \delta_{i,j}\cdot I,$$

so that the elements specified in the statement of the theorem do indeed project correctly onto the identity matrices, and are therefore the primitive central idempotent elements. □

While the identity matrix is a primitive *central* idempotent element in the matrix ring $M_n(k)$, where k is a field, it is never a primitive idempotent element if $n > 1$ since it is the sum of the orthogonal (noncentral) primitive idempotent elements $I = E_{1,1} + \cdots + E_{n,n}$. Furthermore, removing the hypothesis of centrality we can no longer say that decompositions of the identity as a sum of primitive idempotent elements are unique; indeed, any conjugate expression by an invertible matrix will also be a sum of orthogonal primitive idempotent elements. Applying these comments to a matrix summand of $\mathbb{C}G$, the primitive idempotent decompositions of 1 will never be unique if we have a non-abelian matrix summand—which, of course, happens precisely when G is non-abelian. It is unfortunately the case that in terms of the group elements there is in general no known formula for primitive idempotent elements of $\mathbb{C}G$ lying in a non-abelian matrix summand.

3.7 Burnside's $p^a q^b$ Theorem

We conclude this chapter with Burnside's remarkable "$p^a q^b$ theorem," which establishes a group-theoretic result using the ideas of representation theory we have so far developed, together with some admirable ingenuity. In the course of the proof, we again make use of the idea of integrality, but this time we also require Galois theory at one point. This is needed to show that if ζ is a field element that is expressible as a sum of roots of unity, then every algebraic conjugate of ζ is again expressible as a sum of roots of unity. We present Burnside's Theorem here because of its importance as a theorem in its own right, not because anything later depends on it. In view of this, the proof (which is fairly long) can be omitted without subsequent loss of understanding.

Recall that a group G is *solvable* if it has a composition series in which all of the composition factors are cyclic. Thus, a group is not solvable precisely if it has a non-abelian composition factor.

Theorem 3.7.1 (Burnside's $p^a q^b$ Theorem). *Let G be a group of order $p^a q^b$ where p and q are primes. Then G is solvable.*

Proof. We suppose the result is false, and consider a group G of minimal order subject to being not solvable and of order $p^a q^b$.

Step 1. The group G is simple, not abelian, and not of prime-power order; for if it were abelian or of prime-power order it would be solvable, and if G had a normal subgroup N then one of N and G/N would be a smaller group of order $p^\alpha q^\beta$ which was not solvable.

Step 2. We show that G contains an element g whose conjugacy class has size q^d for some $d > 0$. Let P be a Sylow p-subgroup, $1 \neq g \in Z(P)$. Then $C_G(g) \supseteq P$ so $|G : C_G(g)| = q^d$ for some $d > 0$, and this is the number of conjugates of g.

Step 3. We show that there is a simple nonidentity character χ of G such that $q \nmid \chi(1)$ and $\chi(g) \neq 0$. To prove this, suppose to the contrary that whenever $\chi \neq 1$ and $q \nmid \chi(1)$ then $\chi(g) = 0$. Let R denote the ring of algebraic integers in $\mathbb{C}$. Consider the orthogonality relation between the column of 1 (consisting of character degrees) and the column of g:

$$1 + \sum_{\chi \neq 1} \chi(1)\chi(g) = 0.$$

Then q divides every term apart from 1 in the sum on the left, and so $1 \in qR$. Thus, $q^{-1} \in R$. But $q^{-1} \in \mathbb{Q}$ and so $q^{-1} \in \mathbb{Z}$ by Theorem 3.5.1, a contradiction. We now fix a nonidentity character χ for which $q \nmid \chi(1)$ and $\chi(g) \neq 0$.

Step 4. Recall that the number of conjugates of g is q^d. We show that

$$\frac{q^d \chi(g)}{\chi(1)}$$

is an algebraic integer. To do this, we use Lemma 3.4.2 and Proposition 3.5.3. These imply that if $\overline{g} = \sum_{h \sim g} h \in \mathbb{C}G$ is the sum of the elements conjugate to g and ρ is a representation affording the character χ then $\overline{g} \in Z(\mathbb{C}G)$ and

$$\begin{aligned}\rho(\overline{g}) &= \frac{1}{\chi(1)} \sum_{h \sim g} \chi(h) \cdot I \\ &= \frac{q^d \chi(g)}{\chi(1)} \cdot I,\end{aligned}$$

where I is the identity matrix. Now by Proposition 3.5.2 this is integral over $\rho(\mathbb{Z}) = \mathbb{Z} \cdot I$, which proves what we want.

Step 5. We deduce that $\frac{\chi(g)}{\chi(1)}$ is an algebraic integer. This arises from the fact that $q \nmid \chi(1)$. We can find $\lambda, \mu \in \mathbb{Z}$ so that $\lambda q^d + \mu\chi(1) = 1$. Now

$$\frac{\chi(g)}{\chi(1)} = \lambda \frac{q^d \chi(g)}{\chi(1)} + \mu\chi(g)$$

is a sum of algebraic integers.

Step 6. We show that $|\chi(g)| = \chi(1)$ and put $\zeta = \chi(g)/\chi(1)$. We consider the algebraic conjugates of ζ, which are the roots of the minimal polynomial of ζ over $\mathbb{Q}$. They are all algebraic integers, since ζ and its algebraic conjugates are all roots of the same polynomials over $\mathbb{Q}$. Thus, the product $N(\zeta)$ of the algebraic conjugates is an algebraic integer. Since it is also $\pm$ the constant term of the minimal polynomial of ζ, it is rational and nonzero. Therefore, $0 \neq N(\zeta) \in \mathbb{Z}$ by Theorem 3.5.1.

Now $\chi(g)$ is the sum of the eigenvalues of $\chi(g)$, of which there are $\chi(1)$, each of which is a root of unity. Hence by the triangle inequality, $|\chi(g)| \leq \chi(1)$. By Galois theory, the same is true and a similar inequality holds for each algebraic conjugate of $\chi(g)$. We conclude that all algebraic conjugates of ζ have absolute value at most 1. Therefore, $|N(\zeta)| \leq 1$. The only possibility is $|N(\zeta)| = 1$ and $|\zeta| = 1$, so that $|\chi(g)| = \chi(1)$, as was to be shown.

Step 7. We will obtain a contradiction by considering the subgroup

$$H = \{h \in G \mid |\chi(h)| = \chi(1)\}$$

where χ is the simple nonidentity character introduced in Step 3. We argue first that H is a normal subgroup. If the eigenvalues of $\rho(h)$ are $\lambda_1, \ldots, \lambda_n$ then, since these are roots of unity, $|\lambda_1 + \cdots + \lambda_n| = n$ if and only if $\lambda_1 = \cdots = \lambda_n$. Thus, $|\chi(h)| = \chi(1)$ if and only if $\rho(h)$ is multiplication by some scalar, and from this, we see immediately that H is a normal subgroup. It also implies that $H/\operatorname{Ker}\rho$ is abelian. From Step 6, we see that H contains the nonidentity element g, so simplicity of G forces $H = G$. Since ρ is not the trivial representation $\operatorname{Ker}\rho \neq G$, so simplicity of G again forces $\operatorname{Ker}\rho = 1$, so that G must be abelian. However, G was seen not to be abelian in Step 1, and this contradiction completes the proof. □

3.8 Summary of Chapter 3

- There is a tensor product operation on RG-modules whose result is an RG-module.
- Characters are class functions.

- The character of a direct sum, tensor product, or dual is the sum, product, or complex conjugate of the characters.
- The characters of the simple representations form an orthonormal basis for class functions with respect to a certain bilinear form.
- The character table is square and satisfies row and column orthogonality relations. The number of rows of the table equals the number of conjugacy classes in the group.
- The conjugacy class sums form a basis for the center of RG.
- The simple character degrees are divisors of $|G|$. The sum of their squares equals $|G|$.
- There is a formula for the primitive central idempotents in $\mathbb{C}G$.
- Every group of order $p^a q^b$ is solvable.

3.9 Exercises for Chapter 3

We assume throughout these exercises that representations are finite-dimensional.

1. (a) By using characters, show that if V and W are $\mathbb{C}G$-modules then we have isomorphisms $(V \otimes_{\mathbb{C}} W)^* \cong V^* \otimes_{\mathbb{C}} W^*$, and $({}_{\mathbb{C}G}\mathbb{C}G)^* \cong {}_{\mathbb{C}G}\mathbb{C}G$ as $\mathbb{C}G$-modules.

(b) If k is any field and V, W are kG-modules, show that there are isomorphisms $(V \otimes_k W)^* \cong V^* \otimes_k W^*$, and $({}_{kG}kG)^* \cong {}_{kG}kG$ as kG-modules.

2. Consider a ring with identity that is the direct sum (as a ring) of nonzero subrings $A = A_1 \oplus \cdots \oplus A_r$. Suppose that A has exactly n isomorphism types of simple modules. Show that $r \leq n$.

3. Let g be any nonidentity element of a group G. Show that G has a simple complex character χ for which $\chi(g)$ has negative real part.

4. Suppose that V is a representation of G over $\mathbb{C}$ for which $\chi_V(g) = 0$ if $g \neq 1$. Show that $\dim V$ is a multiple of $|G|$. Deduce that $V \cong \mathbb{C}G^n$ for some n. Show that if W is any representation of G over $\mathbb{C}$ then $\mathbb{C}G \otimes_{\mathbb{C}} W \cong \mathbb{C}G^{\dim W}$ as $\mathbb{C}G$-modules.

5. Show that if every element of a finite group G is conjugate to its inverse, then every character of G is real-valued.

Conversely, show that if every character of G is real-valued, then every element of G is conjugate to its inverse.

[The quaternion group of order 8 in its action on the algebra of quaternions provides an example of a complex representation that is not equivalent to a real representation, but whose character is real-valued (see Chapter 2, Exercise 12).

In this example, the representation has complex dimension 2, but there is no basis over $\mathbb{C}$ for the representation space such that the group acts by matrices with real entries. A real-valued character does not necessarily come from a real representation.]

6. (a) Let A be a finite-dimensional semisimple algebra over a field. Show that the center $Z(A)$ is a semisimple algebra.

(b) Let G be a finite group and k a field in which $|G|$ is invertible. Show that the number of simple representations of kG is at most the number of conjugacy classes of G.

7. Let G permute a set Ω and let $R\Omega$ denote the permutation representation of G over R determined by Ω. This means $R\Omega$ has a basis in bijection with Ω and each element $g \in G$ acts on $R\Omega$ by permuting the basis elements in the same way that g permutes Ω.

(a) Show that when H is a subgroup of G and $\Omega = G/H$ is the set of left cosets of H in G, the kernel of G in its action on $R\Omega$ is H if and only if H is normal in G.

(b) Show that the normal subgroups of G are precisely the subgroups of the form $\operatorname{Ker} \chi_{i_1} \cap \cdots \cap \operatorname{Ker} \chi_{i_t}$ where $\chi_1, \ldots, \chi_n$ are the simple characters of G. Use Proposition 3.1.1 to deduce that the normal subgroups of G are determined by the character table of G.

(c) Show that G is a simple group if and only if for every nontrivial simple character χ and for every nonidentity element $g \in G$, we have $\chi(g) \neq \chi(1)$.

8. (Exercise due to Jozsef Pelikan) While walking down the street you find a scrap of paper with the following character table on it:

$\cdots$	1	$\cdots$	1	$\cdots$
	1		-1	
	2		-1	
	3		1	
	3		-1	

All except two of the columns are obscured, and while it is clear that there are five rows you cannot read anything of the other columns, including their position. Prove that there is an error in the table. Given that there is exactly *one* error, determine where it is, and what the correct entry should be.

9. A finite group has seven conjugacy classes with representatives $c_1, \ldots, c_7$ (where $c_1 = 1$), and the values of five of its irreducible characters are given by the following table:

c_1	c_2	c_3	c_4	c_5	c_6	c_7
1	1	1	1	1	1	1
1	1	1	1	-1	-1	-1
4	1	-1	0	2	-1	0
4	1	-1	0	-2	1	0
5	-1	0	1	1	1	-1

Calculate the numbers of elements in the various conjugacy classes and the remaining simple characters.

10. Let $g \in G$.

(a) Use Proposition 3.1.1 to prove that g lies in the center of G if and only if $|\chi(g)| = |\chi(1)|$ for every simple complex character χ of G.
(b) Show that if G has a faithful simple complex character (one whose kernel is 1) then the center of G is cyclic.

11. Here is a column of a character table:

g
1
-1
0
-1
-1
$\frac{-1+i\sqrt{11}}{2}$
$\frac{-1-i\sqrt{11}}{2}$
0
1
0

(a) Find the order of g.
(b) Prove that $g \notin Z(G)$.
(c) Show that there exists an element $h \in G$ with the same order as g but not conjugate to g.
(d) Show that there exist two distinct simple characters of G of the same degree.

12. Let A be a semisimple finite-dimensional algebra over a field and let $1 = e_1 + \cdots + e_n$ be a sum of primitive central idempotents in A.

(a) If $f \in A$ is a primitive idempotent (not necessarily central), show that there is a unique i so that $e_i f \neq 0$, and that for this i we have $e_i f = f$.
(b) Show that for all A-modules V and for all i, $e_i V$ is an A-submodule of V and $V = \bigoplus_{i=1}^{n} e_i V$.

(c) Show that if S is a simple A-module there is a unique i so that $e_i S \neq 0$, and that for this i we have $e_i S = S$.
(d) Show that for each i there is a simple module S_i so that $e_i S \neq 0$, and that if A is assumed to be semisimple then S_i is unique up to isomorphism.
(e) Assuming that A is semisimple and $f \in A$ is a primitive idempotent (not necessarily central), deduce that there is a simple module S which is unique up to isomorphism, such that $fS \neq 0$.

13. (a) Let $1 = e_1 + \cdots + e_n$ be a sum of primitive central idempotents in $\mathbb{C}G$ and let V be a $\mathbb{C}G$-module. Show that $e_i V$ is the largest submodule of V that is a direct sum of copies of the simple module S_i identified by $e_i S_i \neq 0$.
(b) Let V be any representation of S_3 over $\mathbb{C}$. Show that the subset

$$V_2 = \{2v - (1, 2, 3)v - (1, 3, 2)v \mid v \in V\}$$

of V is the unique largest $\mathbb{C}S_3$-submodule of V that is a direct sum of copies of the simple 2-dimensional $\mathbb{C}S_3$-module.

14. Let $G = \langle x \rangle$ be cyclic of order n.

(a) Write down a complete set of primitive (central) idempotents in $\mathbb{C}G$.
(b) Let $\mathbb{C}^n$ be an n-dimensional space with basis $v_1, \ldots, v_n$. Let $g : \mathbb{C}^n \to \mathbb{C}^n$ be the linear map of order n specified by $gv_i = v_{i+1}$, $1 \leq i \leq n-1$, $gv_n = v_1$, so that g has matrix

$$T = \begin{bmatrix} 0 & \cdots & 0 & 1 \\ 1 & \ddots & & 0 \\ \vdots & & \ddots & \vdots \\ 0 & \cdots & 1 & 0 \end{bmatrix}.$$

Let $\zeta = e^{2\pi i/n}$ be a primitive nth root of unity. Show that for each d the vector

$$e_d = \sum_{s=1}^{n} \zeta^{-ds} v_s$$

is an eigenvector of T, and that $e_1, \ldots, e_n$ is a basis of $\mathbb{C}^n$.

15. Let x be an element of G.

(a) If χ is a character of G, show that the function $g \mapsto \chi(xg)$ need not be a class function on G.
(b) Show that the fact that the elements of $\mathbb{C}G$ specified in Theorem 3.6.2 are orthogonal idempotents is equivalent to the validity of the following

formulas, for all $x \in G$ and for all of the simple characters χ_i of G:

$$\frac{d_i}{|G|} \sum_{g \in G} \chi_i(g^{-1})\chi_j(xg) = \begin{cases} 0 & \text{if } i \neq j \\ \chi_i(x) & \text{if } i = j. \end{cases}$$

16. Show that the only group G which has character table

g $\lvert C_G(g)\rvert$	1 20	a 5	b 4	b^2 4	b^3 4
χ_0	1	1	1	1	1
χ_1	1	1	i	-1	$-i$
χ_2	1	1	-1	1	-1
χ_3	1	1	$-i$	-1	i
χ_4	4	-1	0	0	0

is

$$G = C_5 \rtimes C_4 = \langle a, b \mid a^5 = b^4 = 1,\ bab^{-1} = a^2 \rangle.$$

4
The Construction of Modules and Characters

In this chapter, we describe the most important methods for constructing representations and character tables of groups. We describe the characters of cyclic groups, and show also that simple characters of direct products are products of characters of the factors. This allows us to construct the simple characters of abelian groups and, in fact, all degree 1 characters of any finite group. We go on to describe the construction of representations by induction from subgroups, that includes the special case of permutation representations. Having obtained new characters of a group in these ways, we break them apart using orthogonality relations so as to obtain characters of smaller representations. An important tool in the process is Frobenius reciprocity. We conclude the section with a description of symmetric and exterior powers.

4.1 Cyclic Groups and Direct Products

We start with the particular case of complex characters of cyclic groups, noting that some properties of simple representations of cyclic groups over more general fields were already explored in Exercise 11 from Chapter 2. For representations of cyclic groups over fields where the characteristic divides the order of the group, see Proposition 6.1.1, Theorem 6.1.2, Example 8.2.1, and Corollary 11.2.2.

Proposition 4.1.1. *Let $G = \langle x \mid x^n = 1\rangle$ be a cyclic group of order n, and let $\zeta_n \in \mathbb{C}$ be a primitive nth root of unity. Then the simple complex characters of G are the n functions*

$$\chi_r(x^s) = \zeta_n^{rs}$$

where $0 \le r \le n-1$.

Proof. We merely observe that the mapping

$$x^s \mapsto \zeta_n^{rs}$$

is a group homomorphism

$$G \to GL(1, \mathbb{C}) = \mathbb{C}^*$$

giving a 1-dimensional representation with character χ_r, that must necessarily be simple. These characters are all distinct, and since the number of them equals the group order we have them all. □

We next show how to obtain the simple characters of a product of groups in terms of the characters of the groups in the product. Combining this with the last result, we obtain the character table of any finite abelian group. We describe a construction that works over any ring R. Suppose that $\rho_1 : G_1 \to GL(V_1)$ and $\rho_2 : G_2 \to GL(V_2)$ are representations of groups G_1 and G_2. We may define an action of $G_1 \times G_2$ on $V_1 \otimes_R V_2$ by the formula

$$(g_1, g_2)(v_1 \otimes v_2) = g_1 v_1 \otimes g_2 v_2,$$

where $g_i \in G_i$ and $v_i \in V_i$. When R is a field we may choose bases for V_1 and V_2, and now (g_1, g_2) acts via the tensor product of the matrices by which g_1 and g_2 act. It follows when $R = \mathbb{C}$ that

$$\chi_{V_1 \otimes V_2}(g_1, g_2) = \chi_{V_1}(g_1)\chi_{V_2}(g_2).$$

Theorem 4.1.2. *Let $V_1, \ldots, V_m$ and $W_1, \ldots, W_n$ be complete lists of the simple complex representations of groups G_1 and G_2. Then the representations $V_i \otimes W_j$ with $1 \leq i \leq m$ and $1 \leq j \leq n$ form a complete list of the simple complex $G_1 \times G_2$ representations.*

Theorem 4.1.2 is false in general when the field over which we are working is not algebraically closed (see Exercise 10 at the end of this chapter). The theorem is an instance of a more general fact to do with representations of finite-dimensional algebras A and B over an algebraically closed field k: the simple representations of $A \otimes_k B$ are precisely the modules $S \otimes_k T$, where S is a simple A-module and T is a simple B-module. This is proved in [10, Theorem 10.38]. The connection between this result for abstract algebras and group representations is that the group algebra $R[G_1 \times G_2]$ over any commutative ring R is isomorphic to $RG_1 \otimes_R RG_2$, which we may see by observing that this tensor product has a basis consisting of elements $g_1 \otimes g_2$ with $g_i \in G_i$ that multiply in the same way as the elements of the group $G_1 \times G_2$.

Proof. We first verify that the representations $V_i \otimes W_j$ are simple using the criterion of 3.10:

$$\begin{aligned}\langle \chi_{V_i\otimes W_j}, \chi_{V_i\otimes W_j}\rangle &= \frac{1}{|G_1 \times G_2|} \sum_{(g_1,g_2)\in G_1\times G_2} \overline{\chi_{V_i\otimes W_j}(g_1, g_2)}\chi_{V_i\otimes W_j}(g_1, g_2) \\ &= \frac{1}{|G_1||G_2|} \sum_{(g_1,g_2)\in G_1\times G_2} \overline{\chi_{V_i}(g_1)}\chi_{V_i}(g_1)\overline{\chi_{V_j}(g_2)}\chi_{V_j}(g_2) \\ &= \frac{1}{|G_1|}\sum_{g_1\in G_1} \overline{\chi_{V_i}(g_1)}\chi_{V_i}(g_1) \cdot \frac{1}{|G_2|}\sum_{g_2\in G_2} \overline{\chi_{V_j}(g_2)}\chi_{V_j}(g_2) \\ &= 1.\end{aligned}$$

The characters of these representations are distinct, since by a similar calculation if $(i, j) \neq (r, s)$ then $\langle \chi_{V_i\otimes W_j}, \chi_{V_r\otimes W_s}\rangle = 0$. To show that we have the complete list, we observe that if $\dim V_i = d_i$ and $\dim W_j = e_j$ then $V_i \otimes W_j$ is a representation of degree d_ie_j and

$$\sum_{i=1}^{m}\sum_{j=1}^{n}(d_ie_j)^2 = \sum_{i=1}^{m} d_i^2 \cdot \sum_{j=1}^{n} e_j^2 = |G_1||G_2|.$$

This establishes what we need, using 2.5 or 3.13. □

Example 4.1.3. Putting the last two results together enables us to compute the character table of any finite abelian group. To give a very small example, let

$$G = \langle x, y \mid x^2 = y^2 = [x, y] = 1\rangle \cong C_2 \times C_2.$$

The character tables of $\langle x\rangle$ and $\langle y\rangle$ are

$\langle x\rangle$

g	1	x
$\lvert C_G(g)\rvert$	2	2
χ_1	1	1
χ_2	1	−1

$\langle y\rangle$

g	1	y
$\lvert C_G(g)\rvert$	2	2
ψ_1	1	1
ψ_2	1	−1

and the character table of $C_2 \times C_2$ is

$\langle x\rangle \times \langle y\rangle$

g	1	x	y	xy
$\lvert C_G(g)\rvert$	4	4	4	4
$\chi_1\psi_1$	1	1	1	1
$\chi_2\psi_1$	1	−1	1	−1
$\chi_1\psi_2$	1	1	−1	−1
$\chi_2\psi_2$	1	−1	−1	1

We immediately notice that this identifies the character table of $C_2 \times C_2$ as the tensor product of the character tables of C_2 and C_2, and evidently this is true in general.

Corollary 4.1.4. *The character table of a direct product $G_1 \times G_2$ is the tensor product of the character tables of G_1 and G_2.*

We may see from this theory that all simple complex characters of an abelian group have degree 1 (a result already shown in Corollary 2.1.7), and that in fact this property characterizes abelian groups. We will give a different argument later on in Theorem 5.3.3 that shows that over any algebraically closed field the simple representations of abelian groups all have degree 1.

Theorem 4.1.5. *The following are equivalent for a finite group G:*

(1) G is abelian,
(2) all simple complex representations of G have degree 1.

Proof. Since the simple representations over $\mathbb{C}$ of every finite cyclic group all have degree 1, and since every finite abelian group is a direct product of cyclic groups, the last result shows that all simple representations of a finite abelian group have degree 1.

Conversely, we may use the fact that $|G| = \sum_{i=1}^{r} d_i^2$ where $d_1, \ldots, d_r$ are the degrees of the simple representations. We deduce that $d_i = 1$ for all $i \Leftrightarrow r = |G| \Leftrightarrow$ every conjugacy class has size $1 \Leftrightarrow$ every element is central $\Leftrightarrow G$ is abelian.

Another proof of this result may be obtained from the fact that $\mathbb{C}G$ is a direct sum of matrix algebras over $\mathbb{C}$, a summand $M_n(\mathbb{C})$ appearing precisely if there is a simple module of dimension n. The group and hence the group ring are abelian if and only if n is always 1. □

4.2 Lifting (or Inflating) from a Quotient Group

Whenever we have a group homomorphism $G \to H$ and a representation of H we get a representation of G: regarding the representation of H as a group homomorphism $H \to GL(V)$ we simply compose the two homomorphisms. The resulting representation of G is called the *lift* or *inflation* of the representation of H. By this means, copies of the character tables of quotient groups of G all appear in the character table of G, because the lift of a simple representation is always simple. This observation, although straightforward, allows us to fill out the character table of a group very rapidly, provided the group has normal subgroups.

As an instance of this, we may construct the part of the character table of any finite group that consists of characters of degree 1 by combining the previous results with the next one, which is formulated so as to be true over any field.

Proposition 4.2.1. *The degree 1 representations of any finite group G over any field are precisely the degree 1 representations of G/G', lifted to G via the homomorphism $G \to G/G'$.*

Proof. We only have to observe that a degree 1 representation of G over a field k is a homomorphism $G \to GL(1, k) = k^\times$ that takes values in an abelian group, and so has kernel containing G'. Thus, such a homomorphism is always a composite $G \to G/G' \to GL(1, k)$ obtained from a degree 1 representation of G/G'. □

Example 4.2.2. Neither implication of Theorem 4.1.5 holds if we do not assume that our representations are defined over an algebraically closed field of characteristic prime to $|G|$, such as $\mathbb{C}$. We have seen in examples before now that over $\mathbb{R}$ the 2-dimensional representation of the cyclic group $\langle x \mid x^3 = 1\rangle$, in which x acts as rotation through $\frac{2\pi}{3}$, is simple since there is no 1-dimensional subspace stable under the group action. We need to pass to $\mathbb{C}$ to split it as a sum of two representations of degree 1. It is also possible to find a non-abelian group all of whose simple representations do have degree 1: we shall see in Proposition 6.3 that this happens whenever G is a p-group and we consider representations in characteristic p.

4.3 Induction and Restriction

We now consider how to construct representations of a group from representations of its subgroups. Let H be a subgroup of G and V an RH-module where R is a commutative ring with 1. We define an RG-module

$$V \uparrow_H^G = RG \otimes_{RH} V$$

with the action of G coming from the left module action on RG:

$$x \cdot \left(\sum_{g \in G} a_g g \otimes v \right) = \left(x \sum_{g \in G} a_g g \right) \otimes v,$$

where $x, g \in G$, $a_g \in R$, and $v \in V$. We refer to this module as V *induced* from H to G, and say that $V \uparrow_H^G$ is an *induced module*. In many books, the notation V^G is used for this induced module, but for us this conflicts with the notation for fixed points. The operation $\uparrow_H^G$ is called *induction*.

We analyze the structure of an induced module, denoting the set of left cosets $\{gH \mid g \in G\}$ by G/H.

Proposition 4.3.1. *Let H be a subgroup of G, let V be an RH-module, and let $g_1H, \ldots, g_{|G:H|}H$ be a list of the left cosets G/H. Then*

$$V \uparrow_H^G = \bigoplus_{i=1}^{|G:H|} g_i \otimes V$$

as R-modules, where $g_i \otimes V = \{g_i \otimes v \mid v \in V\} \subseteq RG \otimes_{RH} V$. Each $g_i \otimes V$ is isomorphic to V as an R-module, and in case V is free as an R-module, we have

$$\operatorname{rank}_R V \uparrow_H^G = |G : H| \operatorname{rank}_R V.$$

If $x \in G$ then $x(g_i \otimes V) = g_j \otimes V$ where $xg_i = g_jh$ for some $h \in H$. Thus, the R-submodules $g_i \otimes V$ of $V \uparrow_H^G$ are permuted under the action of G. This action is transitive, and if $g_1 \in H$ then $\operatorname{Stab}_G(g_1 \otimes V) = H$.

Proof. We have $RG_{RH} = \bigoplus_{i=1}^{|G:H|} g_iRH \cong RH^{|G:H|}$ as right RH-modules. This is because in its right action H permutes the group element basis of RG with orbits $g_1H, \ldots, g_{|G:H|}H$. Each orbit spans a right RH-submodule $R[g_iH] = g_iRH$ of RG and so RG is their direct sum. Each of these submodules is isomorphic to RH_{RH} as right RH-modules via the isomorphism specified by $g_ih \mapsto h$ for each $h \in H$. Now

$$\begin{aligned} RG \otimes_{RH} V &= \Big(\bigoplus_{i=1}^{|G:H|} g_iRH\Big) \otimes_{RH} V \\ &= \bigoplus_{i=1}^{|G:H|} (g_iRH \otimes_{RH} V) \\ &= \bigoplus_{i=1}^{|G:H|} g_i \otimes_{RH} V \end{aligned}$$

and as R-modules $g_iRH \otimes_{RH} V \cong RH \otimes_{RH} V \cong V$.

We next show that with its left action on $RG \otimes_{RH} V$ coming from the left action on RG, G permutes these R-submodules. If $x \in G$ and $xg_i = g_jh$ with $h \in H$ then

$$\begin{aligned} x(g_i \otimes v) &= xg_i \otimes v \\ &= g_jh \otimes v \\ &= g_j \otimes hv, \end{aligned}$$

so that $x(g_i \otimes v) \subseteq g_j \otimes V$. We argue that we have equality using the invertibility of x. For, by a similar argument to the preceding one, we have $x^{-1}g_j \otimes V \subseteq g_i \otimes V$, and so $g_j \otimes V = xx^{-1}(g_j \otimes V) \subseteq x(g_i \otimes V)$. This action of G on the subspaces is transitive since given two subspaces $g_i \otimes V$ and $g_j \otimes V$ we have $(g_j g_i^{-1})g_i \otimes V = g_j \otimes V$.

Now to compute the stabilizer of $g_1 \otimes V$ where $g_1 \in H$, if $x \in H$ then

$$x(g_1 \otimes V) = g_1(g_1^{-1}xg_1) \otimes V = g_1 \otimes V,$$

and if $x \notin H$ then $x \in g_i H$ for some $i \neq 1$ and so $x(g_1 \otimes V) = g_i \otimes V$. Thus, $\mathrm{Stab}_G(g_1 \otimes V) = H$. □

The structure of induced modules described in the last result in fact characterizes these modules, giving an extremely useful criterion for a module to be of this form that we will use several times later on.

Proposition 4.3.2. *Let M be an RG-module that has an R-submodule V with the property that M is the direct sum of the R-submodules $\{gV \mid g \in G\}$. Let $H = \{g \in G \mid gV = V\}$. Then $M \cong V \uparrow_H^G$.*

We are using the notation $\{gV \mid g \in G\}$ to indicate the set of distinct possibilities for gV, so that if $gV = hV$ with $g \neq h$ we do not count gV twice.

Proof. We define a map of R-modules

$$RG \otimes_{RH} V \to M$$
$$g \otimes v \mapsto gv,$$

extending this specification from the generators to the whole of $RG \otimes_{RH} V$ by R-linearity. This is in fact a map of RG-modules. The R-submodules gV of M are in bijection with the cosets G/H, since G permutes them transitively, and the stabilizer of one of them is H. Thus, each of $RG \otimes_{RH} V$ and M is the direct sum of $|G : H|$ R-submodules $g \otimes V$ and gV, respectively, each isomorphic to V via isomorphisms $g \otimes v \leftrightarrow v$ and $gv \leftrightarrow v$. Thus, on each summand the map specified by $g \otimes v \mapsto gv$ is an isomorphism, and so $RG \otimes_{RH} V \to M$ is itself an isomorphism. □

Example 4.3.3. Immediately from the definitions, we have

$$R \uparrow_1^G \cong RG \otimes_R R \cong RG,$$

so that RG is induced from the identity subgroup.

Example 4.3.4. Permutation modules: suppose that Ω is a G-set; that is, a set with an action of G by permutations. We may form $R\Omega$, the free R-module with the elements of Ω as a basis, and it acquires the structure of an RG-module via the permutation action of G on this basis. This is the *permutation module* (or *permutation representation*) determined by Ω. Now $R\Omega = \bigoplus_{\omega \in \Omega} R\omega$ is the direct sum of rank 1 R-submodules, each generated by a basis vector. In case, G acts transitively on Ω these are permuted transitively by G. If we pick any element $\omega \in \Omega$ and let $H = \mathrm{Stab}_G\, \omega$ then H is also the stabilizer of the space $R\omega$, and $R\Omega \cong R \uparrow_H^G$ by Proposition 4.3.2. This shows that permutation modules on transitive G-sets are exactly the modules that are induced from the trivial module on some subgroup.

Characters of permutation modules over $\mathbb{C}$ are easily computed. Over any ring R, each element $g \in G$ acts on a permutation module $R\Omega$ by means of a permutation matrix with respect to the basis Ω. Such a matrix has trace equal to the number of entries 1 on the diagonal, the other diagonal entries being 0. A 1 on the diagonal is produced each time the corresponding basis element is fixed by g. We deduce from this, for the permutation module $\mathbb{C}\Omega$, that the value of its character on g equals the number of fixed points of g on Ω:

$$\chi_{\mathbb{C}\Omega}(g) = |\Omega^{\langle g \rangle}|.$$

Transitive permutation modules may be realized as quotient modules of RG, and also as submodules of RG. The induced module $R \uparrow_H^G = RG \otimes_{RH} R$ is a quotient of $RG = RG \otimes_R R$ from the definition of the tensor product. To realize this permutation module also as a submodule of RG, write $\overline{H} := \sum_{h \in H} h \in RG$. Then $\overline{H}$ generates a submodule of RG isomorphic to $R \uparrow_H^G$. This may be proved as an application of Proposition 4.3.2, and it is Exercise 11(a) at the end of this chapter.

The main feature of induction is that it is one of the two main operations (the other being restriction) that relate the representations of a group to those of a subgroup. When working over $\mathbb{C}$ it provides a way of constructing new characters. With this in mind, we give the formula for the character of an induced representation. If χ is the character of a representation V of a subgroup H, we write simply $\chi \uparrow_H^G$ for the character of $V \uparrow_H^G$. We will also write $[G/H]$ to denote some (arbitrary) set of representatives of the left cosets of H in G. With this convention, $[G/H]$ is a set of elements of G, not a set of cosets.

Proposition 4.3.5. *Let H be a subgroup of G and let V be a $\mathbb{C}H$-module with character χ. Then the character of $V \uparrow_H^G$ is*

$$\chi \uparrow_H^G (g) = \frac{1}{|H|} \sum_{\substack{t \in G \\ t^{-1}gt \in H}} \chi(t^{-1}gt)$$

$$= \sum_{\substack{t \in [G/H] \\ t^{-1}gt \in H}} \chi(t^{-1}gt).$$

Proof. The two formulas on the right are in fact the same, since if $t^{-1}gt \in H$ and $h \in H$ then $(th)^{-1}gth \in H$ also, and so $\{t \in G \mid t^{-1}gt \in H\}$ is a union of left cosets of H. Since $\chi(t^{-1}gt) = \chi((th)^{-1}gth)$ the terms in the first sum are constant on the cosets of H, and we obtain the second sum by choosing one representative from each coset and multiplying by $|H|$.

Using the vector space decomposition of Proposition 4.3.1, we obtain that the trace of g on $V \uparrow_H^G$ is the sum of the traces of g on the spaces $t \otimes V$ that are invariant under g, where $t \in [G/H]$. This is because if g does not leave $t \otimes V$ invariant, we get a matrix of zeros on the diagonal at that point in the block matrix decomposition for the matrix of g. Thus, we only get a nonzero contribution from subspaces $t \otimes V$ with $gt \otimes V = t \otimes V$. This happens if and only if $t^{-1}gt \otimes V = 1 \otimes V$, that is $t^{-1}gt \in H$. We have

$$\chi \uparrow_H^G (g) = \sum_{\substack{t \in [G/H] \\ t^{-1}gt \in H}} \text{trace of } g \text{ on } t \otimes V.$$

Now g acts on $t \otimes V$ as

$$g(t \otimes v) = t(t^{-1}gt) \otimes v = t \otimes (t^{-1}gt)v,$$

and so the trace of g on this space is $\chi(t^{-1}gt)$. Combining this with the last expression gives the result. □

We see in the preceding proof that g leaves invariant $t \otimes V$ if and only if $t^{-1}gt \in H$, or in other words $g \in tHt^{-1}$. Thus, $\operatorname{Stab}_G(t \otimes V) = tHt^{-1}$. Furthermore, if we identify $t \otimes V$ with V by means of the bijection $t \otimes v \leftrightarrow v$, then g acts on $t \otimes V$ via the composite homomorphism

$$\langle g \rangle \xrightarrow{c_{t^{-1}}} H \xrightarrow{\rho} GL(V),$$

where ρ is the homomorphism associated to V and $c_a(x) = axa^{-1}$ is the automorphism of G that is conjugation by $a \in G$.

Example 4.3.6. To make clearer what the terms in the expression for the induced character are, consider $G = S_3$ and $H = \langle (123) \rangle$, the normal subgroup of order 3. To avoid expressions such as $(())$ we will write the identity element of S_3 as e. We may take the coset representatives $[G/H]$ to be $\{e, (12)\}$. If χ is the trivial character of H then

$$\begin{aligned} \chi \uparrow_H^G (e) &= \chi(e^e) + \chi(e^{(12)}) = 2, \\ \chi \uparrow_H^G ((12)) &= \text{the empty sum} = 0, \\ \chi \uparrow_H^G ((123)) &= \chi((123)^e) + \chi((123)^{(12)}) = 2. \end{aligned}$$

Recalling the character table of S_3, we find that $\chi \uparrow_H^G$ is the sum of the trivial character and the sign character of S_3.

Induced representations can be hard to understand from first principles, so we now develop some formalism that will enable us to compute with them more easily. The companion notion to induction is that of *restriction* of representations. If H is a subgroup of G and W is a representation of G, we denote it by $W \downarrow_H^G$ the representation of H whose representation space is again W, and where the elements of H act the same way on W as they do when regarded as elements of G. In other words, we just forget about the elements of G that are outside H. When W is a representation in characteristic zero with character ψ, we will write $\psi \downarrow_H^G$ for the character of $W \downarrow_H^G$. Its values are the same as those of ψ, but the domain of definition is restricted to H.

Restriction and induction are a particular case of the following more general situation. Whenever we have a (unital) homomorphism of rings $A \to B$, an A-module V and a B-module W, we may form the B-module $B \otimes_A V$ and the A-module $W \downarrow_A^B$. On taking $A = RH$ and $B = RG$, we obtain the induction and restriction we have been studying. There are, in fact, further operations that relate the representations of A and B, and we mention one now, namely, *coinduction*. Given an A-module V we may form the B-module $\mathrm{Hom}_A(B, V)$, which acquires the structure of a left B-module because of the right action of B on itself, and this is the *coinduced* module obtained from V. We will prove in Corollary 4.3.8 that induction and coinduction are the same in the case of group rings, so that we will not need to consider coinduction separately.

Lemma 4.3.7. *Let $A \to B$ be a homomorphism of rings, V an A-module and W a B-module.*

(1) (Left adjoint of restriction) $\mathrm{Hom}_B(B \otimes_A V, W) \cong \mathrm{Hom}_A(V, W \downarrow_A)$.

(2) (Right adjoint of restriction)

$$\mathrm{Hom}_A(W \downarrow_A, V) \cong \mathrm{Hom}_B(W, \mathrm{Hom}_A(B, V)).$$

(3) (Transitivity of induction) If $\phi : B \to C$ is another ring homomorphism then

$$C \otimes_B (B \otimes_A V) \cong C \otimes_A V.$$

Proof. In the case of (1), the mutually inverse isomorphisms are

$$f \mapsto (v \mapsto f(1 \otimes v))$$

and

$$(b \otimes v \mapsto bg(v)) \leftarrow g.$$

In the case of (2), the mutually inverse isomorphisms are

$$f \mapsto (w \mapsto (b \mapsto f(bw)))$$

and

$$(w \mapsto g(w)(1)) \leftarrow g.$$

In the case of (3), the mutually inverse isomorphisms are

$$c \otimes b \otimes v \mapsto c\phi(b) \otimes v$$

and

$$c \otimes 1 \otimes v \leftarrow c \otimes v.$$

There is checking to be done to show that morphisms are indeed well-defined homomorphisms of A-modules and B-modules, and that maps are mutually inverse, but it is all routine. □

It is worth knowing that Lemma 4.3.7 parts (1) and (2) are instances of a single formula to do with bimodules. An *(A, B)-bimodule* T is defined to have the structure of both a left A-module and a right B-module, and such that these two module actions commute: for all $a \in A$, $b \in B$, $t \in T$, we have $(at)b = a(tb)$. The basic adjoint relationship between A-modules V and B-modules W is an isomorphism

$$\mathrm{Hom}_A(T \otimes_B W, V) \cong \mathrm{Hom}_B(W, \mathrm{Hom}_A(T, V))$$

that is given by mutually inverse maps

$$f \mapsto (w \mapsto (t \mapsto f(t \otimes w)))$$

and

$$(t \otimes w \mapsto g(w)(t)) \leftarrow g.$$

Supposing that we have a ring homomorphism $A \to B$, one such bimodule T is the set B with left action of A given by left multiplication after applying the homomorphism to B, and right action of B given by right multiplication. We denote this bimodule ${}_AB_B$. There is a similarly defined (B, A)-bimodule ${}_BB_A$ on which B acts by left multiplication and A acts by right multiplication after applying the homomorphism to B. We have isomorphisms of A-modules

$$\mathrm{Hom}_B({}_BB_A, V) \cong V \downarrow_A^B \cong {}_AB_B \otimes_B V,$$

whereas ${}_BB_A \otimes_A W$ and $\mathrm{Hom}_A({}_AB_B, W)$ are by definition the induction and coinduction of W. Applying the single adjoint isomorphism in the two cases of these bimodules yields the relationships of Lemma 4.3.7 (1) and (2).

Corollary 4.3.8. *Let $H \le K \le G$ be subgroups of G, let V be an RH-module, and let W be an RG-module.*

(1) (Frobenius reciprocity)

$$\mathrm{Hom}_{RG}(V \uparrow_H^G, W) \cong \mathrm{Hom}_{RH}(V, W \downarrow_H^G)$$

and

$$\mathrm{Hom}_{RG}(W, V \uparrow_H^G) \cong \mathrm{Hom}_{RH}(W \downarrow_H^G, V).$$

(2) (Transitivity of induction) $(V \uparrow_H^K) \uparrow_K^G \cong V \uparrow_H^G$ as RG-modules.
(3) (Transitivity of restriction) $(W \downarrow_K^G) \downarrow_H^K = W \downarrow_H^G$ as RH-modules.
(4) $V \uparrow_H^G \otimes_R W \cong (V \otimes_R W \downarrow_H^G) \uparrow_H^G$ as RG-modules. In particular,

$$R \uparrow_H^G \otimes_R W \cong W \downarrow_H^G \uparrow_H^G .$$

(5) (Induced and coinduced are the same) $V \uparrow_H^G \cong \mathrm{Hom}_{RH}(RG, V)$ as RG-modules.

Proof. The first isomorphism of (1) and part (2) follow from the relationships in Lemma 4.3.7, in the case of the ring homomorphism $RH \to RG$. Part (3) also holds in this generality and is immediate. The second isomorphism in (1) as well as (4) and (5) are special for group representations.

Part (4) is the isomorphism

$$(RG \otimes_{RH} V) \otimes_R W \cong RG \otimes_{RH} (V \otimes_R W)$$

and it is not a corollary of Lemma 4.3.7. Here the mutually inverse isomorphisms are

$$(g \otimes v) \otimes w \mapsto g \otimes (v \otimes g^{-1} w)$$

and

$$(g \otimes v) \otimes gw \leftarrow g \otimes (v \otimes w).$$

We prove (5) by exhibiting mutually inverse isomorphisms

$$RG \otimes_{RH} V \cong \mathrm{Hom}_{RH}(RG, V).$$

The first is given by

$$g \otimes v \mapsto \phi_{g,v}$$

where $\phi_{g,v}(x) = (xg)v$ if $x \in Hg^{-1}$ and is 0 otherwise. Here $g, x \in G$ and $v \in V$. In the opposite direction, the map is

$$\sum_{g \in [G/H]} g \otimes \theta(g^{-1}) \leftarrow \theta,$$

where the sum is taken over a set of representatives for the left cosets of H in G. We must check here that $\phi_{g,v}$ is a homomorphism of RH-modules, does specify a map on the tensor product, that θ is well defined, and that the two morphisms are mutually inverse. We have shown that induced and coinduced modules are the same.

Finally, the second isomorphism of part (1) follows from Lemma 4.3.7(2), using (5). □

In the case of representations in characteristic zero, all of these results may be translated into the language of characters. In this setting, the second Frobenius reciprocity formula, which used the fact that coinduced modules and induced modules are the same, becomes much easier. If our interest is only in character theory, we did not need to read about left and right adjoints, coinduction, and bimodules. Recall that if V is a representation of H with character χ and W is a representation of G with character ψ, we write $\chi \uparrow_H^G$ and $\psi \downarrow_H^G$ for the characters of $V \uparrow_H^G$ and $W \downarrow_H^G$.

Corollary 4.3.9. *Let $H \leq K \leq G$ be subgroups of G, let χ be a complex character of H and ψ a complex character of G.*

(1) (Frobenius reciprocity)

$$\langle \chi \uparrow_H^G, \psi \rangle_G = \langle \chi, \psi \downarrow_H^G \rangle_H$$

and

$$\langle \psi, \chi \uparrow_H^G \rangle_G = \langle \psi \downarrow_H^G, \chi \rangle_H.$$

In fact, all four numbers are equal.

(2) (Transitivity of induction) $(\chi \uparrow_H^K) \uparrow_K^G = \chi \uparrow_H^G$.

(3) (Transitivity of restriction) $(\psi \downarrow_K^G) \downarrow_H^K = \psi \downarrow_H^G$.

(4) $\chi \uparrow_H^G \cdot \psi = (\chi \cdot \psi \downarrow_H^G) \uparrow_H^G$.

Proof. In (1), we write $\langle\ ,\ \rangle_G$ and $\langle\ ,\ \rangle_H$ to denote the inner product of characters of G and H, respectively. The four parts are translations of the first four parts of Corollary 4.3.8 into the language of characters. In part (1), we use the fact that the inner products are the dimensions of the Hom groups in Corollary 4.3.8(1). We may deduce the second formula from the first in this context because the Hermitian inner product takes real values on characters, and so is symmetric on them, as observed before Theorem 3.2.3. □

Frobenius reciprocity for complex characters is equivalent to saying that if ψ and χ are simple characters of G and H, respectively, then the multiplicity of ψ as a summand of $\chi \uparrow_H^G$ equals the multiplicity of χ as a summand of $\psi \downarrow_H^G$.

At a more sophisticated level, we may interpret induction, restriction, and Frobenius reciprocity in terms of the space $\mathbb{C}^{\mathrm{cc}(G)}$ of class functions introduced in Chapter 3, that is, the vector space of functions $\mathrm{cc}(G) \to \mathbb{C}$ where $\mathrm{cc}(G)$ is the set of conjugacy classes of G. Since each conjugacy class of H is contained in a unique conjugacy class of G we have a mapping $\mathrm{cc}(H) \to \mathrm{cc}(G)$ and this gives rise by composition to a linear map $\downarrow_H^G : \mathbb{C}^{\mathrm{cc}(G)} \to \mathbb{C}^{\mathrm{cc}(H)}$ that on characters is the restriction operation we have already defined. We may also define a linear map $\uparrow_H^G : \mathbb{C}^{\mathrm{cc}(H)} \to \mathbb{C}^{\mathrm{cc}(G)}$ that, on characters, sends a character χ of H to the character $\chi \uparrow_H^G$. It is possible to define this on arbitrary class functions of H by means of the explicit formula given in Proposition 4.3.5. With this approach the transitivity of induction is not entirely obvious. It is easier to observe that the characters of simple representations of H form a basis of $\mathbb{C}^{\mathrm{cc}(H)}$ and to define $\chi \uparrow_H^G$ in the first instance on these basis elements. We then extend the definition of $\uparrow_H^G$ to arbitrary class functions so that it is a linear map.

With these definitions the formulas of Corollary 4.3.9 hold for arbitrary class functions – except that in part (1) the numbers from the different equations are not equal (as stated there), but are complex conjugates of each other. Frobenius reciprocity becomes an adjoint relationship between $\uparrow_H^G$ and $\downarrow_H^G$, regarded as linear maps between the spaces of class functions. The characters of the simple representations of H and of G form orthonormal bases of $\mathbb{C}^{\mathrm{cc}(H)}$ and of $\mathbb{C}^{\mathrm{cc}(G)}$ and, taking matrices with respect to these bases, Frobenius reciprocity states that the matrix of the induction map is the transpose of the matrix of the restriction map.

Example 4.3.10. Frobenius reciprocity is a most useful tool in calculating with induced characters. In the special case that V and W are simple representations over $\mathbb{C}$ of H and G, respectively, where $H \leq G$, it says that the multiplicity of W as a summand of $V \uparrow_H^G$ equals the multiplicity of V as a summand of $W \downarrow_H^G$. As an example, we may take both V and W to be the trivial representations of their respective groups. As explained in Example 4.3.4, $\mathbb{C} \uparrow_H^G$ is a permutation module. We deduce from Frobenius reciprocity that as representations of G, $\mathbb{C}$ is a direct summand of $\mathbb{C} \uparrow_H^G$ with multiplicity one.

Example 4.3.11. Let $G = \langle x, y \mid x^n = y^2 = 1, yxy^{-1} = x^{-1} \rangle = D_{2n}$, the dihedral group of order $2n$. Suppose that n is odd. We compute that the commutator $[y, x] = x^{n-2}$, and since n is odd we have $G' = \langle x^{n-2} \rangle = \langle x \rangle \cong C_n$ and $G/G' \cong C_2$. Thus, G has two complex characters of degree 1 that we denote 1 and -1.

Let $\chi_{\zeta_n^s}$ denote the degree 1 character of $\langle x\rangle$ specified by $\chi_{\zeta_n^s}(x^r) = \zeta_n^{rs}$ where $\zeta_n = e^{\frac{2\pi i}{n}}$. Then $\chi_{\zeta_n^s} \uparrow_{\langle x\rangle}^G$ has values given in the following table:

D_{2n}, n odd
Ordinary characters

g $\lvert C_G(g)\rvert$	1 $2n$	x n	x^2 n	$\cdots$	$x^{\frac{n-1}{2}}$ n	y 2
χ_1	1	1	1	$\cdots$	1	1
χ_{1u}	1	1	1	$\cdots$	1	-1
$\chi_{\zeta_n^s} \uparrow_{\langle x\rangle}^G$ $(1 \le s \le \frac{n-1}{2})$	2	$\zeta_n^s + \overline{\zeta}_n^s$	$\zeta_n^{2s} + \overline{\zeta}_n^{2s}$	$\cdots$	$\zeta_n^{\frac{n-1}{2}s} + \overline{\zeta}_n^{\frac{n-1}{2}s}$	0

We verify that

$$\langle \chi_s \uparrow_{\langle x\rangle}^G, \pm 1\rangle_G = \langle \chi_s, \pm 1 \downarrow_{\langle x\rangle}^G\rangle_{\langle x\rangle} = 0$$

if $n \nmid s$, using Frobenius reciprocity (or a direct calculation), and hence the characters χ_s must be simple when $n \nmid s$, because otherwise they would be a sum of two characters of degree 1, that must be 1 or -1, and evidently this would not give the correct character values. For $1 \le s \le \frac{n-1}{2}$ they are distinct, and so we have constructed $\frac{n-1}{2} + 2 = \frac{n+3}{2}$ simple characters. This equals the number of conjugacy classes of G, so we have the complete character table.

4.4 Symmetric and Exterior Powers

As further ways of constructing new representations from old ones, we describe the symmetric powers and exterior powers of a representation. If V is a vector space over a field k, its *nth symmetric power* is the vector space

$$S^n(V) = V^{\otimes n}/I,$$

where $V^{\otimes n} = V \otimes \cdots \otimes V$ with n factors, and I is the subspace spanned by tensors of the form $((\cdots \otimes v_i \otimes \cdots \otimes v_j \otimes \cdots) - (\cdots \otimes v_j \otimes \cdots \otimes v_i \otimes \cdots))$, where all places in the two basic tensors are the same except for two of them, where the elements $v_i, v_j \in V$ are interchanged. We write the image of the tensor $v_1 \otimes \cdots \otimes v_n$ in $S^n(V)$ as a (commutative) product $v_1 \cdots v_n$, noting that in $S^n(V)$ it does not matter in which order we write the terms. A good way to think of $S^n(V)$ is as the space of homogeneous polynomials of degree n in a polynomial ring. Indeed, if $u_1, \ldots, u_r$ is any basis of V and we let $k[u_1, \ldots, u_r]_n$ denote the vector space of homogeneous polynomials of degree n in the u_i as indeterminates, there is a surjective linear map

$$V^{\otimes n} \to k[u_1, \ldots, u_r]_n$$
$$u_{i_1} \otimes \cdots \otimes u_{i_n} \mapsto u_{i_1} \cdots u_{i_n}$$

(extended by linearity to the whole of $V^{\otimes n}$). This map contains I in its kernel, so there is induced a map

$$S^n(V) \to k[u_1, \ldots, u_r]_n.$$

This is now an isomorphism since, modulo I, the tensors

$$u_1^{\otimes a_1} \otimes u_2^{\otimes a_2} \otimes \cdots \otimes u_r^{\otimes a_r}$$

where $\sum_{i=1}^{r} a_i = n$ span $V^{\otimes n}$, and they map to the monomials that form a basis of $\dim_k k[u_1, \ldots, u_r]_n$. As is well known, $\dim_k k[u_1, \ldots, u_r]_n = \binom{n+r-1}{n}$.

The *nth exterior power* of V is the vector space

$$\Lambda^n(V) = V^{\otimes n}/J,$$

where J is the subspace spanned by tensors

$$((\cdots \otimes v_i \otimes \cdots \otimes v_j \otimes \cdots) + (\cdots \otimes v_j \otimes \cdots \otimes v_i \otimes \cdots))$$

and

$$(\cdots \otimes v_i \otimes \cdots \otimes v_i \otimes \cdots)$$

where $v_i, v_j \in V$. We write the image of $v_1 \otimes \cdots \otimes v_n$ in $\Lambda^n(V)$ as $v_1 \wedge \cdots \wedge v_n$, so that interchanging v_i and v_j changes the sign of the symbol, and if two of v_i and v_j are equal the symbol is zero. If the characteristic of k is not 2, the second of these properties follows from the first, but for the sake of characteristic 2, we impose it anyway. By an argument similar to the one used for symmetric powers, we see that $\Lambda^n(V)$ has as a basis $\{u_{i_1} \wedge \cdots \wedge u_{i_n} \mid 1 \leq i_1 < \cdots < i_n \leq r\}$, and its dimension is $\binom{r}{n}$. In particular, $\Lambda^n(V) = 0$ if $n > \dim V$.

Suppose now that a group G acts on V and consider the diagonal action of G on $V^{\otimes n}$. The subspaces of relations I and J are preserved by this action, and so there arise actions of G on $S^n(V)$ and $\Lambda^n(V)$:

$$g \cdot (v_1 v_2 \cdots v_n) = (gv_1)(gv_2) \cdots (gv_n)$$
$$g \cdot (v_1 \wedge \cdots \wedge v_n) = (gv_1) \wedge \cdots \wedge (gv_n).$$

Because we substitute the expressions for gv_i into the monomials that form the bases of $S^n(V)$ and $\Lambda^n(V)$, we say that G acts on these spaces by *linear substitutions*. With these actions, we have described the symmetric and exterior powers of the representation V.

Example 4.4.1. Consider the representation of $G = \langle x \mid x^3 = 1 \rangle$ on the vector space V with basis $\{u_1, u_2\}$ given by

$$xu_1 = u_2$$
$$xu_2 = -u_1 - u_2.$$

Then $S^2(V)$ has a basis $\{u_1^2, u_1u_2, u_2^2\}$ and

$$\begin{aligned} x \cdot u_1^2 &= u_2^2, \\ x \cdot (u_1u_2) &= u_2(-u_1 - u_2) = -u_1u_2 - u_2^2, \\ x \cdot u_2^2 &= (-u_1 - u_2)^2 = u_1^2 + 2u_1u_2 + u_2^2. \end{aligned}$$

Similarly, $\Lambda^2(V)$ has basis $\{u_1 \wedge u_2\}$ and

$$x \cdot (u_1 \wedge u_2) = u_2 \wedge (-u_1 - u_2) = u_1 \wedge u_2.$$

The symmetric and exterior powers fit into a more general framework where we consider tensors with different symmetry properties. There is an action of the symmetric group S_n on the n-fold tensor power $V^{\otimes n}$ given by permuting the positions of vectors in a tensor, so that for example, if α, β, γ are vectors in V then

$$\begin{aligned} (1,2)(\alpha \otimes \beta \otimes \gamma) &= \beta \otimes \alpha \otimes \gamma \\ (1,3)(\beta \otimes \alpha \otimes \gamma) &= \gamma \otimes \alpha \otimes \beta. \end{aligned}$$

From the preceding very convincing formulas and the fact that $(1,2,3) = (1,3)(1,2)$, we deduce that

$$(1,2,3)(\alpha \otimes \beta \otimes \gamma) = \gamma \otimes \alpha \otimes \beta,$$

which is evidence that if $\sigma \in S_n$ then

$$\sigma(v_1 \otimes \cdots \otimes v_n) = v_{\sigma^{-1}(1)} \otimes v_{\sigma^{-1}(2)} \otimes \cdots \otimes v_{\sigma^{-1}(n)},$$

a formula that is not quite so obvious. With this action, it is evident that $S^n(V)$ is the largest quotient of $V^{\otimes n}$ on which S_n acts trivially, and when $\text{char}(k) \neq 2$, $\Lambda^n(V)$ is the largest quotient of $V^{\otimes n}$ on which S_n acts as a sum of copies of the sign representation.

We define the *symmetric tensors* or *divided powers* to be the fixed points $(V^{\otimes n})^{S_n}$, and when char $k \neq 2$, we define the *skew-symmetric tensors* to be the largest kS_n-submodule of $V^{\otimes n}$ that is a sum of modules isomorphic to the sign representation. Thus,

$$\begin{aligned} \text{symmetric tensors} &= \{w \in V^{\otimes n} \mid \sigma(w) = w \text{ for all } \sigma \in S_n\} \\ \text{skew-symmetric tensors} &= \{w \in V^{\otimes n} \mid \sigma(w) = \text{sign}(\sigma)w \text{ for all } \sigma \in S_n\}. \end{aligned}$$

When we let G act diagonally on $V^{\otimes n}$ the symmetric tensors, the skew-symmetric tensors, as well as the subspaces I and J defined earlier remain invariant for the action of G. We easily see this directly, but at a more theoretical level the reason is that the actions of G and S_n on $V^{\otimes n}$ commute with each other (as is easily verified), so that $V^{\otimes n}$ acquires the structure of a $k[G \times S_n]$-module, and elements of G act as endomorphisms of $V^{\otimes n}$ as a kS_n-module, and vice-versa. Every endomorphism of the kS_n-module $V^{\otimes n}$ must send the S_n-fixed points to themselves, for example, and so the symmetric tensors are invariant under the action of G. One sees similarly that the other subspaces are also invariant under the action of G.

We remark that, in general, the symmetric power $S^n(V)$ and the symmetric tensors provide nonisomorphic representations of G, as do $\Lambda^n(V)$ and the skew-symmetric tensors. This phenomenon is investigated in Exercises 15 and 16 at the end of this chapter. However, these pairs of kG-modules are isomorphic in characteristic zero, and we now consider in detail the case of the symmetric and exterior square. Suppose that k is a field whose characteristic is not 2. In this situation, the only tensor that is both symmetric and skew-symmetric is 0. Any degree 2 tensor may be written as the sum of a symmetric tensor and a skew-symmetric tensor in the following way:

$$\sum \lambda_{ij} v_i \otimes v_j = \frac{1}{2} \sum \lambda_{ij}(v_i \otimes v_j + v_j \otimes v_i) + \frac{1}{2} \sum \lambda_{ij}(v_i \otimes v_j - v_j \otimes v_i).$$

We deduce from this that

$$V \otimes V = \text{symmetric tensors} \oplus \text{skew-symmetric tensors}$$

as kG-modules. The subspace I that appeared in the definition

$$S^2(V) = (V \otimes V)/I$$

is contained in the space of skew-symmetric tensors, and the subspace J for which $\Lambda^2(V) = (V \otimes V)/J$ is contained in the space of symmetric tensors. By counting dimensions, we see that $\dim I + \dim J = \dim V \otimes V$ and putting this together, we see that

$$\begin{aligned} I =& \text{ skew-symmetric tensors,} \\ J =& \text{ symmetric tensors, and} \\ V \otimes V =& I \oplus J. \end{aligned}$$

From this information, we see on factoring out I and J that

$$\begin{aligned} S^2(V) \cong& \text{ symmetric tensors} \\ \Lambda^2(V) \cong& \text{ skew-symmetric tensors,} \end{aligned}$$

and we have proved the following result.

Proposition 4.4.2. *Suppose V is a representation for G over a field k whose characteristic is not 2. Then*

$$V \otimes V \cong S^2(V) \oplus \Lambda^2(V)$$

as kG-modules, where G acts diagonally on $V \otimes V$ and by linear substitutions on $S^2(V)$ and $\Lambda^2(V)$.

One application of this is that when constructing new representations from an existing representation V by taking tensor products, the tensor square will always decompose (away from characteristic 2) giving two smaller representations.

Suppose now that $k = \mathbb{C}$. If χ is the character of a representation V, we write $S^2\chi$ and $\Lambda^2\chi$ for the characters of $S^2(V)$ and $\Lambda^2(V)$.

Proposition 4.4.3. *Let χ be the character of a representation V of G over $\mathbb{C}$. Then*

$$S^2\chi(g) = \frac{1}{2}(\chi(g)^2 + \chi(g^2))$$
$$\Lambda^2\chi(g) = \frac{1}{2}(\chi(g)^2 - \chi(g^2)).$$

Proof. For each $g \in G$, $V\downarrow^G_{\langle g\rangle}$ is the direct sum of 1-dimensional representations of the cyclic group $\langle g\rangle$, and so we may choose a basis $u_1, \dots, u_r$ for V such that $g \cdot u_i = \lambda_i u_i$ for scalars λ_i. The monomials u_i^2 with $1 \le i \le r$ and $u_i u_j$ with $1 \le i < j \le r$ form a basis for S^2V, and so the eigenvalues of g on this space are λ_i^2 with $1 \le i \le r$ and $\lambda_i\lambda_j$ with $1 \le i < j \le r$. Therefore,

$$\begin{aligned} S^2\chi(g) &= \sum_{i=1}^{r} \lambda_i^2 + \sum_{1\le i<j\le r} \lambda_i\lambda_j \\ &= \frac{1}{2}((\lambda_1 + \cdots + \lambda_r)^2 + (\lambda_1^2 + \cdots + \lambda_r^2)) \\ &= \frac{1}{2}(\chi(g)^2 + \chi(g^2)). \end{aligned}$$

Similarly, Λ^2V has a basis $u_i \wedge u_j$ with $1 \le i < j \le r$, so the eigenvalues of g on Λ^2V are $\lambda_i\lambda_j$ with $1 \le i < j \le r$ and

$$\begin{aligned} \Lambda^2\chi(g) &= \sum_{1\le i<j\le r} \lambda_i\lambda_j \\ &= \frac{1}{2}((\lambda_1 + \cdots + \lambda_r)^2 - (\lambda_1^2 + \cdots + \lambda_r^2)) \\ &= \frac{1}{2}(\chi(g)^2 - \chi(g^2)). \end{aligned}$$ □

There is a formula due to Molien for the generating function of characters of the symmetric powers of V. We present Molien's Theorem in Exercise 19.

4.5 The Construction of Character Tables

We may now summarize some major techniques used in constructing complex character tables. The first things to do are to determine

- the conjugacy classes in G,
- the abelianization G/G',
- the 1-dimensional characters of G.

We construct characters of degree larger than 1 as

- natural representations of G,
- representations lifted from quotient groups,
- representations induced from subgroups,
- tensor products of other representations,
- symmetric and exterior powers of other representations,
- contragredients of other representations.

As a special case of the induced representations, we have permutation representations, which are induced from the trivial module. The representations obtained by these methods might not be simple, so we test them for simplicity and subtract off known character summands using the

- orthogonality relations,

which are assisted in the case of induced characters, by

- Frobenius reciprocity.

The orthogonality relations provide a check on the accuracy of our calculations, and also enable us to complete the final row of the character table. The facts that the character degrees divide $|G|$ and that the sum of the squares of the degrees equals $|G|$ also help in this.

4.6 Summary of Chapter 4

- The characters of representations obtained by induction, tensor product, symmetric, and exterior powers are all useful in constructing character tables and there are formulas for these characters.
- The character table of $G_1 \times G_2$ is the tensor product of the character tables of G_1 and G_2.
- The degree 1 characters of G are precisely the characters of the simple representations of G/G'.

- The simple complex characters of a cyclic group of order n are the n homomorphisms to the group of nth roots of unity.
- Induced characters may be decomposed using Frobenius reciprocity.
- An induced module may be identified by the fact that it is a direct sum of subspaces that are permuted by G.

4.7 Exercises for Chapter 4

In these exercises, every module is supposed to be a finite-dimensional vector space over the ground field.

1. (a) Compute the character table of the dihedral group D_{2n} when n is even.

(b) Compute the character table of the quaternion group Q_8 that was described in Chapter 2, Exercise 12.

(c) Use this information to show that the posets of normal subgroups of D_8 and Q_8 are isomorphic.

[The groups D_8 and Q_8 provide an example of nonisomorphic groups whose character tables are "the same."]

2. Let G be the non-abelian group of order 21:

$$G = \langle x, y \mid x^7 = y^3 = 1, yxy^{-1} = x^2 \rangle.$$

Show that G has 5 conjugacy classes, and find its character table.

3. Find the character table of the following group of order 36:

$$G = \langle a, b, c \mid a^3 = b^3 = c^4 = 1, ab = ba, cac^{-1} = b, cbc^{-1} = a^2 \rangle.$$

[It follows from these relations that $\langle a, b \rangle$ is a normal subgroup of G of order 9.]

4. Compute the character table of the symmetric group S_5 by the methods of this chapter. To help in doing this, consider especially the decomposition of the permutation representation on 5 symbols, the symmetric and exterior square of the summands, as well as tensor product with the sign representation.

5. Compute the character tables of the alternating groups A_4 and A_5 using the following procedure. You may assume that A_5 is a simple group that is isomorphic to the group of rotations of a regular icosahedron, and that A_4 is isomorphic to the group of rotations of a regular tetrahedron.

(a) Compute the conjugacy classes by observing that each conjugacy class of even permutations in S_n is either a single class in A_n or the union of two classes of A_n, and that this can be determined by computing centralizers of elements in A_n and comparing them with the centralizers in S_n.

(b) Compute the abelianization of each group, and hence the 1-dimensional representations.

(c) Obtain further representations using the methods of this section. We have natural 3-dimensional representations in each case. It is also helpful to consider induced representations from the Sylow 2-subgroup in the case of A_4, and from the subgroup A_4 in the case of A_5.

6. Let k be any field, H a subgroup of G, and V a representation of H over k. Show that $V^* \uparrow_H^G \cong (V \uparrow_H^G)^*$. Deduce from this that $kG \cong (kG)^*$ and (more generally) that permutation modules are self-dual (i.e., isomorphic to their dual).

7. Let k be any field, and V any representation of G over k. Prove that $V \otimes kG$ is isomorphic to a direct sum of copies of kG.

8. The tensor product $V = \mathbb{R}^3 \otimes_{\mathbb{R}} \mathbb{R}^3 \otimes_{\mathbb{R}} \mathbb{R}^3$ is a vector space of dimension 27 with basis the tensors $e_i \otimes e_j \otimes e_k$ where e_1, e_2, e_3 is a standerd basis for $\mathbb{R}^3$. The symmetric group S_3 acts on V by permuting the positions of the suffixes, so for instance $(1, 2) \cdot (e_3 \otimes e_1 \otimes e_2) = e_1 \otimes e_3 \otimes e_2$.

(a) Find the multiplicity of each simple representation of S_3 in a decomposition of V as a direct sum of simple representations. [Observe that V is a permutation representation.]

(b) Give also the decomposition of V as a direct sum of three subspaces consisting of tensors with different symmetry properties under the action of S_3. What are the dimensions of these subspaces? Find a basis for each subspace. [Use the result of Chapter 3, Exercise 13.]

(c) The Schur algebra $S_{\mathbb{C}}(3,3)$ may be defined to be the endomorphism ring $\mathrm{Hom}_{\mathbb{C}S_3}(V, V)$. Show that $S_{\mathbb{C}}(3,3)$ is semisimple and find the dimensions of its simple representations.

9. Let V be a representation of G over a field k of characteristic zero. Prove that the symmetric power $S^n(V)$ is isomorphic as a kG-module to the space of symmetric tensors in $V^{\otimes n}$.

10. Show that every simple representation of $C_3 \times C_3$ over $\mathbb{R}$ has dimension 1 or 2. Deduce that if V is a simple 2-dimensional representation of C_3 over $\mathbb{R}$ then $V \otimes V$ is not a simple $\mathbb{R}[C_3 \times C_3]$-module.

11. Let H be a subgroup of G.

(a) Write $\overline{H} = \sum_{h \in H} h$ for the sum of the elements of H, as an element of RG. Show that $RG \cdot \overline{H} \cong R \uparrow_H^G$ as left RG-modules. Show also that $RG \cdot \overline{H}$ equals the fixed points of H in its action on RG from the *right*.

(b) More generally, let $\rho : H \to R^\times$ be a 1-dimensional representation of H (that is, a group homomorphism to the units of R). Write $\tilde{H} := \sum_{h \in H} \rho(h)h \in RG$. Show that $RG \cdot \tilde{H} \cong \rho^* \uparrow_H^G$ as RG-modules.

12. Let H be a subgroup of G and V an RH-module. Show that if V can be generated by d elements as an RH-module then $V \uparrow_H^G$ can be generated by d elements as an RG-module.

13. Let U, V be kG-modules where k is a field, and suppose we are given a non-degenerate bilinear pairing

$$\langle \ , \ \rangle : U \times V \to k$$

that is G-invariant, that is, $\langle u, v \rangle = \langle gu, gv \rangle$ for all $u \in U$, $v \in V$, $g \in G$. If U_1 is a subspace of U, let $U_1^\perp = \{v \in V \mid \langle u, v \rangle = 0 \text{ for all } u \in U_1\}$ and if V_1 is a subspace of V, let $V_1^\perp = \{u \in U \mid \langle u, v \rangle = 0 \text{ for all } v \in V_1\}$.

(a) Show that $V \cong U^*$ as kG-modules, and that there is an identification of V with U^* so that $\langle \ , \ \rangle$ can be identified with the canonical pairing $U \times U^* \to k$.

(b) Show that if U_1 and V_1 are kG-submodules, then so are $U_1^\perp$ and $V_1^\perp$.

(c) Show that if $U_1 \subseteq U_2$ are kG-submodules of U then

$$U_1^\perp / U_2^\perp \cong (U_2/U_1)^*$$

as kG-modules.

(d) Show that the composition factors of U^* are the duals of the composition factors of U.

14. Let Ω be a finite G-set and $k\Omega$ the corresponding permutation module, where k is a field. Let $\langle \ , \ \rangle : k\Omega \times k\Omega \to k$ be the symmetric bilinear form specified on the elements of Ω as

$$\langle \omega_1, \omega_2 \rangle = \begin{cases} 1 & \text{if } \omega_1 = \omega_2 \\ 0 & \text{otherwise.} \end{cases}$$

(a) Show that this bilinear form is G-invariant, that is, $\langle \omega_1, \omega_2 \rangle = \langle g\omega_1, g\omega_2 \rangle$ for all $g \in G$.

(b) Deduce from this that $k\Omega$ is self-dual, that is, $k\Omega \cong (k\Omega)^*$. [Compare with Exercise 6.]

15. Let V be a kG-module where k is a field, and let $\langle\ ,\ \rangle : V \times V^* \to k$ be the canonical pairing between V and its dual, so $\langle v, f\rangle = f(v)$.

(a) Show that the specification

$$\langle v_1 \otimes \cdots \otimes v_n, f_1 \otimes \cdots \otimes f_n\rangle = f_1(v_1)\cdots f_n(v_n)$$

determines a nondegenerate bilinear pairing

$$\langle\ ,\ \rangle : V^{\otimes n} \times (V^*)^{\otimes n} \to k$$

that is invariant both for the diagonal action of G and the action of S_n given by permuting the positions of the tensors.

(b) Let I and J be the subspaces of $V^{\otimes n}$ that appear in the definitions of the symmetric and exterior powers, so $S^n(V) = V^{\otimes n}/I$ and $\Lambda^{\otimes n} = V^{\otimes n}/J$. Show that $I^\perp$ (defined in Exercise 13) equals the space of symmetric tensors in $(V^*)^{\otimes n}$, and that $J^\perp$ equals the space of skew-symmetric tensors in $(V^*)^{\otimes n}$ (at least, when $\operatorname{char} k \neq 2$).

(c) Show that $(S^n(V))^* \cong \mathrm{ST}^n(V^*)$, and that $(\Lambda^n(V))^* \cong \mathrm{SST}^n(V^*)$, where ST^n denotes the symmetric tensors, and in general, we define the skew-symmetric tensors $\mathrm{SST}^n(V^*)$ to be $J^\perp$.

16. Let $G = C_2 \times C_2$ be the Klein four group with generators a and b, and $k = \mathbb{F}_2$ the field of two elements. Let V be a 3-dimensional space on which a and b act via the matrices

$$\begin{bmatrix} 1 & 0 & 0 \\ 1 & 1 & 0 \\ 0 & 0 & 1 \end{bmatrix} \quad \text{and} \quad \begin{bmatrix} 1 & 0 & 0 \\ 0 & 1 & 0 \\ 1 & 0 & 1 \end{bmatrix}.$$

Show that $S^2(V)$ is not isomorphic to either $ST^2(V)$ or $ST^2(V)^*$, where ST denotes the symmetric tensors. [Hint: Compute the dimensions of the spaces of fixed points of these representations.]

17. (Cauchy–Frobenius Lemma) A lemma often attributed to Burnside states that if a finite group G permutes a finite set Ω then the number of orbits of G on Ω equals the average number of fixed points of elements of G on Ω:

$$|G\backslash\Omega| = \frac{1}{|G|}\sum_{g\in G} |\Omega^{\langle g\rangle}|.$$

Prove this lemma by showing that both sides of the equation are equal to $\langle \chi_{\mathbb{C}}, \chi_{\mathbb{C}\Omega}\rangle$. [In the first edition of his book, Burnside attributed this result to Frobenius, who first stated and proved it. Frobenius, in turn, credited Cauchy with the transitive case of the result.]

18. (Artin's Induction Theorem) Let $\mathbb{C}^{cc(G)}$ denote the vector space of class functions on G and let $\mathcal{C}$ be a set of subgroups of G that contains a representative of each conjugacy class of cyclic subgroups of G. Consider the linear mappings

$$\mathrm{res}_{\mathcal{C}} : \mathbb{C}^{cc(G)} \to \bigoplus_{H \in \mathcal{C}} \mathbb{C}^{cc(H)}$$

and

$$\mathrm{ind}_{\mathcal{C}} : \bigoplus_{H \in \mathcal{C}} \mathbb{C}^{cc(H)} \to \mathbb{C}^{cc(G)}$$

whose component homomorphisms are the linear mappings given by restriction

$$\downarrow_H^G : \mathbb{C}^{cc(G)} \to \mathbb{C}^{cc(H)}$$

and induction

$$\uparrow_H^G : \mathbb{C}^{cc(H)} \to \mathbb{C}^{cc(G)}.$$

(a) With respect to the usual inner product $\langle\ ,\ \rangle_G$ on $\mathbb{C}^{cc(G)}$ and the inner product on $\bigoplus_{H \in \mathcal{C}} \mathbb{C}^{cc(H)}$ that is the orthogonal sum of the $\langle\ ,\ \rangle_H$, show that $\mathrm{res}_{\mathcal{C}}$ and $\mathrm{ind}_{\mathcal{C}}$ are the transpose of each other.

(b) Show that $\mathrm{res}_{\mathcal{C}}$ is injective.
[Use the fact that $\mathbb{C}^{cc(G)}$ has a basis consisting of characters, that take their information from cyclic subgroups.]

(c) Prove Artin's Induction Theorem: in $\mathbb{C}^{cc(G)}$ every character χ can be written as a rational linear combination

$$\chi = \sum a_{H,\psi} \psi \uparrow_H^G,$$

where the sum is taken over cyclic subgroups H of G, ψ ranges over characters of H and $a_{H,\psi} \in \mathbb{Q}$.
[Deduce this from surjectivity of $\mathrm{ind}_{\mathcal{C}}$ and the fact that it is given by a matrix with integer entries. A stronger version of Artin's Theorem is possible: there is a proof due to Brauer which gives an explicit formula for the coefficients $a_{H,\psi}$; from this, we may deduce that when χ is the character of a $\mathbb{Q}G$-module the ψ that arise may all be taken to be the trivial character.]

(d) Show that if U is any $\mathbb{C}G$-module then there are $\mathbb{C}G$-modules P and Q, each a direct sum of modules of the form $V \uparrow_H^G$ where H is cyclic, for various V and H, so that $U^n \oplus P \cong Q$ for some n, where U^n is the direct sum of n copies of U.

19. (Molien's Theorem) (a) Let $\rho : G \to GL(V)$ be a complex representation of G, so that V is a $\mathbb{C}G$-module, and for each n let $\chi_{S^n(V)}$ be the character of the nth symmetric power of V. Show that for each $g \in G$, there is an equality of formal power series

$$\sum_{n=0}^{\infty} \chi_{S^n(V)}(g)t^n = \frac{1}{\det(1 - t\rho(g))}.$$

Here t is an indeterminate, and the determinant that appears in this expression is of a matrix with entries in the polynomial ring $\mathbb{C}[t]$, so that the determinant is a polynomial in t. On expanding the rational function on the right, we obtain a formal power series that is asserted to be equal to the formal power series on the left.

[Choose a basis for V so that g acts diagonally, with eigenvalues $\xi_1, \dots, \xi_d$. Show that on both sides of the equation the coefficient of t^n is equal to $\sum_{i_1+\cdots+i_d=n} \xi_1^{i_1} \cdots \xi_d^{i_d}$.]

(b) If W is a simple $\mathbb{C}G$-module, we may write the multiplicity of W as a summand of $S^n(V)$ as $\langle \chi_{S^n(V)}, \chi_W \rangle$ and consider the formal power series

$$M_V(W) = \sum_{i=0}^{\infty} \langle \chi_{S^n(V)}, \chi_W \rangle t^n.$$

Show that

$$M_V(W) = \frac{1}{|G|} \sum_{g \in G} \frac{\chi_W(g^{-1})}{\det(1 - t\rho(g))}.$$

(c) When $G = S_3$ and V is the 2-dimensional simple $\mathbb{C}S_3$-module show that

$$\begin{aligned}
M_V(\mathbb{C}) &= \frac{1}{(1-t^2)(1-t^3)} \\
&= 1 + t^2 + t^3 + t^4 + t^5 + 2t^6 + t^7 + 2t^8 + 2t^9 + 2t^{10} + \cdots, \\
M_V(\epsilon) &= \frac{t^3}{(1-t^2)(1-t^3)} \\
&= t^3 + t^5 + t^6 + t^7 + t^8 + 2t^9 + t^{10} + \cdots, \\
M_V(V) &= \frac{t(1+t)}{(1-t^2)(1-t^3)} \\
&= t + t^2 + t^3 + 2t^4 + 2t^5 + 2t^6 + 3t^7 + 3t^8 + 3t^9 + 4t^{10} + \cdots,
\end{aligned}$$

where $\mathbb{C}$ denotes the trivial module and ϵ the sign representation. Deduce, for example, that the eighth symmetric power $S^8(V) \cong \mathbb{C}^2 \oplus \epsilon \oplus V^3$.

5

More on Induction and Restriction: Theorems of Mackey and Clifford

The results in this chapter go more deeply into the theory. They apply over fields of arbitrary characteristic, and even over arbitrary rings in the case of Mackey's decomposition formula. We start with this formula, which is a relationship between induction and restriction. After that, we explain Clifford's Theorem, which shows what happens when a simple representation is restricted to a normal subgroup. These results will have many consequences later on. At the end of the chapter, we will see the consequence of Clifford's Theorem that simple representations of p-groups are induced from 1-dimensional representations of subgroups.

5.1 Double Cosets

For Mackey's Theorem, we need to consider double cosets. Given subgroups H and K of G, we define for each $g \in G$ the *(H, K)-double coset*

$$HgK = \{hgk \mid h \in H, k \in K\}.$$

If Ω is a left G-set, we use the notation $G\backslash\Omega$ for the set of orbits of G on Ω, and denote a set of representatives for the orbits by $[G\backslash\Omega]$. Similarly if Ω is a right G-set, we write Ω/G and $[\Omega/G]$. We will use all the time the fact that if Ω is a transitive G-set and $\omega \in \Omega$ then $\Omega \cong G/\operatorname{Stab}_G(\omega)$, the set of left cosets of the stabilizer of ω in G.

Proposition 5.1.1. *Let $H, K \leq G$.*

(1) Each (H, K)-double coset is a disjoint union of right cosets of H and a disjoint union of left cosets of K.

(2) Any two (H, K)-double cosets either coincide or are disjoint. The (H, K)-double cosets partition G.

(3) *The set of* (H, K)*-double cosets is in bijection with the orbits* $H\backslash(G/K)$*, and also with the orbits* $(H\backslash G)/K$ *under the mappings*

$$HgK \mapsto H(gK) \in H\backslash(G/K)$$
$$HgK \mapsto (Hg)K \in (H\backslash G)/K.$$

Proof. (1) If $hgk \in HgK$ and $k_1 \in K$ then $hgk \cdot k_1 = hg(kk_1) \in HgK$ so that the entire left coset of K that contains hgk is contained in HgK. This shows that HgK is a union of left cosets of K, and similarly it is a union of right cosets of H.

(2) If $h_1g_1k_1 = h_2g_2k_2 \in Hg_1K \cap Hg_2K$ then $g_1 = h_1^{-1}h_2g_2k_2k_1^{-1} \in Hg_2K$ so that $Hg_1K \subseteq Hg_2K$, and similarly $Hg_2K \subseteq Hg_1K$. Thus, if two double cosets are not disjoint, they coincide.

(3) In this statement, G acts from the left on the left cosets G/K, hence so does H by restriction of the action. We denote the set of H-orbits on G/K by $H\backslash(G/K)$. The mapping

$$\{\text{double cosets}\} \to H\backslash(G/K)$$
$$HgK \mapsto H(gK)$$

is evidently well defined and surjective. If $H(g_1K) = H(g_2K)$ then $g_2K = hg_1K$ for some $h \in H$, so $g_2 \in Hg_1K$ and $Hg_1K = Hg_2K$ by (2). Hence the mapping is injective.

The proof that double cosets biject with $(H\backslash G)/K$ is similar. □

In view of (3), we denote the set of (H, K)-double cosets in G by $H\backslash G/K$. We denote a set of representatives for these double cosets by $[H\backslash G/K]$.

Example 5.1.2. Consider $S_2 = \{(), (12)\}$ as a subgroup of S_3. We have

$$S_2\backslash S_3/S_2 = \{\{(), (12)\}, \{(123), (132), (13), (23)\}\},$$

while, for example,

$$[S_2\backslash S_3/S_2] = \{(), (123)\}.$$

S_3 acts transitively on $\{1, 2, 3\}$ with $\mathrm{Stab}_{S_3}(3) = S_2$, so as S_3-sets we have

$$S_3/S_2 \cong \{1, 2, 3\}.$$

Thus, the set of orbits on this set under the action of S_2 is

$$S_2\backslash(S_3/S_2) \leftrightarrow \{\{1, 2\}, \{3\}\}.$$

We observe that these orbits are indeed in bijection with the double cosets $S_2\backslash S_3/S_2$.

This example illustrates the point that when computing double cosets it may be advantageous to identify G/K as some naturally occurring G-set, rather than as the set of left cosets.

In the next result, we distinguish between conjugation on the left and on the right: ${}^g x = gxg^{-1}$ and $x^g = g^{-1}xg$. Later on, we will write $c_g(x) = {}^g x$, so that $c_g : H \to {}^g H$ is the homomorphism that is left conjugation by g, and $c_{g^{-1}}(x) = x^g$.

Proposition 5.1.3. *Let H, K be subgroups of G and $g \in G$ an element. We have isomorphisms*

$$HgK/K \cong H/(H \cap {}^g K) \quad \textit{as left } H\textit{-sets}$$

and

$$H\backslash HgK \cong (H^g \cap K)\backslash K \quad \textit{as right } K\textit{-sets.}$$

Thus, the double coset HgK is a union of $|H : H \cap {}^g K|$ left K-cosets and also of $|K : H^g \cap K|$ right H-cosets. We have

$$|G : K| = \sum_{g \in [H\backslash G/K]} |H : H \cap {}^g K|$$

and

$$|G : H| = \sum_{g \in [H\backslash G/K]} |K : H^g \cap K|.$$

Proof. HgK is the union of a single H-orbit of left K-cosets. The stabilizer in H of one of these is

$$\begin{aligned} \mathrm{Stab}_H(gK) &= \{h \in H \mid hgK = gK\} \\ &= \{h \in H \mid h^g K = K\} \\ &= \{h \in H \mid h^g \in K\} \\ &= H \cap {}^g K. \end{aligned}$$

Thus, $HgK/K \cong H/(H \cap {}^g K)$ as left H-sets and the number of left K-cosets in HgK equals $|H : H \cap {}^g K|$. By summing these numbers over all double cosets, we obtain the total number of left K-cosets $|G : K|$.

The argument with right H-cosets is similar. □

5.2 Mackey's Theorem

We introduce *conjugation* of representations, a concept we have in fact already met with induced representations. Suppose H is a subgroup of G, $g \in G$, and V is a representation of H. We define a representation ${}^g V$ of ${}^g H$ by specifying that ${}^gV = V$ as a set, and if ${}^gh \in {}^gH$ then ${}^gh \cdot v = hv$. Thus, if $\rho : H \to GL(V)$ was the original representation, the conjugate representation is the composite homomorphism ${}^gH \xrightarrow{c_{g^{-1}}} H \xrightarrow{\rho} GL(V)$ where $c_{g^{-1}}({}^gh) = h$.

When studying the structure of induced representations

$$V \uparrow_H^G = \bigoplus_{g \in [G/H]} g \otimes V,$$

the subspace $g \otimes V$ is in fact a representation for ${}^gH$; for

$$ghg^{-1} \cdot (g \otimes v) = ghg^{-1}g \otimes v = gh \otimes v = g \otimes hv.$$

When $g \otimes V$ is identified with V via the linear isomorphism $g \otimes v \mapsto v$, the action of ${}^gH$ on V that arises coincides with the action we have just described on ${}^gV$.

Theorem 5.2.1 (Mackey decomposition formula). *Let H, K be subgroups of G and V a representation for K over a commutative ring R. Then*

$$(V \uparrow_K^G) \downarrow_H^G \cong \bigoplus_{g \in [H\backslash G/K]} ({}^g(V \downarrow_{H^g \cap K}^K)) \uparrow_{H \cap {}^gK}^H$$

as RH-modules.

Proof. We have $V \uparrow_K^G = \bigoplus_{x \in [G/K]} x \otimes V$. Consider a particular double coset HgK. The terms

$$\bigoplus_{\substack{x \in [G/K] \\ x \in HgK}} x \otimes V$$

form an R-submodule invariant under the action of H, since it is the direct sum of an orbit of R-submodules permuted by H. Now

$$\begin{aligned} \mathrm{Stab}_H(g \otimes V) &= \{h \in H \mid hg \otimes V = g \otimes V\} \\ &= \{h \in H \mid g^{-1}hg \in \mathrm{Stab}_G(1 \otimes V) = K\} \\ &= H \cap {}^gK. \end{aligned}$$

Therefore, as a representation for H, this subspace is $(g \otimes V) \uparrow_{H \cap {}^gK}^H$ by Proposition 4.3.2. As observed before the statement of this theorem, we have $g \otimes V \cong {}^g(V \downarrow_{H^g \cap K}^K)$ as a representation of $H \cap {}^gK$. Putting these expressions together gives the result. □

As an application of Mackey's Theorem, we consider permutation modules arising from multiply transitive G-sets. We say that a G-set Ω is *n-transitive* (or, more properly, the action of G on Ω is n-transitive) if Ω has at least n elements and for every pair of n-tuples $(a_1, \ldots, a_n)$ and $(b_1, \ldots, b_n)$ in which the a_i are distinct elements of Ω and the b_i are distinct elements of Ω, there exists $g \in G$ with $ga_i = b_i$ for every i. For example, S_n acts n-transitively on $\{1, \ldots, n\}$, and one may show that A_n acts $(n-2)$-transitively on $\{1, \ldots, n\}$ provided $n \geq 3$. Notice that if G acts n-transitively on Ω then it also acts $(n-1)$-transitively on Ω.

Lemma 5.2.2. *Let Ω be a G-set with at least n elements (where $n \geq 1$) and let $\omega \in \Omega$. Then G acts n-transitively on Ω if and only if G acts transitively on Ω and* $\mathrm{Stab}_G(\omega)$ *acts $(n-1)$-transitively on $\Omega - \{\omega\}$.*

Proof. If G acts n-transitively then G also acts transitively, and if $a_2, \ldots, a_n$ and $b_2, \ldots, b_n$ are two lists of $n-1$ distinct points of Ω, none of them equal to ω, then there exists $g \in G$ so that $g(\omega) = (\omega)$ and $g(a_i) = b_i$ for all i. This shows that $\mathrm{Stab}_G(\omega)$ acts $(n-1)$-transitively on $\Omega - \{\omega\}$.

Conversely, suppose G acts transitively on Ω and $\mathrm{Stab}_G(\omega)$ acts $(n-1)$-transitively on $\Omega - \{\omega\}$. Let $a_1, \ldots, a_n$ and $b_1, \ldots, b_n$ be two lists of n distinct points of Ω. We may find $u, v \in G$ so that $ua_1 = \omega$ and $v\omega = b_1$, by transitivity of G on Ω. The elements $\omega, ua_2, \ldots, ua_n$ are distinct, as are the points $\omega, v^{-1}b_2, \ldots, v^{-1}b_n$, so we can find $g \in \mathrm{Stab}_G\,\omega$ so that $gua_i = v^{-1}b_i$ when $2 \leq i \leq n$. Now $vgua_i = b_i$ for $1 \leq i \leq n$ and this shows that G acts n-transitively on Ω. □

Proposition 5.2.3. *Whenever Ω is a G-set the permutation module $\mathbb{C}\Omega$ may be written as a direct sum of $\mathbb{C}G$-modules*

$$\mathbb{C}\Omega = \mathbb{C} \oplus V$$

for some module V. Suppose that $|\Omega| \geq 2$, so $V \neq 0$. The representation V is simple if and only if G acts 2-transitively on Ω. In that case, V is not the trivial representation.

Proof. Pick any orbit of G on Ω. It is isomorphic as a G-set to G/H for some subgroup $H \leq G$ and so $\mathbb{C}[G/H]$ is a direct summand of $\mathbb{C}\Omega$, with character $1 \uparrow_H^G$. Since

$$\langle 1, 1 \uparrow_H^G \rangle_G = \langle 1, 1 \rangle_H = 1$$

by Frobenius reciprocity, we deduce that $\mathbb{C}$ is a summand of $\mathbb{C}[G/H]$ and hence of $\mathbb{C}\Omega$.

In the equivalence of statements that forms the third sentence, neither side is true if G has more than one orbit on Ω, so we may assume $\Omega = G/H$. The character of $\mathbb{C}\Omega$ is $1 \uparrow_H^G$, and we compute

$$\begin{aligned}
\langle 1 \uparrow_H^G, 1 \uparrow_H^G \rangle_G &= \langle (1 \uparrow_H^G) \downarrow_H^G, 1 \rangle_H \\
&= \langle \sum_{g \in [H\backslash G/H]} ({}^g 1) \uparrow_{H \cap {}^g H}^H, 1 \rangle_H \\
&= \sum_{g \in [H\backslash G/H]} \langle 1 \uparrow_{H \cap {}^g H}^H, 1 \rangle_H \\
&= \sum_{g \in [H\backslash G/H]} \langle 1, 1 \rangle_{H \cap {}^g H} \\
&= \sum_{g \in [H\backslash G/H]} 1 \\
&= |H\backslash G/H|,
\end{aligned}$$

using Frobenius reciprocity twice and Mackey's formula. Now $|H\backslash G/H|$ is the number of orbits of H (the stabilizer of a point) on G/H. By Lemma 5.2.2, this number is 2 if G acts 2-transitively on Ω, and otherwise, it is greater than 2 (since $|\Omega| \geq 2$ was a hypothesis). Writing $\mathbb{C}[G/H] = S_1 \oplus \cdots \oplus S_n$ as a direct sum of simple representations, we have

$$\langle 1 \uparrow_H^G, 1 \uparrow_H^G \rangle_G \geq n,$$

and we get the value 2 for the inner product if and only if there are 2 simple representations in this expression, and they are nonisomorphic. This is equivalent to requiring that V is simple, because it could only be the trivial representation if G acts trivially on G/H, which our hypotheses exclude. In any case, we deduce that V is not the trivial representation. □

Example 5.2.4. Let $\Omega = \{1, \ldots, n\}$ acted upon transitively by S_n and also by A_n. Then $\mathbb{C}\Omega \cong \mathbb{C} \oplus V$ where V is a simple representation of S_n, which remains simple on restriction to A_n provided $n \geq 4$.

5.3 Clifford's Theorem

We now turn to Clifford's Theorem, which we present in a weak and a strong form. The weak form is used as a step in proving the strong form a little later, and as a result in its own right it only has force in a situation where $|G|$ is not invertible in the ground ring. In these versions of Clifford's Theorem, we make the hypothesis that the ground ring is a field, but this is no loss of generality in view of Exercise 15 from Chapter 1.

Theorem 5.3.1 (Weak form of Clifford's Theorem). *Let k be any field, U a simple kG-module and N a normal subgroup of G. Then $U\downarrow_N^G$ is semisimple as a kN-module.*

Proof. Let V be any simple kN-submodule of $U\downarrow_N^G$. For every $g\in G$, gV is also a kN-submodule since if $n\in N$ we have $n(gv)=g(g^{-1}ng)v\in gV$, using the fact that N is normal. Evidently gV is also simple, since if W were a kN-submodule of gV then $g^{-1}W$ would be a submodule of V. Now $\sum_{g\in G} gV$ is a nonzero G-invariant subspace of the simple kG-module U, and so $\sum_{g\in G} gV=U$. As a kN-module, we see that $U\downarrow_N^G$ is the sum of simple submodules, and hence $U\downarrow_N^G$ is semisimple by the results of Chapter 1. □

The kN-submodules gV that appear in the proof of Theorem 5.3.1 are isomorphic to modules we have seen before. Since $N\lhd G$, the conjugate module ${}^gV$ is a representation for ${}^gN=N$. The mapping

$$\begin{aligned}{}^gV&\to gV\\ v&\mapsto gv\end{aligned}$$

is an isomorphism of kN-modules, since if $n\in N$ the action on ${}^gV$ is $n\cdot v=g^{-1}ngv$ and the action on gV is $n(gv)=g(g^{-1}ngv)$. Recall also that these modules appeared when we described induced modules. Part of Clifford's Theorem states that the simple module U is in fact an induced module.

Theorem 5.3.2 (Clifford's Theorem). *Let k be any field, U a simple kG-module and N a normal subgroup of G. We may write $U\downarrow_N^G=S_1^{a_1}\oplus\cdots\oplus S_r^{a_r}$ where the S_i are nonisomorphic simple kN-modules, occurring with multiplicities a_i. (We refer to the summands $S_i^{a_i}$ as the homogeneous components.) Then*

(1) G permutes the homogeneous components transitively;
(2) $a_1=a_2=\cdots=a_r$ and $\dim S_1=\dim S_2=\cdots=\dim S_r$; and
(3) if $H=\mathrm{Stab}_G(S_1^{a_1})$ then $U\cong S_1^{a_1}\uparrow_H^G$ as kG-modules.

Proof. The fact that $U\downarrow_N^G$ is semisimple and hence can be written as a direct sum as claimed follows from Theorem 5.3.1. We observe that, by Corollary 1.2.7, the homogeneous component $S_i^{a_i}$ is characterized as the unique largest kN-submodule that is isomorphic to a direct sum of copies of S_i. If $g\in G$ then $g(S_i^{a_i})$ is a direct sum of isomorphic simple modules gS_i, and so by this characterization must be contained in one of the homogeneous components: $g(S_i^{a_i})\subseteq S_j^{a_j}$ for some j. Since $U=g(S_1^{a_1})\oplus\cdots\oplus g(S_r^{a_r})$, by counting dimensions, we have $g(S_i^{a_i})=S_j^{a_j}$. Thus, G permutes the homogeneous components. Since $\sum_{g\in G} g(S_1^{a_1})$ is a nonzero G-invariant submodule of the simple module U, it must equal U, and so the action on the homogeneous components is transitive. This establishes (1), and (2) follows since for any pair (i, j), we can find $g \in G$ with $g(S_i^{a_i}) = S_j^{a_j}$, so $a_i = a_j$ and $\dim S_i = \dim S_j$. Finally, (3) is a direct consequence of Proposition 4.3.2. □

For now, we give just one application of Clifford's Theorem, which is Corollary 5.3.4. In the proof of Corollary 5.3.4, we will need a fact about representations of abelian groups that so far we have only proved when $|G|$ is invertible in Corollary 2.1.7 (and over $\mathbb{C}$ in Theorem 4.1.5).

Theorem 5.3.3. *Let k be any algebraically closed field. If G is abelian then every simple kG-module has dimension 1.*

Proof. Consider a simple kG-module S and let $g \in G$. In its action on S, g has an eigenvalue λ, with nonzero eigenspace S_λ. Since all elements $h \in G$ commute with g, we have $hS_\lambda = S_\lambda$ (by the argument that if $v \in S_\lambda$ then $gv = \lambda v$, so $g(hv) = h(gv) = h\lambda v = \lambda hv$, so $hv \in S_\lambda$; but also the action of h is invertible). Thus, S_λ is a kG-submodule of S, so $S_\lambda = S$ by simplicity of S. It follows that every element $g \in G$ acts by scalar multiplication on S, and such a simple module S must have dimension 1. □

As a consequence of the last result and Proposition 4.2.1, over an algebraically closed field the degree 1 representations of any group G are the same as the representations of G/G', lifted to G via the quotient homomorphism $G \to G'$.

The next result about simple representations of p-groups is true as stated when k has characteristic p, but it has no force in that situation because (as we will see in the next section) the only simple representation of a p-group in characteristic p is the trivial representation. We are thus only really interested in the following result over fields of characteristic other than p, and in particular over fields of characteristic 0.

Corollary 5.3.4. *Let k be any algebraically closed field and G a p-group. Then every simple module for G has the form $U \uparrow_H^G$ where U is a 1-dimensional module for some subgroup H.*

Proof. We proceed by induction on $|G|$. Let $\rho : G \to GL(S)$ be a simple representation of G over k and put $N = \operatorname{Ker} \rho$. Then S is really a representation of G/N. If $N \neq 1$ then G/N is a group of smaller order than G, so by induction S has the claimed structure as a representation of G/N, and hence also as a representation of G. Thus, we may assume $N = 1$ and G embeds in $GL(S)$.

If G is abelian then all simple representations are 1-dimensional, so we are done. Assume now that G is not abelian. Then G has a normal abelian subgroup A that is not central. To construct this subgroup A, let $Z_2(G)$ denote the second center of G, that is, the preimage in G of $Z(G/Z(G))$. If x is any element of $Z_2(G) - Z(G)$ then $A = \langle Z(G), x\rangle$ is a normal abelian subgroup not contained in $Z(G)$.

We apply Clifford's Theorem:

$$S \downarrow_A^G = S_1^{a_1} \oplus \cdots \oplus S_r^{a_r}$$

and $S = V \uparrow_K^G$ where $V = S_1^{a_1}$ and $K = \mathrm{Stab}_G(S_1^{a_1})$. We argue that V must be a simple kK-module, since if it had a proper submodule W then $W \uparrow_K^G$ would be a proper submodule of S, which is simple. If $K \neq G$ then by induction $V = U \uparrow_H^K$ where U is 1-dimensional, and so $S = (U \uparrow_H^K) \uparrow_K^G = U \uparrow_H^G$ has the required form.

We show finally that the case $K = G$ cannot happen. For if it were to happen then $S \downarrow_A^G = S_1^{a_1}$, and since A is abelian $\dim S_1 = 1$. The elements of A must therefore act via scalar multiplication on S. Since such an action would commute with the action of G, which is faithfully represented on S, we deduce that $A \subseteq Z(G)$, a contradiction. □

This result is useful if we are constructing the character table of a p-group, because it says that we need look no further than induced characters. We note that the conclusion of Corollary 5.3.4 also applies to supersolvable groups, which again have the property, if they are not abelian, that they have a noncentral normal abelian subgroup.

A representation of the form $U \uparrow_H^G$ for some subgroup H and with U a 1-dimensional representation of H is said to be *monomial*. A group G all of whose irreducible complex representations are monomial is called an *M-group*. Thus, p-groups (and also supersolvable groups) are M-groups.

5.4 Summary of Chapter 5

- The Mackey formula: induction followed by restriction is a sum over double cosets of restriction followed by conjugation followed by induction.
- A permutation representation is 2-transitive if and only if the complex permutation module has two summands.
- Clifford's Theorem: the restriction of a simple module to a normal subgroup is semisimple, and the module is induced from the stabilizer of a homogeneous component.
- For a p-group over an algebraically closed field, every simple module is induced from a 1-dimensional module.

5.5 Exercises for Chapter 5

1. Let k be any field, and g any element of a finite group G.

(a) If $K \leq H \leq G$ are subgroups of G, V a kH-module, and W a kK-module, show that $({}^gV) \downarrow^{{}^gH}_{{}^gK} \cong {}^g(V \downarrow^H_K)$ and $({}^gW) \uparrow^{{}^gH}_{{}^gK} \cong {}^g(W \uparrow^H_K)$. [This allows us to put conjugation before, between, or after restriction and induction in Mackey's formula.]

(b) If U is any kG-module, show that $U \cong {}^gU$ by showing that one of the two mappings $U \to {}^gU$ specified by $u \mapsto gu$ and $u \mapsto g^{-1}u$ is always an RG-module isomorphism. [Find which one of these it is.]

2. Let H and K be subgroups of G with $HK = G$ and $H \cap K = 1$. Show that for any kH-module U the module $U \uparrow^G_H \downarrow^G_K$ is a direct sum of copies of the regular representation kK.

3. Let H and K be subgroups of G and consider the permutation modules $R \uparrow^G_H$ and $R \uparrow^G_K$ over a commutative ring R. Show that the space of homomorphisms $\mathrm{Hom}_{RG}(R \uparrow^G_H, R \uparrow^G_K)$ is free as an R-module, with a basis in bijection with the double cosets $H\backslash G/K$. Show that if $R \to S$ is a surjective ring homomorphism then the induced homomorphism

$$\mathrm{Hom}_{RG}(R \uparrow^G_H, R \uparrow^G_K) \to \mathrm{Hom}_{SG}(S \uparrow^G_H, S \uparrow^G_K)$$

is surjective. Deduce that every module homomorphism between SG-permutation modules lifts to a homomorphism between RG-permutation modules.

4. Let k be a field. Show by example that it is possible to find a subgroup H of a group G and a simple kG-module U for which $U \downarrow^G_H$ is not semisimple.

5. Find the complete list of subgroups H of the dihedral group D_8 such that the 2-dimensional simple representation over $\mathbb{C}$ can be written $U \uparrow^G_H$ for some 1-dimensional representation U of H. Do the same thing for the quaternion group Q_8.

6. Compute the character tables of the generalized quaternion group of order 16

$$Q_{16} = \langle x, y \mid x^8 = 1,\ x^4 = y^2,\ yxy^{-1} = x^{-1} \rangle$$

and the semidihedral group of order 16:

$$SD_{16} = \langle x, y \mid x^8 = y^2 = 1,\ yxy^{-1} = x^3 \rangle.$$

7. The following statements generalize Lemma 3.2.2 and Maschke's Theorem. Let H be a subgroup of G and suppose that k is a field in which $|G : H|$ is invertible. Let V be a kG-module.

(a) Show that

$$\frac{1}{|G:H|} \sum_{g \in [G/H]} g : V^H \to V^G$$

is a well-defined map that is a projection of the H-fixed points onto the G-fixed points. In particular, this map is surjective.

(b) Show that if $V \downarrow_H^G$ is semisimple as a kH-module then V is semisimple as a kG-module.

8. Let H be a normal subgroup of G and suppose that k is a field of characteristic p.

(a) Let $p \nmid |G : H|$. Show that if U is a semisimple kH-module then $U \uparrow_H^G$ is a semisimple kG-module.

(b) Let $p \mid |G : H|$. Show by example that if U is a semisimple kH-module then it need not be the case that $U \uparrow_H^G$ is a semisimple kG-module.

9. Let H be a subgroup of G of index 2 (so that H is normal in G) and let k be a field whose characteristic is not 2. The homomorphism $G \to \{\pm 1\} \subset k$ with kernel H is a 1-dimensional representation of G that we will call ϵ. Let S, T be simple kG-modules and let U, V be simple kH-modules. You may assume that $U \uparrow_H^G$ and $V \uparrow_H^G$ are semisimple (this is proved as Exercise 8(a)). Let $g \in G - H$.

(a) Show that $S \downarrow_H^G$ is the direct sum of either 1 or 2 simple kH-modules.

(b) Show that $U \uparrow_H^G$ is the direct sum of either 1 or 2 simple kG-modules.

In the following questions, notice that

$$S \downarrow_H^G \uparrow_H^G \cong S \otimes (k \uparrow_H^G) \cong S \otimes (k \oplus \epsilon) \cong S \oplus (S \otimes \epsilon).$$

For some parts of the questions, it may help to consider

$$\mathrm{Hom}_{kH}(S \downarrow_H^G, T \downarrow_H^G) \quad \text{and} \quad \mathrm{Hom}_{kG}(U \uparrow_H^G, V \uparrow_H^G).$$

(c) Show that the following are equivalent:

  (i) S is the induction to G of a kH-module,

  (ii) $S \downarrow_H^G$ is not simple,

  (iii) $S \cong S \otimes \epsilon$.

(d) Show that the following are equivalent:

  (i) U is the restriction to H of a kG-module,

(ii) $U \uparrow_H^G$ is not simple,
(iii) $U \cong {}^gU$.

(e) Show that $S \downarrow_H^G$ and $T \downarrow_H^G$ have a summand in common if and only if $S \cong T$ or $S \cong T \otimes \epsilon$.

(f) Show that $U \uparrow_H^G$ and $V \uparrow_H^G$ have a summand in common if and only if $U \cong V$ or $U \cong {}^gV$.

(g) We place an equivalence relation $\sim_1$ on the simple kG-modules and an equivalence relation $\sim_2$ on the simple kH-modules:

$$S \sim_1 T \Leftrightarrow S \cong T \text{ or } S \cong T \otimes \epsilon$$
$$U \sim_2 V \Leftrightarrow U \cong V \text{ or } U \cong {}^gV.$$

Show that induction $\uparrow_H^G$ and restriction $\downarrow_H^G$ induce mutually inverse bijections between the equivalence classes of simple kG-modules and of simple kH-modules in such a way that an equivalence class of size 1 corresponds to an equivalence class of size 2, and vice-versa.

(h) Show that the simple kG-modules of odd degree restrict to simple kH-modules, and the number of such modules is even.

(i) In the case where $G = S_4$, $H = A_4$ and $k = \mathbb{C}$, show that there are three equivalence classes of simple characters under $\sim_1$ and $\sim_2$. Verify that $\downarrow_{A_4}^{S_4}$ and $\uparrow_{A_4}^{S_4}$ give mutually inverse bijections between the equivalence classes.

10. Let $G = GL(3, 2)$ be the group of 3×3 invertible matrices over $k = \mathbb{F}_2$ and let

$$H = \left\{ \begin{bmatrix} a & b & 0 \\ c & d & 0 \\ e & f & 1 \end{bmatrix} \;\middle|\; a, b, c, d, e, f, 1 \in \mathbb{F}_2,\ (ad - bc) \neq 0 \right\}.$$

You may assume from group theory that $|G| = 168$. Let V be the natural 3-dimensional space of column vectors on which G-acts.

(a) Show that $|H| = 24$, so that $|G : H| = 7$.

(b) Show that V is simple as a kG-module.

(c) Show that as a kH-module $V \downarrow_H^G$ has a simple socle with trivial H action, and such that the quotient of V by the socle is a simple 2-dimensional module.

(d) Show that $\dim \mathrm{Hom}_{kG}(k \uparrow_H^G, V) = 1$ and $\dim \mathrm{Hom}_{kG}(V, k \uparrow_H^G) = 0$.

(e) Show that $k \uparrow_H^G$ is not semisimple, thereby showing that even if $p \nmid |G : H|$ it need not be the case that the induction of a simple module is semisimple when H is not normal.

(f) Show that $\dim \operatorname{Hom}_{kG}(k\uparrow_H^G, V^*) = 0$ and $\dim \operatorname{Hom}_{kG}(V^*, k\uparrow_H^G) = 1$. Show that $V \not\cong V^*$. Show that $k\uparrow_H^G$ is the direct sum of the trivial module k and a 6-dimensional module that has socle V^* and socle quotient V.

6

Representations of p-Groups in Characteristic p and the Radical

The study of representations of a group over a field whose characteristic divides the group order is more delicate than the case of ordinary representation theory. Modules no longer need be semisimple, and we have to do more than count multiplicities of simple direct summands to determine their isomorphism type. In this chapter, we begin the task of assembling techniques specifically aimed at dealing with this. As a first step, we focus on representations of p-groups in characteristic p. Specific things may be said about them with very little background preparation, and they have impact on representations of all groups. In subsequent chapters, we will gradually fill in the rest of the picture. We start by describing completely the representations of cyclic p-groups. We show that p-groups have only one simple module in characteristic p. We introduce the radical and socle series of modules and deduce that the regular representation is indecomposable, identifying its radical as the augmentation ideal. We conclude with a discussion of Jennings's Theorem on the radical series of the group algebra of a p-group in characteristic p.

6.1 Cyclic p-Groups

We describe all representations of cyclic p-groups over a field of characteristic p using elementary methods. In the first proposition, we reduce their study to that of modules for a principal ideal domain. When G is cyclic of order N, we have already made use of an isomorphism between kG and $k[X]/(X^N - 1)$ in Exercise 11 from Chapter 2. When N is a power of p, we can express this slightly differently.

Proposition 6.1.1. *Let k be a field of characteristic p and let $G = \langle g \mid g^{p^n} = 1\rangle$ be cyclic of order p^n. Then there is a ring isomorphism $kG \cong k[X]/(X^{p^n})$, where $k[X]$ is the polynomial ring in an indeterminate X.*

Proof. We define a mapping

$$G \to k[X]/(X^{p^n})$$
$$g^s \mapsto (X+1)^s.$$

Since

$$(X+1)^{p^n} = X^{p^n} + p(\cdots) + 1 \equiv 1 \ (\mathrm{mod}(X^{p^n})),$$

this mapping is a group homomorphism to the unit group of $k[X]/(X^{p^n})$, and hence it extends to a linear map

$$kG \to k[X]/(X^{p^n})$$

that is an algebra homomorphism. Since g^s is sent to X^s plus terms of lower degree, the images of $1, \ldots, g^{p^n-1}$ form a basis of $k[X]/(X^{p^n})$. The mapping therefore gives a bijection between a basis of kG and a basis of $k[X]/(X^{p^n})$, and so is an isomorphism. □

Direct sum decompositions of modules are the first consideration in describing their structure. We say that a module U for a ring A is *indecomposable* if it cannot be expressed as a direct sum of two modules except in a trivial way, that is, if $U \cong V \oplus W$ then either $V = 0$ or $W = 0$. When A is an algebra over a field, by repeatedly expressing summands of a module as further direct sums, we can express any finite-dimensional module as a direct sum of indecomposable direct summands. It is useful to know, but we will not prove it until Theorem 11.1.6, that for each module these summands are determined up to isomorphism, independently of the choice of direct sum decomposition. This is the content of the Krull–Schmidt Theorem. We point out that, since we need not be in characteristic zero when group representations are semisimple, indecomposable modules need not be simple. This is the point of introducing the new terminology! An example of an indecomposable module that is not simple was given Example 1.1.7, and we will see many more examples.

A module over a ring is said to be *cyclic* if it can be generated by one element. We now exploit the structure theorem for finitely generated modules over a principal ideal domain, which says that such modules are direct sums of cyclic modules.

Theorem 6.1.2. *Let k be a field of characteristic p. Every finitely generated $k[X]/(X^{p^n})$-module is a direct sum of cyclic modules $U_r = k[X]/(X^r)$ where $1 \le r \le p^n$. The only simple module is the 1-dimensional module U_1. Each module U_r has a unique composition series, and hence is indecomposable. From this, it follows that if G is cyclic of order p^n then kG has exactly p^n indecomposable modules, one of each dimension i with $1 \le i \le p^n$, each having a unique composition series.*

Proof. The modules for $k[X]/(X^{p^n})$ may be identified with the modules for $k[X]$ on which X^{p^n} acts as zero. Every finitely generated $k[X]$-module is a direct sum of modules $k[X]/I$ where I is an ideal. Hence every $k[X]/(X^{p^n})$-module is a direct sum of modules $k[X]/I$ on which X^{p^n} acts as zero, which is to say $(X^{p^n}) \subseteq I$. The ideals I that satisfy this last condition are the ideals (a) where $a \mid X^{p^n}$. This forces $I = (X^r)$ where $1 \leq r \leq p^n$, and $k[X]/I = U_r$.

The submodules of U_r must have the form $J/(X^r)$ where J is some ideal containing (X^r), and they are precisely the submodules in the chain

$$0 \subset (X^{r-1})/(X^r) \subset (X^{r-2})/(X^r) \subset \cdots \subset (X)/(X^r) \subset U_r.$$

This is a composition series, since each successive quotient has dimension 1, and since it is a complete list of submodules, it is the only one. If we could write $U_r = V \oplus W$ as a nontrivial direct sum, then U_r would have at least 2 composition series, obtained by taking first a composition series for V, then one for W, or vice-versa. Hence each U_r is indecomposable, and we have a complete list of the indecomposable modules. The only U_r that is simple is U_1, which is the trivial module.

The final identification of the indecomposable kG-modules comes from the isomorphism in Proposition 6.1.1. □

A module with a unique composition series is said to be *uniserial*. It is equivalent to say that its submodules are linearly ordered by inclusion, and the equivalence of these and other conditions are explored in Exercises 3 and 6 at the end of this chapter.

Example 6.1.3. We see from the description of $k[X]/(X^{p^n})$-modules that U_r has a basis $1 + (X^r), X + (X^r), \ldots, X^{r-1} + (X^r)$ so that X acts on U_r with matrix

$$\begin{bmatrix} 0 & & & \\ 1 & 0 & & \\ & \ddots & \ddots & \\ & & 1 & 0 \end{bmatrix}.$$

Translating now to modules for kG where G is a cyclic p-group, the generator g acts on U_r as $X + 1$, which has matrix

$$\begin{bmatrix} 1 & & & \\ 1 & 1 & & \\ & \ddots & \ddots & \\ & & 1 & 1 \end{bmatrix}.$$

Thus, we see that the indecomposable kG-modules are exactly given by specifying that the generator g acts via a matrix that is a single Jordan block, of size up to p^n. It is helpful to picture U_r using a diagram

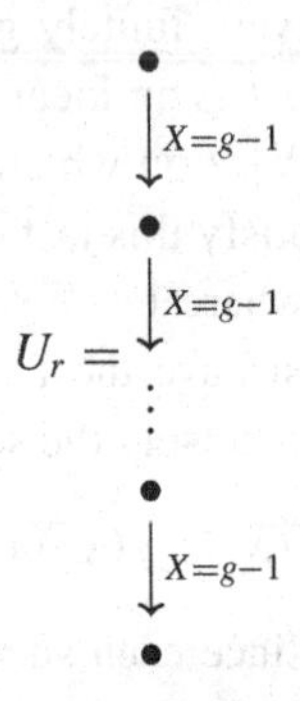

that may be interpreted by saying that the vertices are in bijection with a basis of U_r, and the action of X or $g-1$ is given by the arrows. Where no arrow is shown starting from a particular vertex (as happens in this case only with the bottom vertex), the interpretation is that X and $g-1$ act as zero.

6.2 Simple Modules for Groups with Normal *p*-Subgroups

The following seemingly innocuous result has profound consequences throughout the rest of this book.

Proposition 6.2.1. *Let k be a field of characteristic p and G a p-group. The only simple kG-module is the trivial module.*

Proof. We offer two proofs of this.

Proof 1. We proceed by induction on $|G|$, the induction starting when G is the identity group, for which the result is true. Suppose $G \neq 1$ and the result is true for p-groups of smaller order. There exists a normal subgroup N of G of index p. If S is any simple kG-module then by Clifford's Theorem $S \downarrow_N^G$ is semisimple. By induction, N acts trivially on S. Thus, S is really a representation of G/N that is cyclic of order p. We have just proved that the only simple representation of this group is the trivial representation.

Proof 2. Let S be any simple kG-module and let $0 \neq x \in S$. The subgroup of S generated by the elements $\{gx \mid g \in G\}$ is invariant under the action of G, it is abelian and of exponent p, since it is a subgroup of a vector space in characteristic p. Thus, it is a finite p group acted on by G. Consider the orbits of G on this finite group. Since G is a p-group the orbits all have size a power of p (or 1), because the size of an orbit is the index of the stabilizer of an element in the orbit. The zero element is fixed by G, and we deduce that there must be another element fixed by G since otherwise the other orbits would all have size p^n with $n \geq 1$, and their union would not be a p-group. Thus, there exists $y \in S$ fixed by G, and now $\langle y \rangle$ is a trivial submodule of S. By simplicity, it must equal S. □

As an application of this, we can give some information about the simple representations of arbitrary finite groups in characteristic p. For this, we observe that in every finite group G there is a unique largest normal p-subgroup of G, denoted $O_p(G)$. For if H and K are normal p-subgroups of G then so is HK, and thus the subgroup generated by all normal p-subgroups of G is again a normal p-subgroup, that evidently contains all the others.

Corollary 6.2.2. *Let k be a field of characteristic p and G a finite group. Then the common kernel of the action of G on all the simple kG-modules is $O_p(G)$. Thus, the simple kG-modules are precisely the simple $k[G/O_p(G)]$-modules, made into kG-modules via the quotient homomorphism $G \to G/O_p(G)$.*

Proof. Let H be the kernel of the action of G on all simple kG-modules, that is,

$$H = \{g \in G \mid \text{for all simple } S \text{ and for all } s \in S,\ gs = s\}.$$

By Clifford's Theorem, if S is a simple kG-module then $S \downarrow^G_{O_p(G)}$ is semisimple. Therefore, by Proposition 6.2.1, $O_p(G)$ acts trivially on S, so that $O_p(G) \subseteq H$. We show that H contains no element of order prime to p. For, suppose $h \in H$ were to have order prime to p. Then $kG \downarrow^G_{\langle h \rangle}$ would be a semisimple $k\langle h \rangle$-module that is the direct sum of modules $S \downarrow^G_{\langle h \rangle}$ with S a simple kG-module. Since h acts trivially on all of these, it must act trivially on kG, which is a contradiction. Therefore, H is a p-group, and since it is normal, $O_p(G) \supseteq H$. We therefore have equality. The last sentence is immediate. □

Example 6.2.3. When G has a normal Sylow p-subgroup H it is a semidirect product $G = H \rtimes K$ for some subgroup K of order prime to p, by the Schur–Zassenhaus Theorem. Groups with this structure include nilpotent groups, which are direct products of their Sylow subgroups and, of course, abelian groups. In this situation, $O_p(G) = H$, and so when k has characteristic p the simple kG-modules may be identified with the simple kK-modules, lifted (or inflated) to representations of G via the quotient homomorphism $G \to K$. We obtain the simple modules for many groups in this way, such as for S_3 over $\mathbb{F}_3$, where the two simple modules are the trivial and sign representations in characteristic 3.

For a different example, let k be a field of characteristic 2, and consider the representations of A_4 over k. Since $O_2(A_4) = C_2 \times C_2$, the simple kA_4 modules are the simple $C_3 = A_4/O_2(A_4)$-representations, made into representations of A_4. Now kC_3 is semisimple, and if k contains a primitive cube root of unity ω

(i.e., if $\mathbb{F}_4 \subseteq k$) there are three 1-dimensional simple representations, on which the generator of C_3 acts as 1, ω or ω^2.

6.3 Radicals, Socles, and the Augmentation Ideal

At this point, we examine further the structure of representations that are not semisimple, and we work in the context of modules for a ring A, that is always supposed to have a 1. At the end of Chapter 1, we defined the *socle* of an A-module U to be the sum of all the simple submodules of U, and we showed (at least in the case that U is finite-dimensional) that it is the unique largest semisimple submodule of U. We now work with quotients and define a dual concept, the *radical* of U. We work with quotients instead of submodules, and use the fact that if M is a submodule of U, the quotient U/M is simple if and only if M is a maximal submodule of U. We put

$$\operatorname{Rad} U = \bigcap\{M \mid M \text{ is a maximal submodule of } U\}.$$

In our applications U will always be Noetherian, so provided $U \neq 0$ this intersection will be nonempty and hence $\operatorname{Rad} U \neq U$. If U has no maximal submodules (for example, if $U = 0$, or in more general situations then we consider here where U might not be Noetherian) we set $\operatorname{Rad} U = U$.

Lemma 6.3.1. *Let U be a module for a ring A.*

(1) *Suppose that $M_1, \ldots, M_n$ are maximal submodules of U. Then there is a subset $I \subseteq \{1, \ldots, n\}$ such that*

$$U/(M_1 \cap \cdots \cap M_n) \cong \bigoplus_{i \in I} U/M_i$$

which, in particular, is a semisimple module.

(2) *Suppose further that U has the descending chain condition on submodules. Then $U/\operatorname{Rad} U$ is a semisimple module, and $\operatorname{Rad} U$ is the unique smallest submodule of U with this property.*

Proof. (1) Let I be a subset of $\{1, \ldots, n\}$ maximal with the property that the quotient homomorphisms $U/(\bigcap_{i \in I} M_i) \to U/M_i$ inducc an isomorphism $U/(\bigcap_{i \in I} M_i) \cong \bigoplus_{i \in I} U/M_i$. We show that $\bigcap_{i \in I} M_i = M_1 \cap \cdots \cap M_n$ and argue by contradiction. If it were not the case, there would exist M_j with $\bigcap_{i \in I} M_i \not\subseteq M_j$. Consider the homomorphism

$$f : U \to (\bigoplus_{i \in I} U/M_i) \oplus U/M_j$$

whose components are the quotient homomorphisms $U \to U/M_k$. This has kernel $M_j \cap \bigcap_{i\in I} M_i$, and it will suffice to show that f is surjective, because this will imply that the larger set $I \cup \{j\}$ has the same property as I, thereby contradicting the maximality of I.

To show that f is surjective let $g : U \to U/\bigcap_{i\in I} M_i \oplus U/M_j$ and observe that $(\bigcap_{i\in I} M_i) + M_j = U$ since the left-hand side is strictly larger than M_j, which is maximal in U. Thus, if $x \in U$, we can write $x = y + z$ where $y \in \bigcap_{i\in I} M_i$ and $z \in M_j$. Now $g(y) = (0, x + M_j)$ and $g(z) = (x + \bigcap_{i\in I} M_i, 0)$ so that both summands $U/\bigcap_{i\in I} M_i$ and U/M_j are contained in the image of g, and g is surjective. Since f is obtained by composing g with the isomorphism that identifies $U/\bigcap_{i\in I} M_i$ with $\bigoplus_{i\in I} U/M_i$, we deduce that f is surjective.

(2) By the assumption that U has the descending chain condition on submodules, $\operatorname{Rad} U$ must be the intersection of finitely many maximal submodules. Therefore, $U/\operatorname{Rad} U$ is semisimple by part (1). If V is a submodule such that U/V is semisimple, say $U/V \cong S_1 \oplus \cdots \oplus S_n$ where the S_i are simple modules, let M_i be the kernel of $U \to U/V \xrightarrow{\text{proj.}} S_i$. Then M_i is maximal and $V = M_1 \cap \cdots \cap M_n$. Thus, $V \supseteq \operatorname{Rad} U$, and $\operatorname{Rad} U$ is contained in every submodule V for which U/V is semisimple. □

We define the *radical* of a ring A to be the radical of the regular representation $\operatorname{Rad}{}_A A$ and write simply $\operatorname{Rad} A$. We present some identifications of the radical that are very important theoretically, and also in determining what it is in particular cases.

Proposition 6.3.2. *Let A be a ring. Then,*

(1) $\operatorname{Rad} A = \{a \in A \mid a \cdot S = 0$ *for every simple A-module $S\}$, and*
(2) $\operatorname{Rad} A$ *is a 2-sided ideal of A.*
(3) Suppose further that A is a finite-dimensional algebra over a field. Then
  (a) $\operatorname{Rad} A$ *is the smallest left ideal of A such that $A/\operatorname{Rad} A$ is a semisimple A-module,*
  (b) A is semisimple if and only if $\operatorname{Rad} A = 0$,
  (c) $\operatorname{Rad} A$ *is nilpotent, and is the largest nilpotent ideal of A.*
  (d) $\operatorname{Rad} A$ *is the unique ideal U of A with the property that U is nilpotent and A/U is semisimple.*

Proof. (1) Given a simple module S and $0 \neq s \in S$, the module homomorphism ${}_A A \to S$ given by $a \mapsto as$ is surjective and its kernel is a maximal left ideal M_s. Now if $a \in \operatorname{Rad} A$ then $a \in M_s$ for every S and $s \in S$, so $as = 0$ and a annihilates every simple module. Conversely, if $a \cdot S = 0$ for every simple module S and M

is a maximal left ideal then A/M is a simple module. Therefore, $a \cdot (A/M) = 0$, which means $a \in M$. Hence $a \in \bigcap_{\text{maximal}M} M = \operatorname{Rad} A$.

(2) Being the intersection of left ideals, $\operatorname{Rad} A$ is also a left ideal of A. Suppose that $a \in \operatorname{Rad} A$ and $b \in A$, so $a \cdot S = 0$ for every simple S. Now $a \cdot bS \subseteq a \cdot S = 0$ so ab has the same property that a does.

(3) (a) and (b) are immediate from Lemma 6.3.1. We prove (c). Choose any composition series

$$0 = A_n \subset A_{n-1} \subset \cdots \subset A_1 \subset A_0 = {}_AA$$

of the regular representation. Since each A_i/A_{i+1} is a simple A-module, $\operatorname{Rad} A \cdot A_i \subseteq A_{i+1}$ by part (1). Hence $(\operatorname{Rad} A)^r \cdot A \subseteq A_r$ and $(\operatorname{Rad} A)^n = 0$.

Suppose now that I is a nilpotent ideal of A, say $I^m = 0$, and let S be any simple A-module. Then

$$0 = I^m \cdot S \subseteq I^{m-1} \cdot S \subseteq \cdots \subseteq IS \subseteq S$$

is a chain of A-submodules of S that are either 0 or S since S is simple. There must be some point where $0 = I^r S \neq I^{r-1} S = S$. Then $IS = I \cdot I^{r-1} S = I^r S = 0$, so in fact that point was the very first step. This shows that $I \subseteq \operatorname{Rad} A$ by part (1). Hence $\operatorname{Rad} A$ contains every nilpotent ideal of A, so is the unique largest such ideal.

Finally, (d) follows from (a) and (c): these imply that $\operatorname{Rad} A$ has the properties stated in (d); and, conversely, these conditions on an ideal U imply by (a) that $U \supseteq \operatorname{Rad} A$, and by (c) that $U \subseteq \operatorname{Rad} A$. □

Note that if I is a nilpotent ideal of A then it is always true that $I \subseteq \operatorname{Rad}(A)$ without the assumption that A is a finite-dimensional algebra. The argument given to prove part 3c of Proposition 6.3.2 shows this.

For any group G and commutative ring R with a 1, the ring homomorphism

$$\epsilon : RG \to R$$
$$g \mapsto 1 \quad \text{for all } g \in G$$

is called the *augmentation map*. As well as being a ring homomorphism it as a homomorphism of RG-modules, in which case it expresses the trivial representation as a homomorphic image of the regular representation. The kernel of ϵ is called the *augmentation ideal*, and is denoted IG. Evidently IG consists of those elements $\sum_{g\in G} a_g g \in RG$ such that $\sum_{g\in G} a_g = 0$. We now show that when k is a field of characteristic p and G is a p-group this construction gives the radical of kG.

Proposition 6.3.3. *Let G be a finite group and R a commutative ring with a 1.*

(1) *Let R denote the trivial RG-module. Then $IG = \{x \in RG \mid x \cdot R = 0\}$.*
(2) *IG is free as an R-module with basis $\{g - 1 \mid 1 \neq g \in G\}$.*
(3) *If $R = k$ is a field of characteristic p and G is a p-group then $IG = \mathrm{Rad}(kG)$. It follows that IG is nilpotent in this case.*

Proof. (1) The augmentation map ϵ is none other than the linear extension to RG of the homomorphism $\rho : G \to GL(1, R)$ that is the trivial representation. Thus, each $x \in RG$ acts on R as multiplication by $\epsilon(x)$, and so will act as 0 precisely if $\epsilon(x) = 0$.

(2) The elements $g - 1$ where g ranges through the nonidentity elements of G are linearly independent since the elements g are, and they lie in IG. We show that they span IG. Suppose $\sum_{g\in G} a_g g \in IG$, which means that $\sum_{g\in G} a_g = 0 \in R$. Then

$$\sum_{g\in G} a_g g = \sum_{g\in G} a_g g - \sum_{g\in G} a_g 1 = \sum_{1\neq g\in G} a_g(g - 1)$$

is an expression as a linear combination of elements $g - 1$.

(3) When G is a p-group and $\mathrm{char}(k) = p$ we have seen in Proposition 6.2.1 that k is the only simple kG-module. The result follows by part (1) and Proposition 6.3.2. □

Working in the generality of a finite-dimensional algebra A again, the radical of A allows us to give a further description of the radical and socle of a module. We present this result for finite-dimensional modules, but it is in fact true without this hypothesis. We leave this stronger version to Exercise 27 at the end of this chapter.

Proposition 6.3.4. *Let A be a finite-dimensional algebra over a field k, and U a finite-dimensional A-module.*

(1) *The following are all descriptions of* $\mathrm{Rad}\, U$*:*
   (a) *the intersection of the maximal submodules of U,*
   (b) *the smallest submodule of U with semisimple quotient,*
   (c) $\mathrm{Rad}\, A \cdot U$.
(2) *The following are all descriptions of* $\mathrm{Soc}\, U$*:*
   (a) *the sum of the simple submodules of U,*
   (b) *the largest semisimple submodule of U,*
   (c) $\{u \in U \mid \mathrm{Rad}\, A \cdot u = 0\}$.

Proof. Under the hypothesis that U is finitely generated, we have seen the equivalence of descriptions (a) and (b) in Lemma 6.3.1 and Corollary 1.2.6. The following arguments actually work without the hypothesis of finite generation, provided we assume the results of Exercises 13 and 14 from Chapter 1. The reader who is satisfied with a proof for finitely generated modules can assume that the equivalence of (a) and (b) has already been proved.

Let us show that the submodule $\mathrm{Rad}\, A \cdot U$ in (1)(c) satisfies condition (1)(b). Firstly, $U/(\mathrm{Rad}\, A \cdot U)$ is a module for $A/\mathrm{Rad}\, A$, which is a semisimple algebra. Hence $U/(\mathrm{Rad}\, A \cdot U)$ is a semisimple module and so $\mathrm{Rad}\, A \cdot U$ contains the submodule of (1)(b). On the other hand, if $V \subseteq U$ is a submodule for which U/V is semisimple then $\mathrm{Rad}\, A \cdot (U/V) = 0$ by Proposition 6.3.2, so $V \supseteq \mathrm{Rad}\, A \cdot U$. In particular, the submodule of (1)(b) contains $\mathrm{Rad}\, A \cdot U$. This shows that the descriptions in (1)(b) and (1)(c) are equivalent.

To show that they give the same submodule as (1)(a), observe that if V is any maximal submodule of U, then (by an argument just used, since U/V is simple) $V \supseteq \mathrm{Rad}\, A \cdot U$, so the intersection of maximal submodules of U contains $\mathrm{Rad}\, A \cdot U$. The intersection of maximal submodules of the semisimple module $U/(\mathrm{Rad}\, A \cdot U)$ is zero, so this gives a containment the other way, since they all correspond to maximal submodules of U. We deduce that the intersection of maximal submodules of U equals $\mathrm{Rad}\, A \cdot U$.

For the conditions in (2), observe that $\{u \in U \mid \mathrm{Rad}\, A \cdot u = 0\}$ is the largest submodule of U annihilated by $\mathrm{Rad}\, A$. It is thus an $A/\mathrm{Rad}\, A$-module and hence is semisimple. Since every semisimple submodule of U is annihilated by $\mathrm{Rad}\, A$, it equals the largest such submodule. □

Example 6.3.5. Consider the situation of Theorem 6.1.2 and Proposition 6.1.1 in which G is a cyclic group of order p^n and k is a field of characteristic p. We see that $\mathrm{Rad}\, U_r \cong U_{r-1}$ and $\mathrm{Soc}\, U_r \cong U_1$ for $1 \le r \le p^n$, taking $U_0 = 0$.

We now iterate the notions of socle and radical: for each A-module U, we define inductively

$$\mathrm{Rad}^n(U) = \mathrm{Rad}(\mathrm{Rad}^{n-1}(U))$$
$$\mathrm{Soc}^n(U)/\mathrm{Soc}^{n-1}(U) = \mathrm{Soc}(U/\mathrm{Soc}^{n-1} U).$$

It is immediate from Proposition 6.3.4 that

$$\mathrm{Rad}^n(U) = (\mathrm{Rad}\, A)^n \cdot U$$
$$\mathrm{Soc}^n(U) = \{u \in U \mid (\mathrm{Rad}\, A)^n \cdot u = 0\}$$

and these submodules of U form chains,

$$\cdots \subseteq \mathrm{Rad}^2 U \subseteq \mathrm{Rad}\, U \subseteq U$$
$$0 \subseteq \mathrm{Soc}\, U \subseteq \mathrm{Soc}^2 U \subseteq \cdots,$$

that are called, respectively, the *radical series* and *socle series* of U. The radical series of U is also known as the *Loewy* series of U. The quotients

$\operatorname{Rad}^{n-1}(U)/\operatorname{Rad}^n(U)$ are called the *radical layers*, or *Loewy layers* of U, and the quotients $\operatorname{Soc}^n(U)/\operatorname{Soc}^{n-1}(U)$ are called the *socle layers* of U.

The next corollary is a deduction from Proposition 6.3.4, and again it is true without the hypothesis that the modules be finite-dimensional.

Corollary 6.3.6. *Let A be a finite-dimensional algebra over a field k, and let U and V be finite-dimensional A-modules. Then for each n, we have*

$$\operatorname{Rad}^n(U \oplus V) = \operatorname{Rad}^n(U) \oplus \operatorname{Rad}^n(V)$$

and

$$\operatorname{Soc}^n(U \oplus V) = \operatorname{Soc}^n(U) \oplus \operatorname{Soc}^n(V).$$

Proof. One way to see this is to use the identifications

$$\operatorname{Rad}^n(U \oplus V) = (\operatorname{Rad} A)^n \cdot (U \oplus V)$$

and

$$\operatorname{Soc}^n(U \oplus V) = \{(u, v) \in U \oplus V \mid (\operatorname{Rad} A)^n \cdot (u, v) = 0\}.$$

□

The next result can be proved in various ways; it is also a consequence of Theorem 7.3.9 in the next chapter.

Corollary 6.3.7. *Let k be a field of characteristic p and G a p-group. Then the regular representation kG is indecomposable.*

Proof. If $kG = U \oplus V$ is the direct sum of two nonzero modules then $\operatorname{Rad} kG = \operatorname{Rad} U \oplus \operatorname{Rad} V$ where $\operatorname{Rad} U \neq U$ and $\operatorname{Rad} V \neq V$, so the codimension of $\operatorname{Rad} kG$ in kG must be at least 2. We know from Proposition 6.3.3 that $\operatorname{Rad} kG$ has codimension 1, a contradiction. □

Proposition 6.3.8. *Let A be a finite-dimensional algebra over a field k, and U an A-module. The radical series of U is the fastest descending series of submodules of U with semisimple quotients, and the socle series of U is the fastest ascending series of U with semisimple quotients. The two series terminate, and if m and n are the least integers for which $\operatorname{Rad}^m U = 0$ and $\operatorname{Soc}^n U = U$ then $m = n$.*

Proof. Suppose that $\cdots \subseteq U_2 \subseteq U_1 \subseteq U_0 = U$ is a series of submodules of U with semisimple quotients. We show by induction on r that $\operatorname{Rad}^r(U) \subseteq U_r$. This is true when $r = 0$. Suppose that $r > 0$ and $\operatorname{Rad}^{r-1}(U) \subseteq U_{r-1}$. Then

$$\operatorname{Rad}^{r-1}(U)/(\operatorname{Rad}^{r-1}(U) \cap U_r) \cong (\operatorname{Rad}^{r-1}(U) + U_r)/U_r \subseteq U_{r-1}/U_r$$

is semisimple, so $\mathrm{Rad}^{r-1}(U) \cap U_r \supseteq \mathrm{Rad}(\mathrm{Rad}^{r-1}(U)) = \mathrm{Rad}^r(U)$. Therefore, $\mathrm{Rad}^r(U) \subseteq U_r$. This shows that the radical series descends at least as fast as the series U_i. The argument that the socle series ascends at least as fast is similar.

Since A is a finite-dimensional algebra we have $(\mathrm{Rad}\,A)^r = 0$ for some r. Then $\mathrm{Rad}^r U = (\mathrm{Rad}\,A)^r \cdot U = 0$ and $\mathrm{Soc}^r U = \{u \in U \mid (\mathrm{Rad}\,A)^r u = 0\} = U$, so the two series terminate. By what we have just proved, the radical series descends at least as fast as the socle series and so has equal or shorter length. By a similar argument (using the fact that the socle series is the fastest ascending series with semisimple quotients) the socle series ascends at least as fast as the radical series and so has equal or shorter length. We conclude that the two lengths are equal. □

The common length of the radical series and socle series of U is called the *Loewy length* of the module U, and from the description of the terms of these series, we see it is the least integer n such that $(\mathrm{Rad}\,A)^n \cdot U = 0$.

6.4 Jennings's Theorem

We conclude this chapter by mentioning without proof the theorem of Jennings which gives an explicit description of the radical series of kG when G is a p-group and k is a field of characteristic p. For a proof see Benson's book [3, Theorem 3.14.6]. Jennings considers the series of normal subgroups κ_r of G that is the most rapidly descending central series $\kappa_1 \supseteq \kappa_2 \supseteq \kappa_3 \supseteq \cdots$ with the properties

(1) $\kappa_1 = G$, and
(2) $g^p \in \kappa_{ip}$ for all $g \in \kappa_i$ and $i \geq 1$.

It follows that $[\kappa_r, \kappa_s] \subseteq \kappa_{r+s}$ and κ_r/κ_{2r} is an elementary abelian p-group for all $r, s \geq 1$. Furthermore, we may generate κ_r recursively as

$$\kappa_r = \langle [\kappa_{r-1}, G],\ \kappa^{(p)}_{\lceil r/p \rceil} \rangle, \quad \kappa_1 = G,$$

where $\lceil r/p \rceil$ is the least integer greater than or equal to r/p, and $\kappa_r^{(p)}$ is the set of pth powers of elements of κ_r. After the first term $\kappa_1 = G$, we see that the second term κ_2 is the Frattini subgroup (the smallest normal subgroup of G for which the quotient is elementary abelian), but after that the terms need to be calculated on a case-by-case basis. Note that because of standard properties of p-groups the series eventually terminates at $\kappa_r = 1$ for some r.

For each $i \geq 1$ let d_i be the dimension of the elementary abelian p-group κ_i/κ_{i+1}, regarded as a vector space over $\mathbb{F}_p$. Choose any elements $x_{i,s} \in G$ such that, for each i, the set $\{x_{i,s}\kappa_{i+1} \mid 1 \leq s \leq d_i\}$ forms a basis for κ_i/κ_{i+1}. Let

$\bar{x}_{i,s} = x_{i,s} - 1 \in kG$. We place these elements $\bar{x}_{i,s}$ in some arbitrary fixed order. If $|G| = p^n$, there are n elements $\bar{x}_{i,s}$ and hence $|G|$ products of the form $\prod \bar{x}_{i,s}^{\alpha_{i,s}}$, where $0 \leq \alpha_{i,s} \leq p-1$, and the factors are taken in the fixed order. The weight of such a product is defined to be $\sum i\alpha_{i,s}$.

In the statement of Jennings's Theorem, note from Proposition 6.3.3 that the radical of kG is the augmentation ideal: $\mathrm{Rad}(kG) = IG$.

Theorem 6.4.1 (Jennings). *Let G be a p-group and k a field of characteristic p. We keep the notation just established.*

(1) The Jennings groups have the description

$$\kappa_r(G) = \{g \in G \mid g - 1 \in (IG)^r\}.$$

(2) For each $r \geq 0$, the set of products $\prod \bar{x}_{i,s}^{\alpha_{i,s}}$ of weight at least r is a basis for $(IG)^r$.

(3) The dimension of the rth radical layer of kG is the coefficient of t^r in the expression

$$\sum_{r \geq 0} (\dim(IG)^r / \dim(IG)^{r+1}) t^r = \prod_{i \geq 1} \left(\frac{(1 - t^{ip})}{(1 - t^i)} \right)^{d_i}.$$

(4) The radical series of kG equals the socle series of kG.

The sets $\{g \in G \mid g - 1 \in (IG)^r\}$ which appear in part (1) of this theorem are subgroups which are known as the *dimension subgroups* of G. The series in part (3) is a polynomial, which may be termed the *Poincaré* polynomial. The main work in proving the theorem is in establishing parts (1) and (2), after which (3) is a formality in view of the expansion

$$\prod_{i \geq 1} \left(\frac{(1 - t^{ip})}{(1 - t^i)} \right)^{d_i} = (1 + t + t^2 + \cdots + t^{p-1})^{d_1} (1 + t^2 + \cdots + t^{2(p-1)})^{d_2} \cdots.$$

The deduction of part (4) uses the fact that $kG \cong kG^*$ (see Chapter 8), the fact that the coefficients in this polynomial are the same when read from bottom to top or top to bottom, and Exercise 7 at the end of this chapter.

Example 6.4.2. Let G be dihedral of order 8, which we take to be generated by an element x of order 4 and an element y of order 2:

$$G = \langle x, y \mid x^4 = y^2 = 1,\ yxy = x^{-1} \rangle.$$

We have $\kappa_1 = D_8$, $\kappa_2 = \langle x^2 \rangle$, $\kappa_3 = 1$, so that $\kappa_1/\kappa_2 \cong C_2 \times C_2$ and $\kappa_2/\kappa_3 \cong C_2$ and $d_1 = 2, d_2 = 1, d_3 = d_4 = \cdots = 0$. We may choose $x_{1,1} = x, x_{1,2} = y$, both of weight 1, and $x_{2,1} = x^2$ of weight 2. Note that $\bar{x}^2 = \overline{x^2}$. Now the products

$\overline{x}^{\alpha_{1,1}}\overline{x}^{2\alpha_{2,1}}\overline{y}^{\alpha_{1,2}} = \overline{x}^{\alpha_{1,1}+2\alpha_{2,1}}\overline{y}^{\alpha_{1,2}}$, where $0 \leq \alpha_{i,s} \leq 1$, form a basis of kG that is compatible with the powers of the radical. These elements may be simplified, but we choose not to do this. The Jennings basis of kG consists of the elements

$$1 \text{ (weight 0)}$$
$$\overline{x}, \overline{y} \text{ (weight 1)}$$
$$\overline{xy}, \overline{x^2} \text{ (weight 2)}$$
$$\overline{x}\overline{x^2}, \overline{x^2}\overline{y} \text{ (weight 3)}$$
$$\overline{x}\overline{x^2}\overline{y} \text{ (weight 4)}$$

The Poincaré polynomial is

$$1 + 2t + 2t^2 + 2t^3 + t^4 = (1+t)^2(1+t^2),$$

showing that the radical series of kG (also the socle series) has layers of dimensions $1, 2, 2, 2, 1$.

In this particular example, it is at least as quick to calculate the powers of the augmentation ideal more directly, and this is done in Exercise 19 of this chapter, but it is also interesting to see the general formalism provided by Jennings's Theorem.

6.5 Summary of Chapter 6

- The group ring of C_{p^n} over a field k of characteristic p is isomorphic to $k[X]/(X^{p^n})$. The indecomposable modules are all cyclic and uniserial.
- The only simple module for a p-group in characteristic p is the trivial module.
- When k is a field of characteristic p, simple kG modules are the same as simple $k[G/O_p(G)]$-modules.
- The radical of a finite-dimensional algebra over a field is the largest nilpotent ideal of the algebra.
- The radical series of a module is the fastest descending series with semisimple factors, and the socle series is the fast ascending such series.
- When G is a p-group and k is a field of characteristic p the radical of kG is the augmentation ideal. The radical series and socle series of kG coincide and the ranks of the factors are given in terms of the Jennings series of G.

6.6 Exercises for Chapter 6

In these exercises, k denotes a field and R is a commutative ring with a 1.

1. Let A be a ring. Prove that for each n, $\text{Soc}^n {}_A A$ is a 2-sided ideal of A.

2. Let $\overline{G} = \sum_{g \in G} g$ as an element of kG, where k is a field.

(a) Show that the subspace $k\overline{G}$ of kG spanned by $\overline{G}$ is an ideal.
(b) Show that this ideal is nilpotent if and only if the characteristic of k divides $|G|$.
(c) Deduce that if kG is semisimple then $\text{char}(k) \nmid |G|$.
(d) Assuming instead that G is a p-group and $\text{char}(k) = p$, show that $k\overline{G} = \text{Soc}(kG)$, the socle of the regular representation.

3. Suppose that U is an indecomposable module with just two composition factors. Show that U is uniserial.

4. Show that for each RG-module U, $U/(IG \cdot U)$ is the largest quotient of U on which G acts trivially. Prove also that $U/(IG \cdot U) \cong R \otimes_{RG} U$.
[The first sentence means that G does act trivially on the given quotient; and if V is any submodule of U such that G acts trivially on U/V, then $V \supseteq IG \cdot U$. By analogy with the notation for fixed points, this largest quotient on which G acts trivially is called the *fixed quotient* of U and is denoted $U_G := U/(IG \cdot U)$.]

5. Prove that if N is a normal subgroup of G and k is a field then $\text{Rad}(kN) = kN \cap \text{Rad}(kG)$.
[Use the descriptions of the radical in Proposition 6.3.2 and also Clifford's Theorem.]

6. Show that the following conditions are equivalent for a module U that has a composition series.

(a) U is uniserial (i.e., U has a unique composition series).
(b) The set of all submodules of U is totally ordered by inclusion.
(c) $\text{Rad}^r U/\text{Rad}^{r+1} U$ is simple for all r.
(d) $\text{Soc}^{r+1} U/\text{Soc}^r U$ is simple for all r.

7. Let U be a finitely generated kG-module and U^* its dual. Show that for each n

$$\text{Soc}^n(U^*) = \{f \in U^* \mid f(\text{Rad}^n(U)) = 0\}$$

and

$$\text{Rad}^n(U^*) = \{f \in U^* \mid f(\text{Soc}^n(U)) = 0\}.$$

Deduce that $\text{Soc}^{n+1}(U^*)/\text{Soc}^n(U^*) \cong (\text{Rad}^n(U)/\text{Rad}^{n+1}(U))^*$ as kG-modules. [Hint: recall Exercise 13 of Chapter 4.]

8. Let Ω be a transitive G-set and $R\Omega$ be the corresponding permutation module. Thus, if $H = \text{Stab}_G(\omega)$ for some $\omega \in \Omega$ then $R\Omega \cong R\uparrow_H^G$, as explained in

Example 4.3.4. There is a homomorphism of RG-modules $\epsilon : R\Omega \to R$ defined as $\epsilon(\sum_{\omega\in\Omega} a_\omega \omega) = \sum_{\omega\in\Omega} a_\omega$. Let $\overline{\Omega} = \sum_{\omega\in\Omega} \omega \in R\Omega$.

(a) Show that every RG-module homomorphism $R\Omega \to R$ is a scalar multiple of ϵ.
(b) Show that the fixed points of G on $R\Omega$ are $(R\Omega)^G = R\cdot\overline{\Omega}$.
(c) Suppose now that $R = k$ is a field. Show that $\epsilon(\overline{\Omega}) = 0$ if and only if $\operatorname{char} k \mid |\Omega|$, and that if this happens then $\overline{\Omega} \in \operatorname{Rad} k\Omega$ and the trivial module k occurs as a composition factor of $k\Omega$ with multiplicity ≥ 2.
(d) Again suppose that $R = k$ is a field. Show that if $\epsilon(\overline{\Omega}) \neq 0$ then ϵ is a split epimorphism and $\overline{\Omega} \notin \operatorname{Rad} k\Omega$.
(e) Show that kG is semisimple if and only if the regular representation kG has the trivial module k as a direct summand (i.e., k is a projective module).
(f) Suppose that G is a p-group and k is a field of characteristic p. Show that $\operatorname{Rad}(k\Omega)$ has codimension 1 in $k\Omega$ and that $k\Omega$ is an indecomposable kG-module.

9. Let Ω be a transitive G-set for a possibly infinite group G and let $R\Omega$ be the corresponding permutation module. Show that Ω is infinite if and only if $(R\Omega)^G = 0$ and deduce that G is infinite if and only if $(RG)^G = 0$.

10. Let g be an endomorphism of a finite-dimensional vector space V over a field k of characteristic p, and suppose that g has finite order p^d for some d.

(a) Show that as a $k\langle g\rangle$-module, V has an indecomposable direct summand of dimension at least $p^{d-1}+1$.

[You may assume the classification of indecomposable modules for cyclic p-groups in characteristic p.]

(b) Deduce that if such an endomorphism g fixes pointwise a subspace of V of codimension 1 then g has order p or 1.

[An endomorphism of finite (not necessarily of prime-power) order that fixes a subspace of codimension 1 is sometimes referred to as a *reflection* in a generalized sense.]

11. Let A be a finite-dimensional algebra over a field and let U be an A-module. Write $\ell(U)$ for the Loewy length of U.

(a) Suppose V is a submodule of U. Show that $\ell(V) \leq \ell(U)$ and $\ell(U/V) \leq \ell(U)$. Show by example that we can have equality here even when $0 < V < U$.

(b) Suppose that $U_1, \ldots, U_n$ are submodules of U with the property that $U = U_1 + \cdots + U_n$. Show that $\ell(U) = \max\{\ell(U_i) \mid 1 \le i \le n\}$.

12. (This exercise extends the theorem of Burnside presented in Chapter 2 Exercise 10 to nonsemisimple algebras.) Let A be a finite-dimensional algebra over a field k, let V be an A-module so that the action of A on V is given by an algebra homomorphism $\rho : A \to \mathrm{End}_k(V)$ and let $I = \mathrm{Ker}\,\rho$.

(a) Show that if V is simple then A/I is a semisimple ring with only one simple module (up to isomorphism).
(b) Assuming that k is algebraically closed, show that ρ is surjective if and only if V is simple.
(c) If $A = kG$ is a group algebra, k is algebraically closed and $\dim V = n$, show that V is simple if and only if there exist n^2 elements $g_1, \ldots, g_{n^2}$ of G so that $\rho(g_1), \ldots, \rho(g_{n^2})$ are linearly independent.

The next five exercises give a direct proof of the result that is part of Proposition 6.3.3, that for a p-group in characteristic p the augmentation ideal is nilpotent.

13. Show that if elements $g_1, \ldots, g_n$ generate G as a group, then $(g_1 - 1), \ldots, (g_n - 1)$ generate the augmentation ideal IG as a left ideal of kG.
[Use the formula $(gh - 1) = g(h - 1) + (g - 1)$.]

14. Suppose that k is a field of characteristic p and G is a p-group. Prove that each element $(g - 1)$ is nilpotent. (More generally, every element of IG is nilpotent.)

15. Show that if N is a normal subgroup of G then the left ideal

$$RG \cdot IN = \{x \cdot y \mid x \in RG, y \in IN\}$$

of RG generated by IN is the kernel of the ring homomorphism $RG \to R[G/N]$ and is in fact a 2-sided ideal in RG.
[One approach to this uses the formula $g(n - 1) = ({}^g n - 1)g$.]
Show that $(RG \cdot IN)^r = RG \cdot (IN)^r$ for all r.

16. Show that if a particular element $(g - 1)$ appears n times in a product

$$(g_1 - 1) \cdots (g_r - 1)$$

then

$$(g_1 - 1) \cdots (g_r - 1) \equiv (g - 1)^n \cdot x \quad \text{modulo}\, kG \cdot (IG')$$

for some $x \in kG$, where G' denotes the commutator subgroup.
[Use the formula $(g-1)(h-1) = (h-1)(g-1) + (ghg^{-1}h^{-1}-1)hg$.]

Show that if G is a p-group and k a field of characteristic p then $IG^r \subseteq kG \cdot IG'$ for some power r.

17. Prove that if G is a p-group and k is a field of characteristic p then $(IG)^r = 0$ for some power r.

18. Let k be a field of characteristic p. Show that the Loewy length of kC_p^n, the group algebra of the direct product of n copies of a cycle of order p, is $n(p-1)+1$.

19. The dihedral group of order $2n$ has a presentation

$$D_{2n} = \langle x, y \mid x^2 = y^2 = (xy)^n = 1 \rangle.$$

Let k be a field of characteristic 2. Show that when n is a power of 2, each power $(ID_{2n})^r$ of the augmentation ideal is spanned modulo $(ID_{2n})^{r+1}$ by the two products $(x-1)(y-1)(x-1)(y-1)\cdots$ and $(y-1)(x-1)(y-1)(x-1)\cdots$ of length r. Hence calculate the Loewy length of kD_{2n} and show that $\mathrm{Rad}(kD_{2n})/\mathrm{Soc}(kD_{2n})$ is the direct sum of two kD_{2n}-modules that are uniserial.

20. When $n \geq 3$, the generalized quaternion group of order 2^n has a presentation

$$Q_{2^n} = \langle x, y \mid x^{2^{n-1}} = 1,\ y^2 = x^{2^{n-2}},\ yxy^{-1} = x^{-1} \rangle.$$

Let k be a field of characteristic 2. Show that when $r \geq 1$ each power $(IQ_{2^n})^r$ of the augmentation ideal is spanned modulo the next higher power $(IQ_{2^n})^{r+1}$ by $(x-1)^r$ and $(x-1)^{r-1}(y-1)$. Hence calculate the Loewy length of kQ_{2^n}.

21. Let H be a subgroup of G and let IH be the augmentation ideal of RH, which we may regard as a subset of RG. Show that $RG \cdot IH \cong IH \uparrow_H^G$ as RG-modules, and that $RG/(RG \cdot IH) \cong R \uparrow_H^G$ as RG-modules. Show also that $RG/(RG \cdot IH)$ is the largest quotient of RG on which H acts trivially when acting from the *right*.

22. (a) Let G be any group and $IG \subset \mathbb{Z}G$ the augmentation ideal over $\mathbb{Z}$. Prove that $IG/(IG)^2 \cong G/G'$ as abelian groups.

[Consider the homomorphism of abelian groups $IG \to G/G'$ given by $g-1 \mapsto gG'$. Use the formula $ab-1 = (a-1)+(b-1)+(a-1)(b-1)$ to show that $(IG)^2$ is contained in the kernel, and that the homomorphism $G/G' \to IG/(IG)^2$ given by $gG' \mapsto g-1+(IG)^2$ is well defined.]

(b) For any group G, write $d(G)$ for the smallest size of a set of generators of G as a group, and if U is a $\mathbb{Z}G$-module write $d(U)$ for the smallest size of a set of generators of U as a $\mathbb{Z}G$-module. Use Exercise 13 to show that $d(G/G') \le d(IG) \le d(G)$ with equality when G is a p-group. [For the final equality use properties of the Frattini subgroup of G.]

(c) If now R is any commutative ring with 1 and $IG \subset RG$ is the augmentation ideal of G over R, show that $IG/(IG)^2 \cong R \otimes_{\mathbb{Z}} G/G'$ as R-modules. When G is a p-group and R is a field of characteristic p, show again that $d(G) = d(IG)$.

23. Let U_r be the indecomposable kC_p-module of dimension r, $1 \le r \le p$, where k is a field of characteristic p. Prove that $U_r \cong S^{r-1}(U_2)$, the $(r-1)$ symmetric power.
[One way to proceed is to show that if $C_p = \langle g \rangle$ then $(g-1)^{r-1}$ does not act as zero on $S^{r-1}(U_2)$ and use the classification of indecomposable kC_p-modules.]

24. Let k be a field of characteristic p and let $G = C_{p^n}$ be a cyclic group of order p^n. Suppose that $H \le G$ is the subgroup of order p^t, for some $t \le n$, and for each r with $1 \le r \le p^t$ write $U_{r,H}$ for the indecomposable kH-module of dimension r.

(a) Show that $U_{r,H} \uparrow_H^G \cong U_{|G:H|r,G}$ as kG-modules, so that indecomposable modules induce to indecomposable modules. [Exploit the fact that if V is a cyclic module then so is $V \uparrow_H^G$, by Chapter 4, Exercise 12.]

(b) Write $r = ap^{n-t} + b$ where $0 \le b < p^{n-t}$. Show that

$$U_{r,G} \downarrow_H^G \cong (U_{a+1,H})^b \oplus (U_{a,H})^{p^{n-t}-b}$$

as kH-modules. [If $G = \langle x \rangle$, $H = \langle x^{p^{n-t}} \rangle$ use the fact that $x^{p^{n-t}} - 1 = (x-1)^{p^{n-t}}$ to show that the largest power of $x^{p^{n-t}} - 1$ that is nonzero on $U_{r,G}$ is the ath power, and use this to identify the summands in a decomposition of $U_{r,G} \downarrow_H^G$.]

25. Let $G = SL(2, p)$, the group of 2×2 matrices over $\mathbb{F}_p$ that have determinant 1, where p is a prime. The subgroups

$$P_1 = \left\{ \begin{bmatrix} 1 & \lambda \\ 0 & 1 \end{bmatrix} \middle| \lambda \in \mathbb{F}_p \right\}, \quad P_2 = \left\{ \begin{bmatrix} 1 & 0 \\ \lambda & 1 \end{bmatrix} \middle| \lambda \in \mathbb{F}_p \right\}$$

have order p. Let U_2 be the natural 2-dimensional module on which G acts.

(a) When $0 \le r \le p-1$, prove that $S^r(U_2)$ is a uniserial $\mathbb{F}_p P_1$-module, and also a uniserial $\mathbb{F}_p P_2$-module, but that the only subspaces of $S^r(U_2)$ that are invariant under both P_1 and P_2 are 0 and $S^r(U_2)$. Deduce that $S^r(U_2)$ is a simple $\mathbb{F}_p G$-module.

(b) Show further, when p is odd, that the matrix $\begin{bmatrix} -1 & 0 \\ 0 & -1 \end{bmatrix}$ acts as the identity on $S^r(U_2)$ if and only if r is even, and hence that we have constructed $(p+1)/2$ simple representations of

$$PSL(2, p) := SL(2, p)/\left\{\pm\begin{bmatrix} 1 & 0 \\ 0 & 1 \end{bmatrix}\right\}.$$

[Background to the question that is not needed to solve it: $|G| = p(p^2 - 1)$; both P_1 and P_2 are Sylow p-subgroups of G. In fact, the simple modules constructed here form a complete list of the simple $\mathbb{F}_p SL(2, p)$-modules.]

26. Let k be a field of characteristic p and suppose that G has a normal Sylow p-subgroup N. Show that $\operatorname{Rad} kG = kG \cdot \operatorname{Rad} kN$.
[Use Exercises 5 and 15, and show that $k[G/N]$ is the largest semisimple quotient of kG.]

27. Let A be a finite-dimensional algebra over a field and let U be an arbitrary A-module. As in the text, we define $\operatorname{Rad} U$ to be the intersection of the maximal submodules of U.

(a) Use Exercises 13 and 14 from Chapter 1 to show that Proposition 6.3.4 holds without the hypothesis of finite generation. That is, show that $\operatorname{Rad} U = \operatorname{Rad} A \cdot U$, and that $\operatorname{Rad} U$ is the smallest submodule of U with semisimple quotient. Show also that $\operatorname{Soc} U$, defined as the sum of the simple submodules of U, is the largest semisimple submodule of U, and it is the set of elements of U annihilated by $\operatorname{Rad} A$. [It is a question of copying the arguments from Proposition 6.3.4. Note that the property of A being used is that $A/\operatorname{Rad} A$ is semisimple.]
(b) Show that if $U \neq 0$ then U has a nonzero finite-dimensional homomorphic image. [Use the fact that $\operatorname{Rad} A$ is nilpotent.]
(c) Show that each proper submodule of U is contained in a maximal submodule.

7
Projective Modules for Finite-Dimensional Algebras

In previous sections, we have seen the start of techniques to describe modules that are not semisimple. The most basic decomposition of such a module is one that expresses it as a direct sum of modules that cannot be decomposed as a direct sum any further. These summands are called indecomposable modules. We have also examined the notions of radical series and socle series of a module, which are series of canonically defined submodules that may shed light on submodule structure. We combine these two notions in the study of projective modules for group rings, working at first in the generality of modules for finite-dimensional algebras over a field. In this situation the indecomposable projective modules are the indecomposable summands of the regular representation. We will see that they are identified by the structure of their radical quotient. The projective modules are important because their structure is part of the structure of the regular representation. Since every module is a homomorphic image of a direct sum of copies of the regular representation, by knowing the structure of the projectives we gain some insight into the structure of all modules.

7.1 Characterizations of Projective and Injective Modules

Recall that a module M over a ring A is said to be *free* if it has a basis; that is, a subset $\{x_i \mid i \in I\}$ that spans M as an A-module, and is linearly independent over A. To say that $\{x_i \mid i \in I\}$ is a basis of M is equivalent to requiring $M = \bigoplus_{i \in I} Ax_i$ with $A \cong Ax_i$ via an isomorphism $a \mapsto ax_i$ for all i. Thus M is a finitely generated free module if and only if $M \cong A^n$ for some n. These conditions are also equivalent to the condition in the following proposition:

Proposition 7.1.1. *Let A be a ring and M an A-module. The following are equivalent for a subset $\{x_i \mid i \in I\}$ of M:*

(1) $\{x_i \mid i \in I\}$ *is a basis of* M,
(2) for every module N *and mapping of sets* $\phi : \{x_i \mid i \in I\} \to N$ *there exists a unique module homomorphism* $\psi : M \to N$ *that extends* ϕ.

Proof. The proof is standard. If $\{x_i \mid i \in I\}$ is a basis, then given ϕ we may define $\psi(\sum_{i \in I} a_i x_i) = \sum_{i \in I} a_i \phi(x_i)$ and this is evidently the unique module homomorphism extending ϕ. This shows that (1) implies (2). Conversely if condition (2) holds we may construct the free module F with $\{x_i \mid i \in I\}$ as a basis and use the condition to construct a homomorphism from $M \to F$ that is the identity on $\{x_i \mid i \in I\}$. The fact just shown that the free module also satisfies condition (2) allows us to construct a homomorphism $F \to M$ that is again the identity on $\{x_i \mid i \in I\}$, and the two homomorphisms have composites in both directions that are the identity, since these are the unique extensions of the identity map on $\{x_i \mid i \in I\}$. They are therefore isomorphisms and from this condition (1) follows. □

We define a module homomorphism $f : M \to N$ to be a *split epimorphism* if and only if there exists a homomorphism $g : N \to M$ so that $fg = 1_N$, the identity map on N. Note that a split epimorphism is necessarily an epimorphism since if $x \in N$ then $x = f(g(x))$ so that x lies in the image of f. We define similarly f to be a *split monomorphism* if there exists a homomorphism $g : N \to M$ so that $gf = 1_M$. Necessarily a split monomorphism is a monomorphism. We are about to show that if f is a split epimorphism then N is (isomorphic to) a direct summand of M. To combine both this and the corresponding result for split monomorphisms it is convenient to introduce short exact sequences. We say that a diagram of modules and module homomorphisms $L \xrightarrow{\alpha} M \xrightarrow{\beta} N$ is *exact* at M if $\operatorname{Im} \alpha = \operatorname{Ker} \beta$. A *short exact sequence* of modules is a diagram $0 \to L \xrightarrow{\alpha} M \xrightarrow{\beta} N \to 0$ that is exact at each of L, M and N. Exactness at L and N means simply that α is a monomorphism and β is an epimorphism.

Proposition 7.1.2. *Let* $0 \to L \xrightarrow{\alpha} M \xrightarrow{\beta} N \to 0$ *be a short exact sequence of modules over a ring. The following are equivalent:*

(1) α *is a split monomorphism,*
(2) β *is a split epimorphism,*
(3) there is a commutative diagram

$$\begin{array}{ccccccccc} 0 & \to & L & \xrightarrow{\alpha} & M & \xrightarrow{\beta} & N & \to & 0 \\ & & \| & & \downarrow{\scriptstyle \gamma} & & \| & & \\ 0 & \to & L & \xrightarrow{\iota_1} & L \oplus N & \xrightarrow{\pi_2} & N & \to & 0 \end{array}$$

where ι_1 and π_2 are inclusion into the first summand and projection onto the second summand,

(4) *for every module U the sequence*

$$0 \to \mathrm{Hom}_A(U,L) \to \mathrm{Hom}_A(U,M) \to \mathrm{Hom}_A(U,N) \to 0$$

is exact,

(4′) *for every module U the sequence*

$$0 \to \mathrm{Hom}_A(N,U) \to \mathrm{Hom}_A(M,U) \to \mathrm{Hom}_A(L,U) \to 0$$

is exact.

In any diagram such as the one in (3) the morphism γ is necessarily an isomorphism. Thus if any of the listed conditions is satisfied it follows that $M \cong L \oplus N$.

Proof. Condition (3) implies the first two, since the existence of such a commutative diagram implies that α is split by $\pi_1\gamma$ and β is split by $\gamma^{-1}\iota_2$, and it also implies the last two conditions because the commutative diagram produces similar commutative diagrams after applying $\mathrm{Hom}_A(U,-)$ and $\mathrm{Hom}_A(\ \ ,U)$.

Conversely if condition (1) is satisfied, so that $\delta\alpha = 1_L$ for some homomorphism $\delta : M \to L$, we obtain a commutative diagram as in (3) on taking the components of γ to be δ and β. If condition (2) is satisfied we obtain a commutative diagram similar to the one in (3) but with a homomorphism $\zeta : L \oplus N \to M$ in the wrong direction, whose components are α and a splitting of β. We obtain the diagram of (3) on showing that in any such diagram the middle vertical homomorphism must be invertible.

The fact that the middle homomorphism in the diagram must be invertible is a consequence of both the "five lemma" and the "snake lemma" in homological algebra. We leave it here as an exercise.

Finally if (4) holds then on taking U to be N we deduce that the identity map on N is the image of a homomorphism $\epsilon : U \to M$, so that $1_N = \beta\epsilon$ and β is split epi, so that (2) holds. Equally if (4′) holds then taking U to be L we see that the identity map on L is the image of a homomorphism $\delta : M \to U$, so that $1_L = \delta\alpha$ and (1) holds. □

In the event that α and β are split, we say that the short exact sequence in Proposition 7.1.2 is *split*. Notice that whenever $\beta : M \to N$ is an epimorphism it is part of the short exact sequence $0 \to \mathrm{Ker}\,\beta \hookrightarrow M \xrightarrow{\beta} N \to 0$, and so we deduce that if β is a split epimorphism then N is a direct summand of M. A similar comment evidently applies to split monomorphisms.

Proposition 7.1.3. *The following are equivalent for an A-module P.*

(1) P is a direct summand of a free module.
(2) Every epimorphism $V \to P$ is split.
(3) For every pair of morphisms

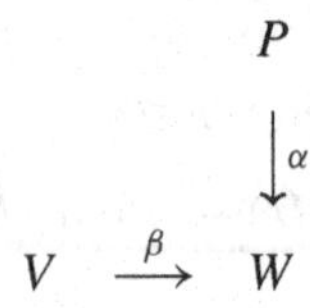

where β is an epimorphism, there exists a morphism $\gamma : P \to V$ with $\beta\gamma = \alpha$.
(4) For every short exact sequence of A-modules $0 \to V \to W \to X \to 0$, the corresponding sequence

$$0 \to \mathrm{Hom}_A(P, V) \to \mathrm{Hom}_A(P, W) \to \mathrm{Hom}_A(P, X) \to 0$$

is exact.

Proof. This result is standard and we do not prove it here. In condition (4) the sequence of homomorphism groups is always exact at the left-hand terms $\mathrm{Hom}_A(P, V)$ and $\mathrm{Hom}_A(P, W)$ without requiring any special property of P (we say that $\mathrm{Hom}_A(P, \ )$ is *left exact*). The force of condition (4) is that the sequence should be exact at the right-hand term. ☐

We say that a module P satisfying any of the four conditions of Proposition 7.1.3 is *projective*. Notice that direct sums and also direct summands of projective modules are projective. An indecomposable module that is projective is an indecomposable projective module, and these modules will be very important in our study. In other texts, the indecomposable projective modules are also known as PIMs, or principal indecomposable modules, but we will not use this terminology here.

We should also mention injective modules, which enjoy properties similar to those of projective modules, but in a dual form. We say that a module I is *injective* if and only if whenever there are morphisms

$$\begin{array}{ccc} & & I \\ & & \uparrow{\scriptstyle\alpha} \\ V & \xleftarrow{\beta} & W \end{array}$$

with β a monomorphism, there exists a morphism $\gamma : V \to I$ so that $\gamma\beta = \alpha$. Dually to Proposition 7.1.3, it is equivalent to require that every monomorphism $I \to V$ be split; and also that $\mathrm{Hom}_A(-, I)$ sends exact sequences to exact sequences. When A is an arbitrary ring we do not have such a nice characterization of injectives analogous to the property that projective modules are direct summands of free modules. However, for group algebras over a field we will show in Corollary 8.13 that injective modules are the same thing as projective modules, so that in this context they are indeed summands of free modules.

7.2 Projectives by Means of Idempotents

One way to obtain projective A-modules is from idempotents of the ring A. If $e^2 = e \in A$ then ${}_AA = Ae \oplus A(1-e)$ as A-modules, and so the submodules Ae and $A(1-e)$ are projective. We formalize this with the next result, which should be compared with Proposition 3.6.1, in which we were dealing with ring summands of A and central idempotents.

Proposition 7.2.1. *Let A be a ring. The decompositions of the regular representation as a direct sum of submodules*

$$ {}_AA = A_1 \oplus \cdots \oplus A_r $$

biject with expressions $1 = e_1 + \cdots + e_r$ for the identity of A as a sum of orthogonal idempotents, in such a way that $A_i = Ae_i$. The summand A_i is indecomposable if and only if the idempotent e_i is primitive.

Proof. Suppose that $1 = e_1 + \cdots + e_r$ is an expression for the identity as a sum of orthogonal idempotents. Then

$$ {}_AA = Ae_1 \oplus \cdots \oplus Ae_r, $$

for the Ae_i are evidently submodules of A, and their sum is A since if $x \in A$ then $x = xe_1 + \cdots + xe_r$. The sum is direct since if $x \in Ae_i \cap \sum_{j \neq i} Ae_j$ then $x = xe_i$ and also $x = \sum_{j\neq i} a_j e_j$ so $x = xe_i = \sum_{j \neq i} a_j e_j e_i = 0$.

Conversely, suppose that ${}_AA = A_1 \oplus \cdots \oplus A_r$ is a direct sum of submodules. We may write $1 = e_1 + \cdots + e_r$ where $e_i \in A_i$ is a uniquely determined element. Now $e_i = e_i 1 = e_i e_1 + \cdots + e_i e_r$ is an expression in which $e_i e_j \in A_j$, and since the only such expression is e_i itself we deduce that

$$ e_i e_j = \begin{cases} e_i & \text{if } i = j \\ 0 & \text{otherwise.} \end{cases} $$

The two constructions just described, in which we associate an expression for 1 as a sum of idempotents to a module direct sum decomposition and vice-versa, are mutually inverse, giving a bijection as claimed.

If a summand A_i decomposes as the direct sum of two other summands, this gives rise to an expression for e_i as a sum of two orthogonal idempotents, and conversely. Thus, A_i is indecomposable if and only if e_i is primitive. □

In Proposition 3.6.1, it was proved that in a decomposition of A as a direct sum of indecomposable rings, the rings are uniquely determined as subsets of A and the corresponding primitive central idempotents are also unique. We point out that the corresponding uniqueness property need not hold with module decompositions of ${}_AA$ that are not ring decompositions. For an example of this, we take $A = M_2(R)$, the ring of 2×2 matrices over a ring R, and consider the two decompositions

$$ {}_AA = A\begin{bmatrix} 1 & 0 \\ 0 & 0 \end{bmatrix} \oplus A\begin{bmatrix} 0 & 0 \\ 0 & 1 \end{bmatrix} = A\begin{bmatrix} 0 & 1 \\ 0 & 1 \end{bmatrix} \oplus A\begin{bmatrix} 1 & -1 \\ 0 & 0 \end{bmatrix}. $$

The submodules here are all different. We will see later that if A is a finite-dimensional algebra over a field then in any two decompositions of ${}_AA$ as a direct sum of indecomposable submodules, the submodules are isomorphic in pairs.

We will also see that when A is a finite-dimensional algebra over a field, every indecomposable projective A-module may be realized as Ae for some primitive idempotent e. For other rings this need not be true: an example is $\mathbb{Z}G$, for which it is the case that the only idempotents are 0 and 1 (see Exercise 1 in Chapter 8). For certain finite groups (an example is the cyclic group of order 23, but this takes us beyond the scope of this book) there exist indecomposable projective $\mathbb{Z}G$-modules that are not free, so such modules will never have the form $\mathbb{Z}Ge$ for any idempotent element e.

Example 7.2.2. We present an example of a decomposition of the regular representation in a situation that is not semisimple. Many of the observations we will make are consequences of theory to be presented in later sections, but it seems worthwhile to show that the calculations can be done by direct arguments.

Consider the group ring $\mathbb{F}_4S_3$, where $\mathbb{F}_4$ is the field of 4 elements. The choice of $\mathbb{F}_4$ is made because at one point it will be useful to have all cube roots of unity available, but in fact many of the observations we are about to make also hold over the field $\mathbb{F}_2$. By Proposition 4.2.1, the 1-dimensional representations of S_3 are the simple representations of $S_3/S_3' \cong C_2$, lifted to S_3. But $\mathbb{F}_4C_2$ has only one simple module, namely, the trivial module, by Proposition 6.2.1, so this is the only 1-dimensional $\mathbb{F}_4S_3$-module. The 2-dimensional representation of S_3 constructed in Chapter 1 over any coefficient ring is now seen to be simple here, since otherwise it would have a trivial submodule; but the eigenvalues of the element $(1, 2, 3)$ on this module are ω and ω^2, where $\omega \in \mathbb{F}_4$ is a primitive cube root of 1, so there is no trivial submodule.

Let $K = \langle(1, 2, 3)\rangle$ be the subgroup of S_3 of order 3. Now $\mathbb{F}_4K$ is semisimple with three 1-dimensional representations on which $(1, 2, 3)$ acts as $1, \omega$, and ω^2, respectively. In fact,

$$\mathbb{F}_4K = \mathbb{F}_4Ke_1 \oplus \mathbb{F}_4Ke_2 \oplus \mathbb{F}_4Ke_3,$$

where

$$\begin{aligned} e_1 &= () + (1, 2, 3) + (1, 3, 2), \\ e_2 &= () + \omega(1, 2, 3) + \omega^2(1, 3, 2), \\ e_3 &= () + \omega^2(1, 2, 3) + \omega(1, 3, 2) \end{aligned}$$

are orthogonal idempotents in $\mathbb{F}_4K$. We may see that these are orthogonal idempotents by direct calculation, but it can also be seen by observing that the corresponding elements of $\mathbb{C}K$ with ω replaced by $e^{\frac{2\pi i}{3}}$ are orthogonal and square to 3 times themselves (Theorem 3.6.2), and lie in $\mathbb{Z}[e^{\frac{2\pi i}{3}}]K$. Reduction modulo 2 gives a ring homomorphism $\mathbb{Z}[e^{\frac{2\pi i}{3}}] \to \mathbb{F}_4$ that maps these elements to e_1, e_2, and e_3, while retaining their properties. Thus,

$$\mathbb{F}_4S_3 = \mathbb{F}_4S_3e_1 \oplus \mathbb{F}_4S_3e_2 \oplus \mathbb{F}_4S_3e_3,$$

and we have constructed modules $\mathbb{F}_4S_3e_i$ that are projective. We have not yet shown that they are indecomposable.

We easily compute that

$$(1, 2, 3)e_1 = e_1, \quad (1, 2, 3)e_2 = \omega^2e_2, \quad (1, 2, 3)e_3 = \omega e_3$$

and from this, we see that $K \cdot \mathbb{F}_4e_i = \mathbb{F}_4e_i$ for all i. Since $S_3 = K \cup (1, 2)K$ we have $\mathbb{F}_4S_3e_i = \mathbb{F}_4e_i \oplus \mathbb{F}_4(1, 2)e_i$, which has dimension 2 for all i. We have already seen that when $i = 2$ or 3, e_i is an eigenvector for $(1, 2, 3)$ with eigenvalue ω or ω^2, and a similar calculation shows that the same is true for $(1, 2)e_i$. Thus, when $i = 2$ or 3, $\mathbb{F}_4S_3e_i$ has no trivial submodule and hence is simple by the observations made at the start of this example. We have an isomorphism of $\mathbb{F}_4S_3$-modules

$$\begin{aligned} \mathbb{F}_4S_3e_2 &\to \mathbb{F}_4S_3e_3, \\ e_2 &\mapsto (1, 2)e_3, \\ (1, 2)e_2 &\mapsto e_3. \end{aligned}$$

On the other hand, $\mathbb{F}_4 S_3 e_1$ has fixed points $\mathbb{F}_4 \sum_{g \in S_3} g$ of dimension 1 and so has two composition factors, which are trivial. On restriction to $\mathbb{F}_4\langle(1,2)\rangle$ it is the regular representation, and it is a uniserial module.

We see from all this that $\mathbb{F}_4 S_3 = \begin{smallmatrix} 1 \\ 1 \end{smallmatrix} \oplus 2 \oplus 2$, in a diagrammatic notation. Thus, the 2-dimensional simple $\mathbb{F}_4 S_3$-module is projective, and the trivial module appears as the unique simple quotient of a projective module of dimension 2 whose socle is also the trivial module. These summands of $\mathbb{F}_4 S_3$ are indecomposable, and so e_1, e_2 and e_3 are primitive idempotents in $\mathbb{F}_4 S_3$. We see also that the radical of $\mathbb{F}_4 S_3$ is the span of $\sum_{g \in S_3} g$.

7.3 Projective Covers, Nakayama's Lemma, and Lifting of Idempotents

We now develop the theory of projective covers. We first make the definition that an *essential epimorphism* is an epimorphism of modules $f : U \to V$ with the property that no proper submodule of U is mapped surjectively onto V by f. An equivalent formulation is that whenever $g : W \to U$ is a map such that fg is an epimorphism, then g is an epimorphism. One immediately asks for examples of essential epimorphisms, but it is probably more instructive to consider epimorphisms that are not essential. If $U \to V$ is any epimorphism and X is a nonzero module then the epimorphism $U \oplus X \to V$ constructed as the given map on U and zero on X can never be essential. This is because U is a submodule of $U \oplus X$ mapped surjectively onto V. Thus, if $U \to V$ is essential then U can have no direct summands that are mapped to zero. One may think of an essential epimorphism as being minimal, in that no unnecessary parts of U are present.

The greatest source of essential epimorphisms is Nakayama's Lemma, given here in a version for modules over noncommutative rings. Over an arbitrary ring a finiteness condition is required, and that is how we state the result here. We will see in Exercise 10 at the end of this chapter that, when the ring is a finite-dimensional algebra over a field, the result is true for arbitrary modules without any finiteness condition.

Theorem 7.3.1 (Nakayama's Lemma). *If U is any Noetherian module, the homomorphism $U \to U/\operatorname{Rad} U$ is essential. Equivalently, if V is a submodule of U with the property that $V + \operatorname{Rad} U = U$, then $V = U$.*

Proof. Suppose V is a submodule of U. If $V \neq U$ then $V \subseteq M \subset U$ where M is a maximal submodule of U. Now $V + \operatorname{Rad} U \subseteq M$ and so the composite $V \to U \to U/\operatorname{Rad} U$ has image contained in $M/\operatorname{Rad} U$, which is not equal to $U/\operatorname{Rad} U$ since $(U/\operatorname{Rad} U)/(M/\operatorname{Rad} U) \cong U/M \neq 0$. □

When U is a module for a finite-dimensional algebra, it is always true that every proper submodule of U is contained in a maximal submodule, even when U is not finitely generated. This was the only point in the proof of Theorem 7.3.1 where the Noetherian hypothesis was used, and so in this situation $U \to U/\operatorname{Rad} U$ is always essential. This is shown in Exercise 10 of this chapter.

The next result is not at all difficult and could also be proved as an exercise.

Proposition 7.3.2. *(1) Suppose that $f : U \to V$ and $g : V \to W$ are two module homomorphsms. If two of f, g, and gf are essential epimorphisms then so is the third.*

(2) Let $f : U \to V$ be a homomorphism of Noetherian modules. Then f is an essential epimorphism if and only if the homomorphism of radical quotients $U/\operatorname{Rad} U \to V/\operatorname{Rad} V$ is an isomorphism.

(3) Let $f_i : U_i \to V_i$ be homomorphisms of Noetherian modules, where $i = 1, \ldots, n$. The f_i are all essential epimorphisms if and only if

$$\oplus f_i : \bigoplus_i U_i \to \bigoplus_i V_i$$

is an essential epimorphism.

Proof. (1) Suppose f and g are essential epimorphisms. Then gf is an epimorphism also, and it is essential because if U_0 is a proper submodule of U then $f(U_0)$ is a proper submodule of V since f is essential, and hence $g(f(U_0))$ is a proper submodule of S since g is essential.

Next suppose f and gf are essential epimorphisms. Since $W = \operatorname{Im}(gf) \subseteq \operatorname{Im}(g)$, it follows that g is an epimorphism. If V_0 is a proper submodule of V then $f^{-1}(V_0)$ is a proper submodule of U since f is an epimorphism, and now $g(V_0) = gf(f^{-1}(V_0))$ is a proper submodule of S since gf is essential.

Suppose that g and gf are essential epimorphisms. If f were not an epimorphism then $f(U)$ would be a proper submodule of V, so $gf(U)$ would be a proper submodule of W since gf is essential. Since $gf(U) = W$, we conclude that f is an epimorphism. If U_0 is a proper submodule of U then $gf(U_0)$ is a proper submodule of W, since gf is essential, so $f(U_0)$ is a proper submodule of V since g is an epimorphism. Hence f is essential.

(2) Consider the commutative square

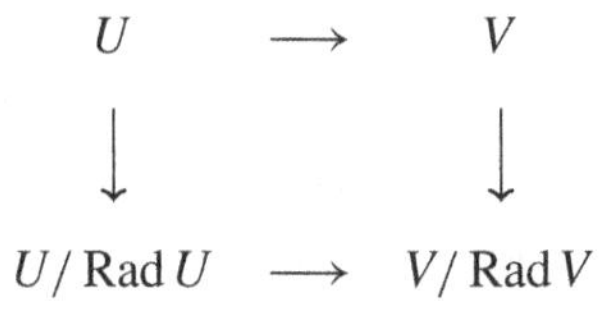

where the vertical homomorphisms are essential epimorphisms by Nakayama's Lemma. Now if either of the horizontal arrows is an essential epimorphism then so is the other, using part (1). The bottom arrow is an essential epimorphism if and only if it is an isomorphism; for $U/\operatorname{Rad} U$ is a semisimple module and so the kernel of the map to $V/\operatorname{Rad} V$ has a direct complement in $U/\operatorname{Rad} U$, which maps onto $V/\operatorname{Rad} V$. Thus, if $U/\operatorname{Rad} U \to V/\operatorname{Rad} V$ is an essential epimorphism its kernel must be zero and hence it must be an isomorphism.

(3) The map

$$(\oplus_i U_i)/\operatorname{Rad}(\oplus_i U_i) \to (\oplus_i V_i)/\operatorname{Rad}(\oplus_i V_i)$$

induced by $\oplus f_i$ may be identified as a map

$$\bigoplus_i (U_i/\operatorname{Rad} U_i) \to \bigoplus_i (V_i/\operatorname{Rad} V_i),$$

and it is an isomorphism if and only if each map $U_i/\operatorname{Rad} U_i \to V_i/\operatorname{Rad} V_i$ is an isomorphism. These conditions hold if and only if $\oplus f_i$ is an essential epimorphism, if and only if each f_i is an essential epimorphism by part (2). □

We define a *projective cover* of a module U to be an essential epimorphism $P \to U$, where P is a projective module. Strictly speaking the projective cover is the homomorphism, but we may also refer to the module P as the projective cover of U. We are justified in calling it *the* projective cover by the second part of the following result, which says that projective covers (if they exist) are unique.

Proposition 7.3.3. *(1) Suppose that $f : P \to U$ is a projective cover of a module U and $g : Q \to U$ is an epimorphism where Q is a projective module. Then we may write $Q = Q_1 \oplus Q_2$ so that g has components $g = (g_1, 0)$ with respect to this direct sum decomposition and $g_1 : Q_1 \to U$ appears in a commutative triangle*

$$\begin{array}{ccc} & & Q_1 \\ & {\scriptstyle\gamma}\swarrow & \downarrow{\scriptstyle g_1} \\ P & \xrightarrow{f} & U \end{array}$$

where γ is an isomorphism.

(2) If any exist, the projective covers of a module U are all isomorphic, by isomorphisms that commute with the essential epimorphisms.

Proof. (1) In the diagram,

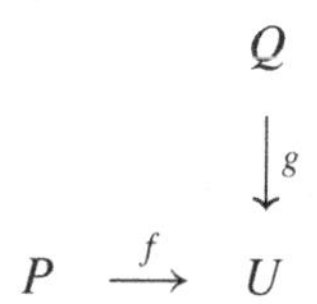

we may lift in both directions to obtain maps $\alpha : P \to Q$ and $\beta : Q \to P$ so that the two triangles commute. Now $f\beta\alpha = g\alpha = f$ is an epimorphism, so $\beta\alpha$ is also an epimorphism since f is essential. Thus, β is an epimorphism. Since P is projective β splits and $Q = Q_1 \oplus Q_2$ where $Q_2 = \operatorname{Ker}\beta$, and β maps Q_1 isomorphically to P. Thus, $g = (f\beta|_{Q_1}, 0)$ is as claimed with $\gamma = \beta|_{Q_1}$.

(2) Supposing that $f : P \to U$ and $g : Q \to U$ are both projective covers, since Q_1 is a submodule of Q that maps onto U and f is essential, we deduce that $Q = Q_1$. Now $\gamma : Q \to P$ is the required isomorphism. □

Corollary 7.3.4. *If P and Q are Noetherian projective modules over a ring then $P \cong Q$ if and only if $P/\operatorname{Rad} P \cong Q/\operatorname{Rad} Q$.*

Proof. By Nakayama's Lemma, P and Q are the projective covers of $P/\operatorname{Rad} P$ and $Q/\operatorname{Rad} Q$. It is clear that if P and Q are isomorphic then so are $P/\operatorname{Rad} P$ and $Q/\operatorname{Rad} Q$, and conversely if these quotients are isomorphic then so are their projective covers, by uniqueness of projective covers. □

If P is a projective module for a finite-dimensional algebra A then Corollary 7.3.4 says that P is determined up to isomorphism by its semisimple quotient $P/\operatorname{Rad} P$. We are going to see that if P is an indecomposable projective A-module, then its radical quotient is simple, and also that every simple A-module arises in this way. Furthermore, every indecomposable projective for a finite-dimensional algebra is isomorphic to a summand of the regular representation (something that is not true in general for projective $\mathbb{Z}G$-modules, for instance). This means that it is isomorphic to a module Af for some primitive idempotent $f \in A$, and the radical quotient $P/\operatorname{Rad} P$ is isomorphic to $(A/\operatorname{Rad} A)e$ where e is a primitive idempotent of $A/\operatorname{Rad} A$ satisfying $e = f + \operatorname{Rad} A$. We will examine this kind of relationship between idempotent elements more closely.

In general, if I is an ideal of a ring A and f is an idempotent of A then clearly $e = f + I$ is an idempotent of A/I, and we say that *f lifts e*. On the other hand, given an idempotent e of A/I it may or may not be possible to lift it to an idempotent of A. If, for every idempotent e in A/I, we can always find an idempotent $f \in A$ such that $e = f + I$ then we say we can *lift idempotents* from A/I to A.

We present the next results about lifting idempotents in the context of a ring with a nilpotent ideal I, but readers familiar with completions will recognize that these results extend to a situation where A is complete with respect to the I-adic topology on A.

Theorem 7.3.5. *Let I be a nilpotent ideal of a ring A and e an idempotent in A/I. Then there exists an idempotent $f \in A$ with $e = f + I$. If e is primitive, so is any lift f.*

Proof. We define idempotents $e_i \in A/I^i$ inductively such that $e_i + I^{i-1}/I^i = e_{i-1}$ for all i, starting with $e_1 = e$. Suppose that e_{i-1} is an idempotent of A/I^{i-1}. Pick any element $a \in A/I^i$ mapping onto e_{i-1}, so that $a^2 - a \in I^{i-1}/I^i$. Since $(I^{i-1})^2 \subseteq I^i$ we have $(a^2 - a)^2 = 0 \in A/I^i$. Put $e_i = 3a^2 - 2a^3$. This does map to $e_{i-1} \in A/I^{i-1}$ and we have

$$\begin{aligned} e_i^2 - e_i &= (3a^2 - 2a^3)(3a^2 - 2a^3 - 1) \\ &= -(3 - 2a)(1 + 2a)(a^2 - a)^2 \\ &= 0. \end{aligned}$$

This completes the inductive definition, and if $I^r = 0$ we put $f = e_r$.

Suppose that e is primitive and that f can be written $f = f_1 + f_2$ where f_1 and f_2 are orthogonal idempotents. Then $e = e_1 + e_2$, where $e_i = f_i + I$, is also a sum of orthogonal idempotents. Therefore one of these is zero, say, $e_1 = 0 \in A/I$. This means that $f_1^2 = f_1 \in I$. But I is nilpotent, and so contains no nonzero idempotent. □

We will very soon see that in the situation of Theorem 7.3.5, if f is primitive, so is e. It depends on the next result, which is a more elaborate version of Theorem 7.3.5.

Corollary 7.3.6. *Let I be a nilpotent ideal of a ring A and let $1 = e_1 + \cdots + e_n$ be a sum of orthogonal idempotents in A/I. Then, we can write $1 = f_1 + \cdots + f_n$ in A, where the f_i are orthogonal idempotents such that $f_i + I = e_i$ for all i. If the e_i are primitive then so are the f_i.*

Proof. We proceed by induction on n, the induction starting when $n = 1$. Suppose that $n > 1$ and the result holds for smaller values of n. We will write $1 = e_1 + E$ in A/I where $E = e_2 + \cdots + e_n$ is an idempotent orthogonal to e_1. By Theorem 7.3.5, we may lift e_1 to an idempotent $f_1 \in A$. Write $F = 1 - f_1$, so that F is an idempotent that lifts E. Now F is the identity element of the ring FAF which has a nilpotent ideal FIF. The composite homomorphism $FAF \hookrightarrow A \to A/I$ has kernel $FAF \cap I$ and this equals FIF, since clearly $FAF \cap I \supseteq FIF$, and if $x \in FAF \cap I$ then $x = FxF \in FIF$, so $FAF \cap I \subseteq FIF$. Inclusion of FAF in A thus induces a monomorphism $FAF/FIF \to A/I$, and its image is $E(A/I)E$. In $E(A/I)E$, the identity element E is the sum of $n - 1$

orthogonal idempotents, and this expression is the image of a similar expression for $F + FIF$ in FAF/FIF. By induction, there is a sum of orthogonal idempotents $F = f_2 + \cdots + f_n$ in FAF that lifts the expression in FAF/FIF and hence also lifts the expression for E in A/I, so we have idempotents $f_i \in A$, $i = 1, \ldots, n$ with $f_i + I = e_i$. These f_i are orthogonal: for $f_2, \ldots, f_n$ are orthogonal in FAF by induction, and if $i > 1$ then $Ff_i = f_i$ so we have $f_1 f_i = f_1 F f_i = 0$.

The final assertion about primitivity is the last part of Theorem 7.3.5. □

Corollary 7.3.7. *Let f be an idempotent in a ring A that has a nilpotent ideal I. Then f is primitive if and only if $f + I$ is primitive.*

Proof. We have seen in Theorem 7.3.5 that if $f + I$ is primitive, then so is f. Conversely, if $f + I$ can be written $f + I = e_1 + e_2$ where the e_i are orthogonal idempotents of A/I, then by applying Corollary 7.3.6 to the ring fAf (of which f is the identity) we may write $f = g_1 + g_2$ where the g_i are orthogonal idempotents of A that lift the e_i. □

We now classify the indecomposable projective modules over a finite-dimensional algebra as the projective covers of the simple modules. We first describe how these projective covers arise, and then show that they exhaust the possibilities for indecomposable projective modules. We postpone explicit examples until the next section, in which we consider group algebras.

Theorem 7.3.8. *Let A be a finite-dimensional algebra over a field and S a simple A-module.*

1. *There is an indecomposable projective module P_S with $P_S / \operatorname{Rad} P_S \cong S$, of the form $P_S = Af$, where f is a primitive idempotent in A.*
2. *The idempotent f has the property that $fS \neq 0$ and if T is any simple module not isomorphic to S then $fT = 0$.*
3. *P_S is the projective cover of S, it is uniquely determined up to isomorphism by this property and has S as its unique simple quotient.*
4. *It is also possible to find an idempotent $f_S \in A$ so that $f_S S = S$ and $f_S T = 0$ for every simple module T not isomorphic to S.*

Proof. Let $e \in A/\operatorname{Rad} A$ be any primitive idempotent such that $eS \neq 0$. It is possible to find such e since we may write 1 as a sum of primitive idempotents and some term in the sum must be nonzero on S. Let f be any lift of e to A, possible by Corollary 7.3.7. Then f is primitive, $fS = eS \neq 0$ and $fT = eT = 0$ if $T \not\cong S$ since a primitive idempotent e in the semisimple ring $A/\operatorname{Rad} A$ is nonzero on a unique isomorphism class of simple modules. We define

$P_S = Af$, an indecomposable projective module. Now

$$P_S/\operatorname{Rad} P_S = Af/(\operatorname{Rad} A \cdot Af) \cong (A/\operatorname{Rad} A)\cdot(f + \operatorname{Rad} A) = S,$$

the isomorphism arising because the map $Af \to (A/\operatorname{Rad} A)\cdot(f+\operatorname{Rad} A)$ defined by $af \mapsto (af + \operatorname{Rad} A)$ has kernel $(\operatorname{Rad} A)\cdot f$. The fact that P_S is the projective cover of S is a consequence of Nakayama's Lemma, and the uniqueness of the projective cover was dealt with in Proposition 7.3.3. Any simple quotient of P_S is a quotient of $P_S/\operatorname{Rad} P_S$, so there is only one of these. Finally, we observe that if we had written 1 as a sum of primitive central idempotents in $A/\operatorname{Rad} A$, the lift of the unique such idempotent that is nonzero on S is the desired idempotent f_S. □

Theorem 7.3.9. *Let A be a finite-dimensional algebra over a field k. Up to isomorphism, the indecomposable projective A-modules are exactly the modules P_S that are the projective covers of the simple modules, and $P_S \cong P_T$ if and only if $S \cong T$. Each projective P_S appears as a direct summand of the regular representation, with multiplicity equal to the multiplicity of S as a summand of $A/\operatorname{Rad} A$. As a left A-module the regular representation decomposes as*

$$A \cong \bigoplus_{\text{simple } S} (P_S)^{n_S},$$

where $n_S = \dim_k S$ if k is algebraically closed, and more generally $n_S = \dim_D S$ where $D = \operatorname{End}_A(S)$.

In what follows we will only prove that finitely generated indecomposable projective modules are isomorphic to P_S, for some simple S. In Exercise 10 at the end of this chapter, it is shown that this accounts for all indecomposable projective modules.

Proof. Let P be an indecomposable projective module and write

$$P/\operatorname{Rad} P \cong S_1 \oplus \cdots \oplus S_n.$$

Then $P \to S_1 \oplus \cdots \oplus S_n$ is a projective cover. Now

$$P_{S_1} \oplus \cdots \oplus P_{S_n} \to S_1 \oplus \cdots \oplus S_n$$

is also a projective cover, and by uniqueness of projective covers we have

$$P \cong P_{S_1} \oplus \cdots \oplus P_{S_n}.$$

Since P is indecomposable we have $n = 1$ and $P \cong P_{S_1}$.

Suppose that each simple A module S occurs with multiplicity n_S as a summand of the semisimple ring $A/\operatorname{Rad} A$. Both A and $\bigoplus_{\text{simple } S} P_S^{n_S}$ are the projective cover of $A/\operatorname{Rad} A$, and so they are isomorphic. We have seen in Corollary 2.1.4 that $n_S = \dim_k S$ when k is algebraically closed, and in Exercise 5 of Chapter 2 that $n_S = \dim_D S$ in general. □

Theorem 7.3.10. *Let A be a finite-dimensional algebra over a field k, and U an A-module. Then U has a projective cover.*

Again, we only give a proof in the case that U is finitely generated, leaving the general case to Exercise 10 of this chapter.

Proof. Since $U/\operatorname{Rad} U$ is semisimple we may write $U/\operatorname{Rad} U = S_1 \oplus \cdots \oplus S_n$, where the S_i are simple modules. Let P_{S_i} be the projective cover of S_i and $h : P_{S_1} \oplus \cdots \oplus P_{S_n} \to U/\operatorname{Rad} U$ the projective cover of $U/\operatorname{Rad} U$. By projectivity, there exists a homomorphism f such that the following diagram commutes:

$$\begin{array}{ccc} & & P_{S_1} \oplus \cdots \oplus P_{S_n} \\ & \overset{f}{\swarrow} & \downarrow h \\ U & \xrightarrow{g} & U/\operatorname{Rad} U \end{array} .$$

Since both g and h are essential epimorphisms, so is f by Proposition 7.3.2. Therefore, f is a projective cover. □

We should really learn more from Theorem 7.3.10 than simply that U has a projective cover: the projective cover of U is the same as the projective cover of $U/\operatorname{Rad} U$.

Example 7.3.11. The arguments that show the existence of projective covers have a sense of inevitability about them and we may get the impression that projective covers always exist in arbitrary situations. In fact, they fail to exist in general for integral group rings. If $G = \{e, g\}$ is a cyclic group of order 2, consider the submodule $3\mathbb{Z} \cdot e + \mathbb{Z} \cdot (e+g)$ of $\mathbb{Z}G$ generated as an abelian group by $3e$ and $e+g$. We rapidly check that this subgroup is invariant under the action of G (so it is a $\mathbb{Z}G$-submodule), and it is not the whole of $\mathbb{Z}G$ since it does not contain e. Applying the augmentation map $\epsilon : \mathbb{Z}G \to \mathbb{Z}$, we have $\epsilon(3e) = 3$ and $\epsilon(e+g) = 2$ so $\epsilon(3\mathbb{Z} \cdot e + \mathbb{Z} \cdot (e+g)) = 3\mathbb{Z} + 2\mathbb{Z} = \mathbb{Z}$. This shows that the epimorphism ϵ is not essential, and so it is not a projective cover of $\mathbb{Z}$. If $\mathbb{Z}$ were to have a projective cover it would be a proper summand of $\mathbb{Z}G$ by Proposition 7.3.3. On reducing modulo 2, we would deduce that $\mathbb{F}_2G$ decomposes, which we know not to be the case by Corollary 6.3.7. This shows that $\mathbb{Z}$ has no projective cover as a $\mathbb{Z}G$-module.

7.4 The Cartan Matrix

Now that we have classified the projective modules for a finite-dimensional algebra we turn to one of their important uses, which is to determine the multiplicity of a simple module S as a composition factor of an arbitrary module U (with a composition series). If

$$0 = U_0 \subset U_1 \subset \cdots \subset U_n = U$$

is any composition series of U, the number of quotients U_i/U_{i-1} isomorphic to S is determined independently of the choice of composition series, by the Jordan–Hölder theorem. We call this number the *(composition factor) multiplicity* of S in U.

Proposition 7.4.1. *Let S be a simple module for a finite-dimensional algebra A with projective cover P_S, and let U be a finite-dimensional A-module.*

(1) If T is a simple A-module then

$$\dim \operatorname{Hom}_A(P_S, T) = \begin{cases} \dim \operatorname{End}_A(S) & \text{if } S \cong T \\ 0 & \text{otherwise.} \end{cases}$$

(2) The multiplicity of S as a composition factor of U is

$$\dim \operatorname{Hom}_A(P_S, U)/\dim \operatorname{End}_A(S).$$

(3) If $e \in A$ is an idempotent then $\dim \operatorname{Hom}_A(Ae, U) = \dim eU$.

We remind the reader that if the ground field k is algebraically closed then $\dim \operatorname{End}_A(S) = 1$ by Schur's Lemma. Thus, the multiplicity of S in U is just $\dim \operatorname{Hom}_A(P_S, U)$ in this case.

Proof. (1) If $P_S \to T$ is any nonzero homomorphism, the kernel must contain $\operatorname{Rad} P_S$, being a maximal submodule of P_S. Since $P_S/\operatorname{Rad} P_S \cong S$ is simple, the kernel must be $\operatorname{Rad} P_S$ and $S \cong T$. Every homomorphism $P_S \to S$ is the composite $P_S \to P_S/\operatorname{Rad} P_S \to S$ of the quotient map and either an isomorphism of $P_S/\operatorname{Rad} P_S$ with S or the zero map. This gives an isomorphism $\operatorname{Hom}_A(P_S, S) \cong \operatorname{End}_A(S)$.

(2) Let

$$0 = U_0 \subset U_1 \subset \cdots \subset U_n = U$$

be a composition series of U. We prove the result by induction on the composition length n, the case $n = 1$ having just been established. Suppose $n > 1$ and that the multiplicity of S in U_{n-1} is $\dim \operatorname{Hom}_A(P_S, U_{n-1})/\dim \operatorname{End}_A(S)$.

The exact sequence

$$0 \to U_{n-1} \to U \to U/U_{n-1} \to 0$$

gives rise to an exact sequence

$$0 \to \operatorname{Hom}_A(P_S, U_{n-1}) \to \operatorname{Hom}_A(P_S, U) \to \operatorname{Hom}_A(P_S, U/U_{n-1}) \to 0$$

by Proposition 7.1.3, so that

$$\dim \operatorname{Hom}_A(P_S, U) = \dim \operatorname{Hom}_A(P_S, U_{n-1}) + \dim \operatorname{Hom}_A(P_S, U/U_{n-1}).$$

Dividing these dimensions by $\dim \operatorname{End}_A(S)$ gives the result, by part (1).

(3) There is an isomorphism of vector spaces $\operatorname{Hom}_A(Ae, U) \cong eU$ specified by $\phi \mapsto \phi(e)$. Note here that since $\phi(e) = \phi(ee) = e\phi(e)$ we must have $\phi(e) \in eU$. This mapping is injective since each A-module homomorphism $\phi : Ae \to U$ is determined by its value on e as $\phi(ae) = a\phi(e)$. It is surjective since the equation just written down does define a module homomorphism for each choice of $\phi(e) \in eU$. □

Again in the context of a finite-dimensional algebra A, we define for each pair of simple A-modules S and T the integer

$$c_{ST} = \text{the composition factor multiplicity of } S \text{ in } P_T.$$

These are called the *Cartan invariants* of A, and they form a matrix $C = (c_{ST})$ with rows and columns indexed by the isomorphism types of simple A-modules, called the *Cartan matrix* of A.

Corollary 7.4.2. *Let A be a finite-dimensional algebra over a field, let S and T be simple A-modules and let e_S, e_T be idempotents so that $P_S = Ae_S$ and $P_T = Ae_T$ are projective covers of S and T. Then,*

$$c_{ST} = \dim \operatorname{Hom}_A(P_S, P_T)/\dim \operatorname{End}_A(S) = \dim e_S A e_T / \dim \operatorname{End}_A(S).$$

If the ground field k is algebraically closed then

$$c_{ST} = \dim \operatorname{Hom}_A(P_S, P_T) = \dim e_S A e_T.$$

While it is rather weak information just to know the composition factors of the projective modules, this is at least a start in describing these modules. We will see later on in the case of group algebras that there is an extremely effective way of computing the Cartan matrix using the decomposition matrix.

7.5 Summary of Chapter 7

- Direct sum decompositions of ${}_AA$ as an A-module (with indecomposable summands) correspond to expressions for 1_A as a sum of orthogonal (primitive) idempotents.
- $U \to U/\operatorname{Rad} U$ is essential.
- Projective covers are unique when they exist. For modules for a finite-dimensional algebra over a field they do exist.
- Idempotents can be lifted through nilpotent ideals.
- The indecomposable projective modules for a finite-dimensional algebra over a field are exactly the projective covers of the simple modules. Each has a unique simple quotient and is a direct summand of the regular representation. Over an algebraically closed field P_S occurs as a summand of the regular representation with multiplicity $\dim S$.

7.6 Exercises for Chapter 7

1. Let A be a finite-dimensional algebra over a field. Show that A is semisimple if and only if all finite-dimensional A-modules are projective.

2. Let P_S be an indecomposable projective module for a finite-dimensional algebra over a field. Show that every nonzero homomorphic image of P_S

(a) has a unique maximal submodule,
(b) is indecomposable, and
(c) has P_S as its projective cover.

3. Let A be a finite-dimensional algebra over a field, and suppose that f, f' are primitive idempotents of A. Show that the indecomposable projective modules Af and Af' are isomorphic if and only if $\dim fS = \dim f'S$ for every simple module S.

4. Let A be a finite-dimensional algebra over a field and $f \in A$ a primitive idempotent. Show that there is a simple A-module S with $fS \neq 0$, and that S is uniquely determined up to isomorphism by this property.

5. Let A be a finite-dimensional algebra over a field, and suppose that Q is a projective A-module. Show that in any expression

$$Q = P_{S_1}^{n_1} \oplus \cdots \oplus P_{S_r}^{n_r},$$

where $S_1, \ldots, S_r$ are nonisomorphic simple modules, we have

$$n_i = \dim \operatorname{Hom}_A(Q, S_i) / \dim \operatorname{End}_A(S_i).$$

6. Let A be a finite-dimensional algebra over a field. Suppose that V is an A-module, and that a certain simple A-module S occurs as a composition factor of V with multiplicity 1. Suppose that there exist nonzero homomorphisms $S \to V$ and $V \to S$. Prove that S is a direct summand of V.

7. Let $G = S_n$, let k be a field of characteristic 2 and let $\Omega = \{1, 2, \ldots, n\}$ permuted transitively by G.

(a) When $n = 3$, show that the permutation module $k\Omega$ is semisimple, being the direct sum of the 1-dimensional trivial module and the 2-dimensional simple module. [Use the information from Example 7.2.2 and Exercise 8 from Chapter 6.]

(b) When $n = 4$, there is a normal subgroup $V \lhd S_4$ with $S_4/V \cong S_3$, where $V = \langle (1, 2)(3, 4), (1, 3)(2, 4) \rangle$. Show that the simple kS_4-modules are precisely the two simple kS_3-modules, made into kS_4-modules via the quotient homomorphism to S_3. Show that $k\Omega$ is uniserial with three composition factors that are the trivial module, the 2-dimensional simple module and the trivial module.

[Use Exercise 8 from Chapter 6.]

8. Show by example that if H is a subgroup of G it need not be true that $\operatorname{Rad} kH \subseteq \operatorname{Rad} kG$.
[Compare this result with Exercise 5 from Chapter 6.]

9. Suppose that we have module homomorphisms $U \xrightarrow{f} V \xrightarrow{g} W$. Show that part of Proposition 7.3.2(1) can be strengthened to say the following: if gf is an essential epimorphism and f is an epimorphism then both f and g are essential epimorphisms.

10. Let U and V be arbitrary (not necessarily Noetherian) modules for a finite-dimensional algebra A. Use the results of Exercise 27 of Chapter 6 to show the following.

(a) Show that the quotient homomorphism $U \to U/\operatorname{Rad} U$ is essential.
(b) Show that a homomorphism $U \to V$ is essential if and only if the homomorphism of radical quotients $U/\operatorname{Rad} U \to V/\operatorname{Rad} V$ is an isomorphism.
(c) Show that U has a projective cover.
(d) Show that every indecomposable projective A-module is finite-dimensional, and hence isomorphic to P_S for some simple module S.
(e) Show that every projective A-module is a direct sum of indecomposable projective modules.

11. In this question, U, V, and W are modules for a finite-dimensional algebra over a field and P_W is the projective cover of W. Assume either that these modules are finite-dimensional, or the results from the last exercise.

(a) Show that $U \to W$ is an essential epimorphism if and only if there is a surjective homomorphism $P_W \to U$ so that the composite $P_W \to U \to W$ is a projective cover of W. In this situation, show that $P_W \to U$ must be a projective cover of U.

(b) Prove the following "extension and converse" to Nakayama's Lemma: let V be any submodule of U. Then $U \to U/V$ is an essential epimorphism $\Leftrightarrow V \subseteq \operatorname{Rad} U$.

8
Projective Modules for Group Algebras

We focus in this chapter on facts about group algebras that are not true for finite-dimensional algebras in general. The results are a mix of general statements and specific examples describing the representations of certain types of groups. At the beginning of the chapter, we summarize the properties of projective modules for p-groups and also the behavior of projective modules under induction and restriction. Toward the end, we show that the Cartan matrix is symmetric (then the field is algebraically closed) and also that projective modules are injective. In the middle, we describe quite explicitly the structure of projective modules for many semidirect products, and we do this by elementary arguments. It shows that the important general theorems are not always necessary to understand specific representations, and it also increases our stock of examples of groups and their representations.

Because of the diversity of topics, it is possible to skip certain results in this chapter without affecting comprehension of what remains. For example, the reader who is more interested in the general results could skip the description of representations of specific groups between Example 8.2.1 and Theorem 8.4.1.

8.1 The Behavior of Projective Modules under Induction, Restriction, and Tensor Product

We start with a basic fact about group algebras of p-groups in characteristic p.

Theorem 8.1.1. *Let k be a field of characteristic p and G a p-group. The regular representation is an indecomposable projective module that is the projective cover of the trivial representation. Every finitely generated projective module is free. The only idempotents in kG are 0 and 1.*

Proof. We have seen in 6.12 that kG is indecomposable and it also follows from 7.14. By Nakayama's Lemma, kG is the projective cover of k. By 7.13 and 6.3, every indecomposable projective is isomorphic to kG. Every finitely generated projective is a direct sum of indecomposable projectives, and so is free. Finally, every idempotent $e \in kG$ gives a module decomposition $kG = kGe \oplus kG(1-e)$. If $e \neq 0$ then we must have $kG = kGe$, so $kG(1-e) = 0$ and $e = 1$. □

The next lemma is a consequence of the fact that induction is both the left and the right adjoint of restriction, as shown in Lemma 4.12. Without using that language we give a direct proof.

Lemma 8.1.2. *Let H be a subgroup of G.*

(1) If P is a projective RG-module then $P \downarrow_H^G$ is a projective RH-module.
(2) If Q is a projective RH-module then $Q \uparrow_H^G$ is a projective RG-module.

Proof. (1) As a RH-module,

$$RG \downarrow_H \cong \bigoplus_{g \in [H\backslash G]} RHg \cong (RH)^{|G:H|},$$

which is a free module. Hence a direct summand of RG^n on restriction to H is a direct summand of $RH^{|G:H|n}$, and so is projective.

(2) We have

$$(RH)\uparrow_H^G \cong (R\uparrow_1^H)\uparrow_H^G \cong R\uparrow_1^G \cong RG$$

so that direct summands of RH^n induce to direct summands of RG^n. □

We now put together Theorem 8.1.1 and Lemma 8.1.2 to obtain an important restriction on projective modules.

Corollary 8.1.3. *Let k be a field of characteristic p and let p^a be the exact power of p that divides $|G|$. If P is a projective kG-module then $p^a \mid \dim P$.*

Proof. Let H be a Sylow p-subgroup of G and P a projective kG-module. Then $P \downarrow_H^G$ is projective by Lemma 8.1.2, hence free as a kH-module by Theorem 8.1.1, and of dimension a multiple of $|H|$. □

We are about to describe in detail the structure of projective modules for some particular groups that are semidirect products, and the next two results will be used in our proofs. The first is valid over any commutative ring R.

Proposition 8.1.4. *Suppose that V is any RG-module that is free as an R-module and P is a projective RG-module. Then $V \otimes_R P$ is projective as an RG-module.*

Proof. If $P \oplus P' \cong RG^n$ then $V \otimes RG^n \cong V \otimes P \oplus V \otimes P'$ and it suffices to show that $V \otimes RG^n$ is free. We offer two proofs of the fact that $V \otimes RG \cong RG^{\operatorname{rank} V}$. The first is that $V \otimes RG \cong V \otimes (R \uparrow_1^G) \cong (V \otimes R) \uparrow_1^G \cong V \uparrow_1^G$, with the middle isomorphism coming from Corollary 4.13. As a module for the identity group, V is just a free R-module and so $V \uparrow_1^G \cong (R \uparrow_1^G)^{\operatorname{rank} V} \cong RG^{\operatorname{rank} V}$.

The second proof is really the same as the first, but we make the isomorphism explicit. Let V^{triv} be the same R-module as V, but with the trivial G-action, so $V^{\mathrm{triv}} \cong R^{\operatorname{rank} V}$ as RG-modules. We define a linear map

$$V \otimes RG \to V^{\mathrm{triv}} \otimes RG$$
$$v \otimes g \mapsto g^{-1}v \otimes g,$$

which has inverse $gw \otimes g \leftarrow w \otimes g$. One checks that these mutually inverse linear maps are RG-module homomorphisms. Finally, $V^{\mathrm{triv}} \otimes RG \cong RG^{\operatorname{rank} V}$. □

In the calculations that follow, we will need to use the fact that for representations over a field, taking the tensor product with a fixed representation preserves exactness.

Lemma 8.1.5. *Let $0 \to U \to V \to W \to 0$ be a short exact sequence of kG-modules and X another kG-module, where k is a field. Then the sequence*

$$0 \to U \otimes_k X \to V \otimes_k X \to W \otimes_k X \to 0$$

is exact. Thus, if U is a submodule of V then

$$(V/U) \otimes_k X \cong (V \otimes_k X)/(U \otimes_k X).$$

Proof. Since the tensor products are taken over k, the question of exactness is independent of the action of G. As a short exact sequence of vector spaces $0 \to U \to V \to W \to 0$ is split, so that $V \cong U \oplus W$ with the morphisms in the sequence as two of the component inclusions and projections. Applying $- \otimes_k X$ to this we get $V \otimes_k X \cong (U \otimes_k X) \oplus (W \otimes_k X)$ with component morphisms given by the morphisms in the sequence $0 \to U \otimes_k X \to V \otimes_k X \to W \otimes_k X \to 0$. This is enough to show that the sequence is (split) exact as a sequence of vector spaces, and hence exact as a sequence of kG-modules.

Another approach to the same thing is to suppose that U is a submodule of V and take a basis $v_1, \dots, v_n$ for V such that $v_1, \dots, v_d$ is a basis for U and let $x_1, \dots, x_m$ be a basis for X. Now the $v_i \otimes_k x_j$ with $1 \le i \le n$ and $1 \le j \le m$

form a basis for $V \otimes_k X$, and the same elements with $1 \le i \le d$ and $1 \le j \le m$ form a basis for $U \otimes_k X$. This shows that $U \otimes_k X$ is a submodule of $V \otimes_k X$, and the quotient has as a basis the images of the $v_i \otimes_k x_j$ with $d+1 \le i \le n$ and $1 \le j \le m$, which is in bijection with a basis of $W \otimes_k X$. □

8.2 Projective and Simple Modules for Direct Products of a p-Group and a p'-Group

Before continuing with the general development of the theory, we describe in detail the projective modules for groups that are a semidirect product with one of the terms a Sylow p-subgroup, since this can be done by direct arguments with the tools we already have at hand. This will be done in general in the next sections, and in this section, we start with the special case of a direct product.

Example 8.2.1. Let k be a field of characteristic p and let $G = H \times K$, where H is a p-group and K has order prime to p. In this example, the only tensor products that appear are tensor products $\otimes_k$ over the field k, and we supress the suffix k from the notation. We make use of the following general isomorphism (not dependent on the particular hypotheses we have here, and seen before in the remark after Theorem 4.1.2):

$$k[H \times K] \cong kH \otimes kK \quad \text{as } k\text{-algebras,}$$

which arises because kG has as a basis the elements (h, k) where $h \in H, k \in K$, and $kH \otimes kK$ has as a basis the corresponding elements $h \otimes k$. These two bases multiply together in the same fashion, and so we have an algebra isomorphism.

Let us write $kK = S_1^{n_1} \oplus \cdots \oplus S_r^{n_r}$, where $S_1, \ldots, S_r$ are the nonisomorphic simple kK-modules, bearing in mind that kK is semisimple since K has order relatively prime to p. Since $H = O_p(G)$, these are also the nonisomorphic simple kG-modules, by Corollary 6.4. We have

$$kG = kH \otimes kK = (kH \otimes S_1)^{n_1} \oplus \cdots \oplus (kH \otimes S_r)^{n_r}$$

as kG-modules, and so the $kH \otimes S_i$ are projective kG-modules. Each does occur with multiplicity equal to the multiplicity of S_i as a summand of $kG/\operatorname{Rad}(kG)$, and so must be indecomposable, using 7.14. We have therefore constructed all the indecomposable projective kG-modules, and they are the modules $P_{S_i} = kH \otimes S_i$.

Suppose that $0 \subset P_1 \subset \cdots \subset P_n = kH$ is a composition series of the regular representation of H. Since H is a p-group, all the composition factors are the trivial representation, k. Because $\_ \otimes_k S_i$ preserves exact sequences, the series

$0 \subset P_1 \otimes S_i \subset \cdots \subset P_n \otimes S_i = P_{S_i}$ has quotients $k \otimes S_i = S_i$, which are simple, and so this is a composition series of P_{S_i}. There is only one isomorphism type of composition factor.

We can also see this from the ring-theoretic structure of kG. Assuming that k is algebraically closed (to make the notation easier), we have $\mathrm{End}_{kG}(S_i) = k$ for each i and $kK \cong \bigoplus_{i=1}^{r} M_{n_i}(k)$ as rings where $n_i = \dim S_i$. Now

$$kG = kH \otimes kK \cong \bigoplus_{i=1}^{r} M_{n_i}(kH)$$

as rings, since $kH \otimes M_n(k) \cong M_n(kH)$. The latter can be proved by observing that the two sides of the isomorphism have bases that multiply together in the same way. The projective kG-modules $P_{S_i} = kH \otimes S_i$ are now identified in this matrix description as column vectors of length n_i with entries in kH.

We next observe that

$$\mathrm{Rad}\, kG = \mathrm{Rad}(kH) \otimes kK \cong \bigoplus_{i=1}^{r} M_{n_i}(\mathrm{Rad}(kH)).$$

The first equality holds because the quotient

$$(kH \otimes kK)/(\mathrm{Rad}(kH) \otimes kK) \cong (kH/\mathrm{Rad}\, kH) \otimes kK \cong k \otimes kK$$

is semisimple, since kK is semisimple, so that

$$\mathrm{Rad}(kH \otimes kK) \subseteq \mathrm{Rad}(kH) \otimes kK.$$

On the other hand, $\mathrm{Rad}(kH) \otimes kK$ is a nilpotent ideal of $kH \otimes kK$, so is contained in the radical, and we have equality. To look at it from the matrix point of view, $\bigoplus_{i=1}^{r} M_{n_i}(\mathrm{Rad}(kH))$ is a nilpotent ideal with quotient $\bigoplus_{i=1}^{r} M_{n_i}(k)$, which is semisimple, so again we have correctly identified the radical. Computing the powers of these ideals, we have

$$\mathrm{Rad}^n\, kG = \mathrm{Rad}^n(kH) \otimes kK \cong \bigoplus_{i=1}^{r} M_{n_i}(\mathrm{Rad}^n(kH))$$

for each n.

Example 8.2.2. As a very specific example, suppose that H is cyclic of order p^s. Then kH has a unique composition series by Theorem 6.2, with terms $P_j = \mathrm{Rad}(kH)^j \cdot kH$. We see from the preceding discussion and the fact that $\mathrm{Rad}\, kG = \mathrm{Rad}(kH) \otimes kK$ that the terms in the composition series of P_{S_i} are $P_j \otimes S_i = \mathrm{Rad}(kG)^j \cdot P_{S_i}$. Thus, the radical series of P_{S_i} is in fact a composition series, and it follows (as in Exercise 6 of Chapter 6) that P_{S_i} also has a unique composition series, there being no more submodules of P_{S_i} other than the ones listed.

8.3 Projective Modules for Groups with a Normal Sylow p-Subgroup

We move on now to describe the projective kG-modules where k is a field of characteristic p and where G is a semidirect product with Sylow p-subgroup H, doing this first in this section when G has the form $G = H \rtimes K$ and afterward in the next section doing it when $G = K \rtimes H$. Before this, we give a module decomposition of the group ring of a semidirect product that holds in all cases.

Lemma 8.3.1. *Let R be a commutative ring and $G = H \rtimes K$ a semidirect product. There is an action of RG on RH such that H acts on RH by left multiplication and K acts by conjugation. With this action, $RH \cong R \uparrow_K^G \cong RG \cdot \overline{K}$ as RG-modules, where $\overline{K} = \sum_{x \in K} x$. We have a tensor decomposition of RG-modules*

$$RG \cong (R \uparrow_K^G) \otimes_R (R \uparrow_H^G),$$

where the structure of $R \uparrow_K^G$ has just been described, and $R \uparrow_H^G \cong RK$ with H acting trivially.

Proof. We may take $R \uparrow_K^G = RG \otimes_{RK} R$ to have as R-basis the tensors $h \otimes 1$ where $h \in H$. The action of an element $x \in H$ on such a basis element is

$$x(h \otimes 1) = xh \otimes 1,$$

and the action of an element $k \in K$ is given by

$$k(h \otimes 1) = kh \otimes 1 = kh \otimes k^{-1}1 = khk^{-1} \otimes 1$$

since the tensor product is taken over RK. Note here that although $kh \otimes 1$ is defined, it is not one of our chosen basis elements (unless $k = 1$), whereas $khk^{-1} \otimes 1$ is one of them. We see from this that the R-linear isomorphism $RH \to R \uparrow_K^G$ given by $h \mapsto h \otimes 1$ is in fact an isomorphism of RG-modules with the specified action on RH, and indeed this specification does give an action.

The identification of $R \uparrow_K^G$ with $RG \cdot \overline{K}$ was seen in 4.9 and in Exercise 11 from Chapter 4, but we give the argument here. The submodule $RG \cdot \overline{K}$ is spanned in RG by the elements $G \cdot \overline{K} = HK \cdot \overline{K} = H \cdot \overline{K}$. However, the elements $h\overline{K}$ are independent in RG as h ranges over H, since they have disjoint supports, so these elements form a basis of $RG \cdot \overline{K}$. The basis elements are permuted transitively by G and the stabilizer of $\overline{K}$ is K, so the $RG \cdot \overline{K} \cong R \uparrow_K^G$.

Since $RH \otimes_R RK$ has an R-basis consisting of the basic tensors $h \otimes k$ here $h \in H, k \in K$, and every element of G is uniquely written in the form hk with $h \in H, k \in K$, the mapping $RH \otimes_R RK \to RG$ specified by $h \otimes k \mapsto hk$ is an R-linear isomorphism. This map commutes with the action of G, since if $uv \in G$

with $u \in H$, $v \in K$ then

$$uv(h \otimes k) = (uv) \cdot h \otimes (uv) \cdot k = u({}^v h) \otimes vk$$

since v acts by conjugation of H and u acts trivially on K. This is mapped to $u({}^v h)vk = (uv)(hk) \in RG$ so the map commutes with the action of G. □

We now suppose that k is a field of characteristic p and $G = H \rtimes K$, where H is a Sylow p-subgroup. We shall see that several of the properties of projective modules that we identified in the case of a direct product $G = H \times K$ still hold for the semidirect product, but not all of them.

Proposition 8.3.2. *Let k be a field of characteristic p and let $G = H \rtimes K$ where H is a p-group and K has order prime to p.*

(1) *Let $e_K = \frac{1}{|K|}\sum_{x \in K} x$. Then the indecomposable projective P_k has the form*

$$P_k \cong kGe_K \cong kH \cong k\uparrow_K^G,$$

where kH is taken to have the kG-module action where H acts by multiplication and K acts by conjugation, as in Lemma 8.3.1.

(2) $\mathrm{Rad}(kG)$ *is the kernel of the ring homomorphism $kG \to kK$ given by the quotient homomorphism $G \to K$, so that $kG/\mathrm{Rad}(kG) \cong kK$ as rings, and also as kG-modules. Furthermore,* $\mathrm{Rad}(kG)$ *is generated both as a left ideal and a right ideal by* $\mathrm{Rad}(kH) = IH$, *the augmentation ideal of kH.*

(3) *The simple kG-modules are precisely the simple kK-modules, regarded as kG-modules via the quotient homomorphism $G \to K$. If S is any simple kG-module with projective cover P_S then $P_S \cong P_k \otimes S \cong S\uparrow_K^G$ as kG-modules.*

(4) *For each simple kG-module S with projective cover P_S, the radical series of P_S has terms* $\mathrm{Rad}^n(P_S) = (IH)^n \otimes S$ *with radical layers $((IH)^{n-1}/(IH)^n) \otimes S$, where IH is taken to have the restriction of the action on RH given in Lemma 8.3.1. Consequently all of the indecomposable projective modules have the same Loewy length.*

(5) *There is an isomorphism of kG-modules $P_k \otimes kK \cong kG$.*

Proof. (1) Since $|K|$ is invertible in k, we have $kGe_K = kG \cdot \overline{K}$ (with the notation of Lemma 8.3.1), and since e_K is an idempotent this is a projective kG-module, and it has the stated identifications by Lemma 8.3.1. This module is indecomposable as a kH-module since H is a p-group, by Theorem 8.1.1, so it is indecomposable as a kG-module. Since its unique simple quotient is represented by $1 \in H$, and K conjugates this trivially, kGe_K is the projective cover of the trivial module.

(2) Since $H = O_p(G)$ the simple kG-modules are precisely the simple kK-modules, by 6.4. Now G acts on these via the ring surjection $kG \to kK$, so the kernel of this map acts as zero on all simple modules and hence is contained in $\operatorname{Rad} kG$. But also kK is a semisimple ring, so the kernel equals $\operatorname{Rad} kG$.

The fact that the kernel of the ring homomorphism $kG \to kK$ (and hence $\operatorname{Rad} kG$) has the description $kG \cdot IH$ is shown in Exercise 15 of Chapter 6, and we also give the argument here. We know that $\operatorname{Rad} kH = IH$ from Proposition 6.3.3. Taking coset representatives $G = \bigsqcup_{g\in[G/H]} gH$ and writing $k[gH]$ for the span in kG of the elements gh, $h \in H$, we have that

$$kG \cdot IH = \left(\sum_{g\in[G/H]} k[gH] \right) \cdot IH = \sum_{g\in[G/H]} gIH = \bigoplus_{g\in[G/H]} gIH,$$

the sum being direct since each term gIH has basis $\{g(h-1) \mid h \in H\}$ with support in the coset gH. The span of the elements of K inside kG is a space complementary to $kG \cdot IH$ that is mapped isomorphically to kK. From this, we see that $kG \cdot IH$ is the kernel of the homomorphism $kG \to kK$. From the fact that H is a normal subgroup, we may calculate directly that $kG \cdot \operatorname{Rad}(kH) = kG \cdot \operatorname{Rad}(kH) \cdot kG = \operatorname{Rad}(kH) \cdot kG$. We may also show this by observing that $kG \cdot \operatorname{Rad}(kH)$ is the kernel of a ring homomorphism, hence is a 2-sided ideal and so $kG \cdot \operatorname{Rad}(kH) = kG \cdot \operatorname{Rad}(kH) \cdot kG$. We could have argued with $\operatorname{Rad}(kH) \cdot kG$ just as well, and so this also is equal to $kG \cdot \operatorname{Rad}(kH) \cdot kG$.

(3) The fact that the simple kG-modules are the same as the simple kK-modules was observed at the start of the proof of (2). If S is a simple module, by Propositon 8.1.4 $P_k \otimes S$ is projective. Tensoring the epimorphism $P_k \to k$ with S gives an epimorphism $P_k \otimes S \to S$, so $P_k \otimes S$ contains P_S as a summand. We show that $P_k \otimes S$ equals P_S. Now

$$\operatorname{Rad}(P_k \otimes S) = \operatorname{Rad}(kG) \cdot (P_k \otimes S) = \operatorname{Rad}(kH) \cdot kG \cdot (P_k \otimes S) = IH \cdot (P_k \otimes S).$$

We show that this equals $(IH \cdot P_k) \otimes S$. The reason for this is that IH is spanned by elements $h - 1$ where $h \in H$, and if $x \otimes s$ is a basic tensor in $P_k \otimes S$ then

$$\begin{aligned}(h-1) \cdot x \otimes s &= h \cdot (x \otimes s) - 1 \cdot (x \otimes s) \\ &= hx \otimes hs - x \otimes s \\ &= hx \otimes (h-1)s + (h-1)x \otimes s \\ &= (h-1)x \otimes s\end{aligned}$$

since H is a normal p-subgroup which thus acts trivially on S. Note that this argument does not depend on the first module in the tensor product being P_k. To continue the argument, we have

$$(IH \cdot P_k) \otimes S = (IH \cdot kG \cdot P_k) \otimes S = (\mathrm{Rad}(kG) \cdot P_k) \otimes S = \mathrm{Rad}(P_k) \otimes S.$$

Now $\mathrm{Rad}(P_k)$ has codimension 1 in P_k, so $\mathrm{Rad}(P_k) \otimes S$ has codimension $\dim S$ in $P_k \otimes S$. From all this, it follows that $P_k \otimes S$ is the projective cover of S, and so is isomorphic to P_S. We have $k \uparrow_K^G \otimes S \cong S \uparrow_K^G$ by 4.13.

(4) We have seen in the proof of part (3) that $\mathrm{Rad}(P_S) = (IH \cdot P_k) \otimes S$. It was observed there that the validity of the equation $\mathrm{Rad}(P_k \otimes S) = (IH \cdot P_k) \otimes S$ did not depend on upon the particular structure of P_k. Thus, we see by induction that $\mathrm{Rad}^n(P_S) = ((IH)^n \cdot P_k) \otimes S$. Using the identification of P_k as kH that we established in (1) gives the result.

(5) This is immediate from Lemma 8.3.1 and part (1). □

Note in Proposition 8.3.2(1) that when P_k is identified with kH using the module action from Lemma 8.3.1, it is not being identifed with the subring kH of kG, which does not immediately have the structure of a kG-module.

Proposition 8.3.2 allows us to give very specific information in the case of groups with a normal Sylow p-subgroup that is cyclic. This extends the description of the projectives that we already gave in Example 8.2.1 for the case when the Sylow p-subgroup is a direct factor (as happens, for instance, when G is cyclic). We recall from Chapter 6 that a *uniserial* module is one with a unique composition series, and that some equivalent conditions to this were explored in Exercise 6 of Chapter 6.

Proposition 8.3.3. *Let k be a field of characteristic p and let $G = H \rtimes K$, where $H = \langle x \rangle \cong C_{p^n}$ is cyclic of order p^n and K is a group of order relatively prime to p. Let $\phi : G \to GL(W)$ be the 1-dimensional representation of kG on which H acts trivially and K acts via its conjugation action on $H/\langle x^p \rangle$. Thus, if $y \in K$ conjugates x as ${}^y x = x^r$ then $\phi(y)$ is multiplication by r. If S is any simple kG-module then the projective cover P_S is uniserial with radical quotients* $\mathrm{Rad}^i P_S / \mathrm{Rad}^{i+1} P_S$ *given as $S, W \otimes S, W^{\otimes 2} \otimes S, \dots, W^{\otimes p^n - 1} \otimes S = S$.*

Proof. We know from 6.2 that the powers IH^s are a complete list of the kH-submodules of P_k, and since they are also kG-submodules in the action described in Lemma 8.3.1 we have a complete list of the kG-submodules of P_k. We thus have a unique composition series for P_k as a kG-module.

For each element $y \in K$ as in the statement of the theorem the action of y on $IH/(IH)^2$ is multiplication by r, as the following calculation shows:

$$\begin{aligned} y(x-1) &= {}^y x - 1 \\ &= x^r - 1 \\ &= (x-1)(x^{r-1} + \cdots + x + 1 - r) + r(x-1) \\ &\equiv r(x-1) \pmod{IH^2}. \end{aligned}$$

More generally for some $\alpha \in IH^2$,

$$\begin{aligned} y(x-1)^s &= (x^r - 1)^s \\ &= (r(x-1) + \alpha)^s \\ &= r^s(x-1)^s + sr(x-1)^{s-1}\alpha + \cdots \\ &\equiv r^s(x-1)^s \pmod{IH^{s+1}}, \end{aligned}$$

and so y acts on the quotient IH^s/IH^{s+1} as multiplication by r^s. One way to describe this is that $IH^s/IH^{s+1} = W^{\otimes s}$, the s-fold tensor power. Thus, by Proposition 8.3.2(4), the radical layers of P_S are of the form $W^{\otimes s} \otimes S$. These are simple because W has dimension 1, so the radical series of P_S is its unique composition series, by Exercise 6 from Chapter 6. □

An algebra for which all indecomposable projective and injective modules are uniserial is called a *Nakayama algebra*, so that we have just shown that $k[H \rtimes K]$ is a Nakayama algebra when k has characteristic p, H is a cyclic p-group and K has order prime to p. At least, we have shown that the projectives are uniserial, and the injectives are uniserial because they are the duals of the projectives. We will see in Corollary 8.5.3 that for group algebras over a field, injective modules and projective modules are, in fact, the same thing. In Proposition 11.8, we will give a complete description of the indecomposable modules for a Nakayama algebra. It turns out they are all uniserial and there are finitely many of them.

Let us continue further with our analysis of the composition factors of the indecomposable projective module P_S in the situation of Proposition 8.3.3. Observe that the isomorphism types of these composition factors occur in a cycle that repeats itself: for each element $y \in K$, the map $x \mapsto {}^y x = x^r$ is an automorphism of C_{p^n}, and there is a least positive integer f_y such that $r^{f_y} \equiv 1 \pmod{p^n}$. This f_y divides $p^n - 1$, and letting f be the l.c.m. of all f_y with $y \in K$ we put $p^n - 1 = ef$. Then the modules $k, W, W^{\otimes 2}, W^{\otimes 3}, \ldots$ give rise to f different representations. They repeat e times in P_k, except for k which appears $e+1$ times. A similar repetition occurs with the composition factors $W^{\otimes i} \otimes S$ of P_S.

Example 8.3.4. As a specific example of this, consider the non-abelian group

$$G = \langle x, y \mid x^7 = y^3 = 1,\ yxy^{-1} = x^2 \rangle = C_7 \rtimes C_3$$

of order 21 over the field $\mathbb{F}_7$. For the element y, we have $r = 2$ and $f_y = 3$, so that $f = 3$ and $e = 2$. There are three simple $\mathbb{F}_7G$-modules by Proposition 8.3.2, all 1-dimensional, which we will label k_1, k_2, k_4. The element x acts trivially on all of these modules, and y acts trivially on k_1, as multiplication by 2 on k_2 and as multiplication by 4 on k_4. In the previous notation, $W = k_2$, $W^{\otimes 2} = k_4$, and $W^{\otimes 3} = k_1$. The three projective covers are uniserial with composition factors as shown.

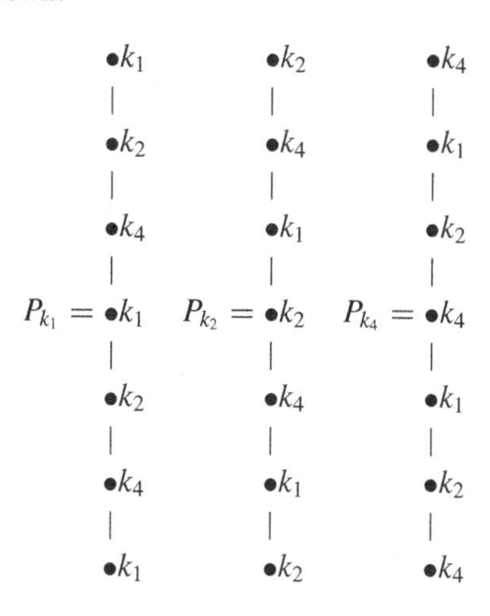

Indecomposable projectives for $\mathbb{F}_7G$, where G is non-abelian of order 21.

8.4 Projective Modules for Groups with a Normal p-Complement

We next consider the projective modules for groups that are a semidirect product of a p-group and a group of order prime to p, but with the roles of these groups the opposite of what they were in the last example. We say that a group G has a *normal p-complement* if and only if it has a normal subgroup $K \lhd G$ of order prime to p with $|G : K|$ a power of p. Necessarily in this situation, if H is a Sylow p-subgroup of G then $G = K \rtimes H$ by the Schur–Zassenhaus Theorem. In this situation, we also say that G is a *p-nilpotent* group, a term that has exactly the same meaning as saying that G has a normal p-complement. To set this property in a context, we mention that a famous theorem of Frobenius characterizes p-nilpotent groups as those groups G with the property that for every p-subgroup Q, $N_G(Q)/C_G(Q)$ is a p-group. This is not a result that we will use here, and we refer to standard texts on group theory for this and other criteria that guarantee the existence of a normal p-complement.

Theorem 8.4.1. *Let G be a finite group and k a field of characteristic p. The following are equivalent.*

(1) G has a normal p-complement.
(2) For every simple kG-module S, the composition factors of the projective cover P_S are all isomorphic to S.
(3) The composition factors of P_k are all isomorphic to k.

Proof. (1) $\Rightarrow$ (2): Let $G = K \rtimes H$ where $p \nmid |K|$ and H is a Sylow p-subgroup of G. We show that kH, regarded as a kG-module via the homomorphism $G \to H$, is a projective module. In fact, since kK is semisimple we may write $kK = k \oplus U$ for some kK-module U, and now $kG = kK \uparrow_K^G = k \uparrow_K^G \oplus U \uparrow_K^G$. Here $k \uparrow_K^G \cong kH$ as kG-modules (they are permutation modules with stabilizer K) and so kH is projective, being a summand of kG.

Now if S is any simple kG-module then $S \otimes_k kH$ is also projective by Proposition 8.1.4, and all its composition factors are copies of $S = S \otimes_k k$ since the composition factors of kH are all k (using Proposition 6.2.1 and Lemma 8.1.5). The indecomposable summands of $S \otimes_k kH$ are all copies of P_S, and their composition factors are all copies of S.

(2) $\Rightarrow$ (3) is immediate.

(3) $\Rightarrow$ (1): Suppose that the composition factors of P_k are all trivial. If $g \in G$ is any element of order prime to p (we say such an element is *p-regular*) then $P_k \downarrow_{\langle g \rangle}^G \cong k^t$ for some t, since $k\langle g \rangle$ is semisimple. Thus, g lies in the kernel of the action on P_k, and if we put

$$K = \langle g \in G \mid g \text{ is } p\text{-regular} \rangle$$

then K is a normal subgroup of G, G/K is a p-group and K acts trivially on P_k. We show that K contains no element of order p: if $g \in K$ were such an element, then as $P_k \downarrow_{\langle g \rangle}^G$ is a projective $k\langle g \rangle$-module, it is isomorphic to a direct sum of copies of $k\langle g \rangle$ by Theorem 8.1.1, and so g does not act trivially on P_k. It follows that $p \nmid |K|$, thus completing the proof. □

In Example 7.2.2, we have already seen an instance of the situation described in Theorem 8.4.1. In that example, we took $G = S_3 = K \rtimes H$ where $H = \langle (1, 2) \rangle$ and $K = \langle (1, 2, 3) \rangle$, and we worked with a field k of characteristic 2, which for a technical reason was $\mathbb{F}_4$. Note that if V is the 2-dimensional simple kS_3-module then $V \otimes kH \cong V \oplus V$ since V is projective. We see from this that the module $S \otimes kH$ that appeared in the proof of Theorem 8.4.1 need not be indecomposable.

8.5 Symmetry of the Group Algebra

Group algebras are symmetric, a term that will be defined before Theorem 8.5.5, and this has a number of important consequences for group representation theory. Some of these consequences may also be deduced in a direct fashion from weaker conditions than symmetry, and we present these direct arguments first. Over a field k, the following properties of the dual $U^* = \mathrm{Hom}_k(U, k)$ are either well known or immediate.

Proposition 8.5.1. *Let k be a field and U a finite-dimensional kG-module. Then*

(1) *$U^{**} \cong U$ as kG-modules,*
(2) *U is semisimple if and only if U^* is semisimple,*
(3) *U is indecomposable if and only if U^* is indecomposable, and*
(4) *a morphism $f : U \to V$ is a monomorphism (epimorphism) if and only if $f^* : V^* \to U^*$ is an epimorphism (monomorphism).*

The first part of the next proposition has already been proved in two different ways in Exercises 6 and 14 from Chapter 4. The proof given here is really the same as one of those earlier proofs, but it is presented a little differently.

Proposition 8.5.2. *Let k be a field. Then*

(1) *$kG^* \cong kG$ as kG-modules, and*
(2) *a finitely generated kG-module P is projective if and only if P^* is projective as a kG-module.*

Proof. (1) We denote the elements of kG^* dual to the basis elements $\{g \mid g \in G\}$ by $\hat{g}$, so that $\hat{g}(h) = \delta_{g,h} \in k$, the Kronecker δ. We define an isomorphism of vector spaces

$$kG \to kG^*$$
$$\sum_{g \in G} a_g g \mapsto \sum_{g \in G} a_g \hat{g}.$$

To see that this is a kG-module homomorphism, we observe that if $x \in G$ then

$$(x\hat{g})(h) = \hat{g}(x^{-1}h) = \delta_{g,x^{-1}h} = \delta_{xg,h} = \widehat{xg}(h)$$

for $g, h \in G$, so that $x\hat{g} = \widehat{xg}$.

(2) Since $P^{**} \cong P$ as kG-modules it suffices to prove one implication. If P is a summand of kG^n then P^* is a summand of $(kG^n)^* \cong kG^n$, and so is also projective. □

From part (1) of Proposition 8.5.2, we might be led to suppose that $P \cong P^*$ whenever P is finitely generated projective, but this is not so in general. We will see in Corollary 8.5.6 that if P_S is an indecomposable projective with simple quotient S then $P_S \cong (P_S)^*$ if and only if $S \cong S^*$.

We now come to a very important property of projective modules for group algebras over a field, which is that they are also injective.

Corollary 8.5.3. *Let k be a field.*

(1) *Finitely generated projective kG-modules are the same as finitely generated injective kG-modules.*
(2) *Each indecomposable projective kG-module has a simple socle.*

The result is true without the hypothesis of finite generation. It may be deduced from the finitely generated case and Exercise 10(e) from Chapter 7, which says that every projective kG-module is a direct sum of indecomposable projectives (and, dually, every injective kG-module is a direct sum of indecomposable injectives).

Proof. (1) Suppose that P is a finitely generated projective kG-module and that there are morphisms

$$\begin{array}{ccc} & & P \\ & & \uparrow \alpha \\ V & \xleftarrow{\beta} & W \end{array}$$

with β injective. Then in the diagram

$$\begin{array}{ccc} & & P^* \\ & & \downarrow \alpha^* \\ V^* & \xrightarrow{\beta^*} & W^* \end{array}$$

β^* is surjective, and so by projectivity of P^* there exists $f : P^* \to V^*$ such that $\beta^* f = \alpha^*$. Since $f^* \beta = \alpha$, we see that P is injective.

To see that all finitely generated injectives are projective, a similar argument shows that their duals are projective, hence injective, whence the original modules are projective, being the duals of injectives.

(2) One way to proceed is to quote Exercise 7 of Chapter 6, which implies that $\mathrm{Soc}(P) \cong (P^* / \mathrm{Rad}\, P^*)^*$. If P is an indecomposable projective module then so is P^* and $P^* / \mathrm{Rad}\, P^*$ is simple. Thus, so is $\mathrm{Soc}(P)$.

Alternatively, since homomorphisms $S \to P$ are in bijection (via duality) with homomorphisms $P^* \to S^*$, if P is indecomposable projective and S is simple then P^* is also indecomposable projective and

$$\dim \mathrm{Hom}_{kG}(S, P) = \dim \mathrm{Hom}_{kG}(P^*, S^*)$$
$$= \begin{cases} \dim \mathrm{End}(S^*) & \text{if } P^* \text{ is the projective cover of } S^* \\ 0 & \text{otherwise.} \end{cases}$$

Since $\dim \mathrm{End}(S^*) = \dim \mathrm{End}(S)$ this implies that P has a unique simple submodule and $\mathrm{Soc}(P)$ is simple. □

An algebra for which injective modules and projective modules coincide is called *self-injective* or *quasi-Frobenius*, so we have just shown that group rings of finite groups over a field are self-injective. An equivalent condition on a finite-dimensional algebra A that it should be self-injective is that the regular representation ${}_AA$ should be an injective A-module.

Corollary 8.5.4. *Suppose U is a kG-module, where k is a field, for which there are submodules $U_0 \subseteq U_1 \subseteq U$ with $U_1/U_0 = P$ a projective module. Then $U \cong P \oplus U'$ for some submodule U' of U.*

Proof. The exact sequence $0 \to U_0 \to U_1 \to P \to 0$ splits, and so $U_1 \cong P \oplus U_0$. Thus, P is isomorphic to a submodule of U, and since P is injective the monomorphism $P \to U$ must split. □

We will now sharpen part (2) of Corollary 8.5.3 by showing that $\mathrm{Soc}\, P_S \cong S$ for group algebras, and we will also show that the Cartan matrix for group algebras is symmetric. These are properties that hold for a class of algebras called symmetric algebras, of which group algebras are examples. We say that a finite-dimensional algebra A over a field k is a *symmetric algebra* if there is a nondegenerate bilinear form $(\ ,\ ) : A \times A \to k$ such that

(1) (symmetry) $(a, b) = (b, a)$ for all $a, b \in A$
(2) (associativity) $(ab, c) = (a, bc)$ for all $a, b, c \in A$.

The group algebra kG is a symmetric algebra with the bilinear form defined on the basis elements by

$$(g, h) = \begin{cases} 1 & \text{if } gh = 1 \\ 0 & \text{otherwise,} \end{cases}$$

as is readily verified. Notice that this bilinear form may be described on general elements $a, b \in kG$ by $(a, b) =$ coefficient of 1 in ab. Having learned that group algebras are symmetric it will be no surprise to learn that matrix algebras are symmetric. When $A = M_n(k)$ is the algebra of $n \times n$ matrices over a field k, the trace bilinear form $(A, B) = \mathrm{tr}(AB)$ gives the structure of a symmetric algebra.

We will use the bilinear form on kG in the proof of the next result. Although we only state it for group algebras, it is valid for symmetric algebras in general.

Theorem 8.5.5. *Let P be an indecomposable projective module for a group algebra kG. Then $P/\operatorname{Rad} P \cong \operatorname{Soc} P$.*

Proof. We may choose a primitive idempotent $e \in kG$ so that $P \cong kGe$ as kG-modules. We claim that $\operatorname{Soc}(kGe) = \operatorname{Soc}(kG) \cdot e$, since $\operatorname{Soc}(kG) \cdot e \subseteq \operatorname{Soc}(kG)$ and $\operatorname{Soc}(kG) \cdot e \subseteq kGe$ so $\operatorname{Soc}(kG) \cdot e \subseteq kGe \cap \operatorname{Soc}(kG) = \operatorname{Soc}(kGe)$, since the last intersection is the largest semisimple submodule of kGe. On the other hand, $\operatorname{Soc}(kGe) \subseteq \operatorname{Soc}(kG)$ since $\operatorname{Soc}(kGe)$ is semisimple so $\operatorname{Soc}(kGe) = \operatorname{Soc}(kGe) \cdot e \subseteq \operatorname{Soc}(kG) \cdot e$.

Next, $\operatorname{Hom}(kGe, \operatorname{Soc}(kG)e)$ and $e \operatorname{Soc}(kG)e$ have the same dimension by 7.17(3), and since $\operatorname{Soc}(kG)e$ is simple, by Corollary 8.5.3, this is nonzero if and only if $\operatorname{Soc}(kG)e \cong kGe/\operatorname{Rad}(kGe)$ by Theorem 7.13. We show that $e \operatorname{Soc}(kG)e \neq 0$.

If $e \operatorname{Soc}(kG)e = 0$ then

$$\begin{aligned} 0 &= (1, e \operatorname{Soc}(kG)e) \\ &= (e, \operatorname{Soc}(kG)e) \\ &= (\operatorname{Soc}(kG)e, e) \\ &= (kG \cdot \operatorname{Soc}(kG)e, e) \\ &= (kG, \operatorname{Soc}(kG)e \cdot e) \\ &= (kG, \operatorname{Soc}(kGe)). \end{aligned}$$

Since the bilinear form is nondegenerate this implies that $\operatorname{Soc}(kGe) = 0$, a contradiction. □

Recall that for any RG-module U, we have defined the *fixed points* of G on U to be $U^G := \{u \in U \mid gu = u \text{ for all } g \in G\}$. We also define the *fixed quotient* of G on U to be $U_G := U/\{(g-1)u \mid 1 \neq g \in G\}$. Then U^G is the largest submodule of U on which G acts trivially and U_G is the largest quotient of U on which G acts trivially.

Corollary 8.5.6. *Let k be a field.*

(1) *If P is any projective kG-module and S is a simple kG-module, the multiplicity of S in P/* Rad *P equals the multiplicity of S in* Soc *P. In particular,*

$$\dim P^G = \dim P_G = \dim(P^*)^G = \dim(P^*)_G.$$

(2) *For every simple kG-module S,* $(P_S)^* \cong P_{S^*}$.

Proof. (1) This is true for every indecomposable projective module, hence also for every projective module. For the middle equality, we may use an argument similar to the one that appeared in the proof of Corollary 8.5.3(2).

(2) We have seen in the proof of Corollary 8.5.3 that $(P_S)^*$ is the projective cover of $(\operatorname{Soc} P_S)^*$, and because of Theorem 8.5.5, we may identify the latter module as S^*. □

From this last observation, we are able to deduce that, over a large enough field, the Cartan matrix of kG is symmetric. We recall that the Cartan invariants are the numbers

$$c_{ST} = \text{multiplicity of } S \text{ as a composition factor of } P_T$$

where S and T are simple. The precise condition we require on the size of the field is that it should be a splitting field, and this is something that is discussed in the next chapter.

Theorem 8.5.7. *Let k be a field and let S, T be simple kG-modules. The Cartan invariants satisfy*

$$c_{ST} \cdot \dim \operatorname{End}_{kG}(T) = c_{TS} \cdot \dim \operatorname{End}_{kG}(S).$$

If $\dim \operatorname{End}_{kG}(S) = 1$ *for all simple modules S (e.g., if k is algebraically closed) then the Cartan matrix* $C = (c_{ST})$ *is symmetric.*

Proof. We recall from 7.18 that

$$c_{ST} = \dim \operatorname{Hom}_{kG}(P_S, P_T)/\dim \operatorname{End}_{kG}(S),$$

and in view of this we must show that

$$\dim \operatorname{Hom}_{kG}(P_S, P_T) = \dim \operatorname{Hom}_{kG}(P_T, P_S).$$

Now

$$\operatorname{Hom}_{kG}(P_S, P_T) = \operatorname{Hom}_k(P_S, P_T)^G \cong (P_S^* \otimes_k P_T)^G$$

by 3.3 and 3.4. Since $P_S^* \otimes_k P_T$ is projective by Proposition 8.1.4, this has the same dimension as

$$(P_S^* \otimes_k P_T)^{*G} \cong (P_S \otimes_k P_T^*)^G \cong \mathrm{Hom}_{kG}(P_T, P_S),$$

using Corollary 8.5.6. □

We conclude this chapter by summarizing some further aspects of injective modules. We define an *essential monomorphism* to be a monomorphism of modules $f : V \to U$ with the property that whenever $g : U \to W$ is a map such that gf is a monomorphism then g is a monomorphism. An *injective hull* (or *injective envelope*) of U is an essential monomorphism $U \to I$ where I is an injective module. By direct arguments, or by taking the corresponding results for essential epimorphisms and projective covers and applying the duality $U \mapsto U^*$, we may establish the following properties for finitely-generated kG-modules.

- The inclusion $\mathrm{Soc}\, U \to U$ is an essential monomorphism.
- Given homomorphisms $W \xrightarrow{g} V \xrightarrow{f} U$, if two of f, g, and fg are essential monomorphisms then so is the third.
- A homomorphism $f : V \to U$ is an essential monomorphism if and only if $f|_{\mathrm{Soc}\, V} : \mathrm{Soc}\, V \to \mathrm{Soc}\, U$ is an isomorphism.
- $U \to I$ is an injective hull if and only if $I^* \to U^*$ is a projective cover. Injective hulls always exist and are unique. From Theorem 8.5.5, we see that $S \to P_S$ is the injective hull of the simple module S.
- The multiplicity of a simple module S as a composition factor of a module U equals $\dim \mathrm{Hom}(U, P_S)/\dim \mathrm{End}(S)$.

8.6 Summary of Chapter 8

- When G is a p-group and k is a field of characteristic p, the regular representation kG is indecomposable.
- The property of projectivity is preserved under induction and restriction.
- Tensor product with a projective modules gives a projective module.
- When k is a field of characteristic p we have an explicit description of the projective kG-modules when G has a normal Sylow p-subgroup, and also when G has a normal p-complement. When G has a cyclic normal Sylow p-subgroup kG is a Nakayama algebra.
- G has a normal p-complement if and only if for every simple module S the composition factors of P_S are all isomorphic to S.
- Projective kG-modules are the same as injective kG-modules.
- kG is a symmetric algebra. $\mathrm{Soc}\, P_S \cong S$ always. The Cartan matrix is symmetric.

8.7 Exercises for Chapter 8

1. Let G be a finite group.

(a) Prove that if P is a finitely generated $\mathbb{Z}G$-module then $\mathrm{rank}_{\mathbb{Z}} P$ is divisible by $|G|$.
(b) Prove that the only idempotents in $\mathbb{Z}G$ are 0 and 1.

[Observe that the rank of P as a free abelian group is the dimension of the image of P under the homomorphism $\mathbb{Z}G \to \mathbb{F}_pG$ for each prime p dividing $|G|$, and this image is a projective $\mathbb{F}_pG$-module. After deducing (a), let e be an idempotent in $\mathbb{Z}G$ and consider the projective $\mathbb{Z}G$-module $\mathbb{Z}Ge$.]

2. (a) Let $H = C_2 \times C_2$ and let k be a field of characteristic 2. Show that $(IH)^2$ is a 1-dimensional space spanned by $\sum_{h \in H} h$.

(b) Let $G = A_4 = (C_2 \times C_2) \rtimes C_3$ and let $\mathbb{F}_4$ be the field with four elements. Compute the radical series of each of the three indecomposable projectives for $\mathbb{F}_4A_4$ and identify each of the quotients

$$\mathrm{Rad}^n P_S / \mathrm{Rad}^{n+1} P_S.$$

Now do the same for the socle series. Hence determine the Cartan matrix of $\mathbb{F}_4A_4$.

[Start by observing that $\mathbb{F}_4A_4$ has 3 simple modules, all of dimension 1, which one might denote by 1, ω and ω^2. This exercise may be done by applying the kind of calculation that led to Proposition 8.3.3.]

(c) Now consider $\mathbb{F}_2A_4$ where $\mathbb{F}_2$ is the field with two elements. Prove that the 2-dimensional $\mathbb{F}_2$-vector space on which a generator of C_3 acts via $\begin{bmatrix} 0 & 1 \\ 1 & 1 \end{bmatrix}$ is a simple $\mathbb{F}_2C_3$-module. Calculate the radical and socle series for each of the two indecomposable projective modules for $\mathbb{F}_2A_4$ and hence determine the Cartan matrix of $\mathbb{F}_2A_4$.

3. Let $G = H \rtimes K$ where H is a p-group, K is a p'-group, and let k be a field of characteristic p. Regard kH as a kG-module via its isomorphism with P_k, so H acts as usual and K acts by conjugation.

(a) Show that for each n, $(IH)^n$ is a kG-submodule of kH, and that $(IH)^n/(IH)^{n+1}$ is a kG-module on which H acts trivially.
(b) Show that

$$P_k = kH \supseteq IH \supseteq (IH)^2 \supseteq (IH)^3 \cdots$$

is the radical series of P_k as a kG-module.

(c) Show that there is a map

$$IH/(IH)^2 \otimes_k (IH)^n/(IH)^{n+1} \to (IH)^{n+1}/(IH)^{n+2}$$
$$x + (IH)^2 \otimes y + (IH)^{n+1} \mapsto xy + (IH)^{n+2}$$

that is a map of kG-modules. Deduce that $(IH)^n/(IH)^{n+1}$ is a homomorphic image of $(IH/(IH)^2)^{\otimes n}$.

(d) Show that the abelianization H/H' becomes a $\mathbb{Z}G$-module under the action $g \cdot xH' = gxg^{-1}H'$. Show that there is a map

$$IH/(IH)^2 \to k \otimes_{\mathbb{Z}} H/H'$$

specified by the formula $(x-1) + (IH)^2 \mapsto 1 \otimes xH'$ of Chapter 6, Exercise 22, and this map is an isomorphism of kG-modules.

4. The group $SL(2,3)$ is isomorphic to the semidirect product $Q_8 \rtimes C_3$ where the cyclic group C_3 acts on $Q_8 = \{\pm 1, \pm i, \pm j, \pm k\}$ by cycling the three generators i, j, and k. Assuming this structure, compute the radical series of each of the three indecomposable projectives for $\mathbb{F}_4 SL(2,3)$ and identify each of the quotients $\operatorname{Rad}^n P_S / \operatorname{Rad}^{n+1} P_S$.

[Use Chapter 6, Exercise 20.]

5. Let $G = P \rtimes S_3$ be a group that is the semidirect product of a 2-group P and the symmetric group of degree 3. (Examples of such groups are $S_4 = V \rtimes S_3$ where $V = \langle (1,2)(3,4), (1,3)(2,4) \rangle$, and $GL(2,3) \cong Q_8 \rtimes S_3$ where Q_8 is the quaternion group of order 8.)

(a) Let k be a field of characteristic 2. Show that kG has two nonisomorphic simple modules.

(b) Let $e_1, e_2, e_3 \in \mathbb{F}_4 S_3$ be the orthogonal idempotents that appeared in Example 7.2.2. Show that each e_i is primitive in $\mathbb{F}_4 G$ and that $\dim \mathbb{F}_4 G e_i = 2|P|$ for all i.

[Use the fact that the $\mathbb{F}_4 G e_i$ are projective modules.]

(c) Show that if $e_1 = () + (1,2,3) + (1,3,2)$ then $\mathbb{F}_4 S_4 e_1$ is the projective cover of the trivial module and that $\mathbb{F}_4 S_4 e_2$ and $\mathbb{F}_4 S_4 e_3$ are isomorphic, being copies of the projective cover of a 2-dimensional module.

(d) Show that $\mathbb{F}_4 G e_i \cong \mathbb{F}_4 \langle (1,2,3) \rangle e_i \uparrow_{\langle (1,2,3) \rangle}^G$ for each i.

6. Let A be a finite-dimensional algebra over a field k and suppose that the left regular representation ${}_A A$ is injective. Show that every projective module is injective and that every injective module is projective.

7. Let U be an indecomposable kG-module, where k is a field, and let P_k be the projective cover of the trivial module. Prove that

$$\dim((\sum_{g\in G} g)\cdot U) = \begin{cases} 1 & \text{if } U \cong P_k \\ 0 & \text{otherwise.} \end{cases}$$

For an arbitrary finite-dimensional module V, show that $\dim((\sum_{g\in G} g)\cdot V)$ is the multiplicity with which P_k occurs as a direct summand of V.
[Observe that $kG^G = P_k^G = k\cdot\sum_{g\in G} g$. Remember that P_k is injective and has socle isomorphic to k.]

8. Let k be a field of characteristic p and let $G = H \rtimes K$ where H is a p-group and K has order prime to p. Show that $\mathrm{Rad}^n(kG) \cong \mathrm{Rad}^n(kH)\uparrow_H^G$ as kG-modules.

9. Let A be a finite-dimensional algebra over a field k, and let $A^* = \mathrm{Hom}_k(A, k)$ be the vector space dual. Regarding A as a right module A_A gives $(A_A)^*$ the structure of a left A-module via the action $(af)(b) = f(ba)$ where $a \in A, b \in A_A$ and $f \in (A_A)^*$. Similarly, regarding A as a left module ${}_AA$ gives $({}_AA)^*$ the structure of a right A-module. In fact, both A and A^* are (A, A)-bimodules. Consider a k-bilinear form $(\ ,\ ) : A \times A \to k$. It determines, and is determined by, a linear map $\phi : A \to A^*$, where $\phi(b)(a) = (a, b)$. Consider three conditions such a form may satisfy:

(1) (symmetry) $(a, b) = (b, a)$ for all $a, b \in A$,
(2) (associativity) $(ab, c) = (a, bc)$ for all $a, b, c \in A$,
(3) $(ba, c) = (a, cb)$ for all $a, b, c \in A$.

(a) Show that if any two of conditions (1), (2), and (3) hold, then so does the third.
(b) Show that ϕ is a map of left A-modules if and only if (2) holds.
(c) Show that ϕ is a map of right A-modules if and only if (3) holds.
(d) Show that the following are equivalent (an algebra satisfying any of these conditions is called a *Frobenius algebra*):
  (i) there is a nondegenerate form on A satisfying (2),
  (ii) $A \cong A^*$ as left A-modules,
  (iii) $A \cong A^*$ as right A-modules,
  (iv) there is a nondegenerate form on A satisfying (3).
(e) For a Frobenius algebra (as in (d)), show that the left A-module ${}_AA$ is injective.
(f) Show that A is a symmetric algebra (i.e., there is a nondegenerate form on A satisfying both (1) and (2)) if and only if $A \cong A^*$ as (A, A)-bimodules.

10. Let S and T be simple kG-modules, with projective covers P_S and P_T, where k is an algebraically closed field.

(a) For each n, prove that

$$\begin{aligned}\mathrm{Hom}_{kG}(P_T, \mathrm{Soc}^n P_S) &\cong \mathrm{Hom}_{kG}(P_T/\mathrm{Rad}^n P_T, \mathrm{Soc}^n P_S)\\ &\cong \mathrm{Hom}_{kG}(P_T/\mathrm{Rad}^n P_T, P_S).\end{aligned}$$

(b) Deduce Landrock's Theorem: the multiplicity of T in the nth socle layer of P_S equals the multiplicity of S in the nth radical layer of P_T.

(c) Use Exercise 7 of Chapter 6 to show that these multiplicities equal the multiplicity of T^* in the nth radical layer of P_{S^*}, and also the multiplicity of S^* in the nth socle layer of P_{T^*}.

11. Let U be a finite-dimensional kG-module, where k is a field, and let P_S be an indecomposable projective kG-module with simple quotient S. Show that in any decomposition of U as a direct sum of indecomposable modules, the multiplicity with which P_S occurs is equal to

$$\frac{\dim \mathrm{Hom}_{kG}(P_S, U) - \dim \mathrm{Hom}_{kG}(P_S/\mathrm{Soc}\, P_S, U)}{\dim \mathrm{End}_{kG}(S)}$$

and also to

$$\frac{\dim \mathrm{Hom}_{kG}(U, P_S) - \dim \mathrm{Hom}_{kG}(U, \mathrm{Rad}\, P_S)}{\dim \mathrm{End}_{kG}(S)}.$$

12. Let k be an algebraically closed field of characteristic p and suppose that G has a normal p-complement, so that $G = K \rtimes H$ where H is a Sylow p-subgroup of G. Let $S_1, \ldots, S_n$ be the simple kG-modules with projective covers P_{S_i}.

(a) Show that there is a ring isomorphism $kG \cong \bigoplus_{i=1}^n M_{\dim S_i}(\mathrm{End}_{kG}(P_{S_i}))$ where the right-hand side is a direct sum of matrix rings with entries in the endomorphism rings of the indecomposable projectives. [Copy the approach of the proof of Wedderburn's Theorem.]

(b) For each i, show that if $P_{S_i} \cong P_k \otimes S_i$ then $\mathrm{End}_{kG}(P_{S_i}) \cong kH$ as rings. [Show that $\dim \mathrm{End}_{kG}(P_{S_i}) = |H|$. Deduce that the obvious map $\mathrm{End}_{kG}(P_k) \to \mathrm{End}_{kG}(P_k \otimes S_i)$ is an isomorphism.]

(c) Show that if $\dim S_i = 1$ then $P_{S_i} \cong P_k \otimes S_i$, and that if $P_{S_i} \cong P_k \otimes S_i$ then $S_i \downarrow_K^G$ is a simple kK-module. [Use the identification $P_k \cong k \uparrow_K^G$.]

(d) Show that if $P_{S_i} \not\cong P_k \otimes S_i$ then $\mathrm{End}_{kG} P_{S_i}$ has dimension smaller than $|H|$.

13. (a) Show that $\mathbb{F}_3A_4$ has just two isomorphism types of simple modules, of dimensions 1 and 3, and that the simple module of dimension 3 is projective.

[Eliminate modules of dimension 2 by observing that a projective cover of such a module must have dimension at least 6. Assume the results of Exercise 12.]

(b) Show that $\mathbb{F}_3A_4 \cong \mathbb{F}_3C_3 \oplus M_3(\mathbb{F}_3)$ as rings.

14. Let D_{30} be the dihedral group of order 30. Using the fact that $D_{30} \cong C_5 \rtimes D_6$ has a normal Sylow 5-subgroup, show that $\mathbb{F}_5D_{30}$ has three simple modules of dimensions 1, 1, and 2. We will label them k_1, k_ϵ, and U, respectively, with k_1 the trivial module. Use the method of Proposition 8.3.3 to show that the indecomposable projectives have the form

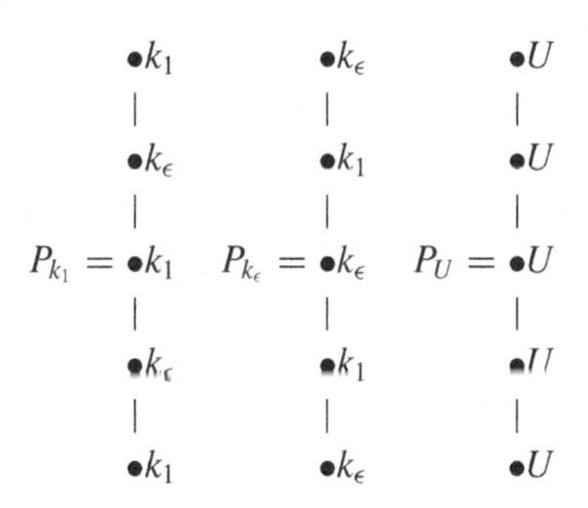

Deduce that the Cartan matrix is $\begin{bmatrix} 3 & 2 & 0 \\ 2 & 3 & 0 \\ 0 & 0 & 5 \end{bmatrix}$ and that $\mathbb{F}_5D_{30} \cong \mathbb{F}_5D_{10} \oplus M_2(\mathbb{F}_5C_5)$ as rings.

9

Changing the Ground Ring: Splitting Fields and the Decomposition Map

We examine the relationship between the representations of a fixed group over different rings. Often we have have assumed that representations are defined over a field that is algebraically closed. What if the field is not algebraically closed? Such a question is significant because representations arise naturally over different fields that might not be algebraically closed, and it is important to know how they change on moving to an extension field, such as the algebraic closure. It is also important to know whether a representation may be defined over some smaller field. We introduce the notion of a splitting field, showing that such a field may always be chosen to be a finite extension of the prime field.

After proving Brauer's Theorem, that over a splitting field of characteristic p the number of nonisomorphic simple representations equals the number of conjugacy classes of elements of order prime to p, we turn to the question of reducing representations from characteristic 0 to characteristic p. The process involves first writing a representation in the valuation ring of a p-local field and then factoring out the maximal ideal of the valuation ring. This gives rise to the decomposition map between the Grothendieck groups of representations in characteristic 0 and characteristic p. We show that this map is well defined and then construct the so-called cde triangle. This provides a very effective way to compute the Cartan matrix in characteristic p from the decomposition map.

In the last part of this chapter, we describe in detail the properties of blocks of defect zero. These are representations in characteristic p that are both simple and projective. They always arise as the reduction modulo p of a simple representations in characteristic zero, and these are also known as blocks of defect 0. The blocks of defect zero have importance in character theory, accounting for many zeroes in character tables, and they are also the subject of some of the deepest investigations in representation theory.

9.1 Some Definitions

Suppose that A is an algebra A over a commutative ring R and that U is an A-module. If $R \to R'$ is a homomorphism to another commutative ring R', we may form the R'-algebra $R' \otimes_R A$, and now $R' \otimes_R U$ becomes an $R' \otimes_R A$-module in an evident way. In this chapter, we study the relationship between the modules U and $R' \otimes_R U$. When we specialize to a group algebra $A = RG$, we will identify $R' \otimes_R RG$ with $R'G$.

We will pay special attention to two particular cases of this construction, the first being when R is a subring of R'. If U is an A-module, we say that the module $V = R' \otimes_R U$ is obtained from U by *extending the scalars from R to R'*; and if an $R' \otimes_R A$-module V has the form $R' \otimes_R U$, we say it can be *written in R*. In this situation, when U is free as an R-module, we may identify U with the subset $1_{R'} \otimes_R U$ of $R' \otimes_R U$, and an R-basis of U becomes an R'-basis of $R' \otimes_R U$ under this identification. In case, $A = RG$ is a group ring, with respect to such a basis of $V = R' \otimes_R U$ the matrices that represent the action of elements $g \in G$ on V have entries in R, and are the same as the matrices representing the action of these elements on U (with respect to the basis of U). Equally, if we can find a basis for an $R'G$-module V so that each $g \in G$ acts by a matrix with elements in R then RG preserves the R-linear span of this basis, and this R-linear span is an RG-module U for which $V = R' \otimes_R U$. Thus, an R'-free module V can be written in R if and only if V has an R'-basis with respect to which G acts via matrices with entries in R.

The second situation to which we will pay particular attention arises when $R' = R/I$ for some ideal I in R. In this case, applying $R' \otimes_R \_$ to a module U is the same as reducing U modulo I. If V is an $R' \otimes_R A$-module of the form $R' \otimes_R U$ for some A-module U, we say that V can be *lifted* to U, and that U is a *lift* of V. Most often we will perform this construction when R is a local ring and I is the maximal ideal of R.

9.2 Splitting Fields

We start by considering the behavior of representations over a field. It is often a help to know that a representation can be written in a small field.

Proposition 9.2.1. *Let $F \subseteq E$ be fields where E is algebraic over F and let A be a finite-dimensional F-algebra. Let V be a finite-dimensional $E \otimes_F A$-module. Then there exists a field K with $F \subseteq K \subseteq E$, of finite degree over F, so that V can be written in K.*

Proof. Let $a^1, \ldots, a^n$ be a basis for A and let a^t act on V with matrix (a^t_{ij}) with respect to some basis of V. Let $K = F[a^t_{ij},\ 1 \le t \le n,\ 1 \le i, j \le d]$. Then

$[K : F]$ is finite since K is an extension of F by finitely many algebraic elements, and A acts by matrices with entries in K. □

Let A be a finite-dimensional F-algebra, where F is a field. A simple A-module U is said to be *absolutely simple* if and only if $E \otimes_F U$ is a simple $E \otimes_F A$-module for all extension fields E of F. We say that an extension field E of F is a *splitting field* for A if and only if every simple $E \otimes_F A$-module is absolutely simple. If A is a group algebra FG, we say that E is a splitting field for G, by extension of the terminology.

Example 9.2.2. The kind of phenomenon that these definitions are designed to address is exemplified by cyclic groups. If $G = \langle g \rangle$ is cyclic of order n then g acts on $\mathbb{C}G$ as a direct sum of 1-dimensional eigenspaces with eigenvalues $e^{\frac{2\pi i}{n}}$. Since these lie outside $\mathbb{Q}$ (if $n \geq 3$), the regular representation of $\mathbb{Q}G$ is a direct sum of simple modules but some of them have dimension greater than 1. On extending scalars to a field containing $e^{\frac{2\pi i}{n}}$ these simple modules decompose as direct sums of 1-dimensional modules. Thus, $\mathbb{Q}$ is not a splitting field for G if $n \geq 3$, but any field containing $\mathbb{Q}(e^{\frac{2\pi i}{n}})$ is a splitting field since the simple modules are now 1-dimensional and remain simple on extension of scalars.

Example 9.2.3. We will use several times the fact that if we let F be any field, then F is a splitting field for the matrix algebra $M_n(F)$. In fact, if $E \supseteq F$ is a field extension then $E \otimes_F M_n(F) \cong M_n(E)$, an isomorphism that is most easily seen by observing that $M_n(F)$ has a basis consisting of the matrices E_{ij} that are nonzero only in position (i, j), where the entry is 1. Thus, $E \otimes_F M_n(F)$ has a basis consisting of the elements $1 \otimes_F E_{ij}$. Since these multiply together in the same fashion as the corresponding basis elements of $M_n(E)$, we obtain the claimed isomorphism. Every simple $M_n(F)$-module is isomorphic to the module of column vectors of length n over F, and on extending the scalars to E, we obtain column vectors of length n over E, which is a simple module for $E \otimes_F M_n(F)$. This shows that F is a splitting field for $M_n(F)$.

Example 9.2.4. As another example, the prime field $\mathbb{F}_p$ is a splitting field for every p-group, since the only simple $\mathbb{F}_pG$-module here is $\mathbb{F}_p$, which is absolutely simple.

Proposition 9.2.5. *Let U be a simple module for a finite-dimensional algebra A over a field F. The following are equivalent.*

(1) *U is absolutely simple.*
(2) $\operatorname{End}_A(U) = F$.
(3) *The matrix algebra summand of $A/\operatorname{Rad} A$ corresponding to U has the form $M_n(F)$, where $n = \dim U$.*

Proof. (2) $\Rightarrow$ (3): If the matrix summand of $A/\operatorname{Rad} A$ corresponding to U is $M_n(D)$ for some division ring D then $D = \operatorname{End}_A(U)$. The hypothesis is that $D = F$ so the matrix summand is $M_n(F)$ and since U identifies as column vectors of length n we have $n = \dim U$.

(3) $\Rightarrow$ (1): The hypothesis is that A acts on U via a surjective ring homomorphism $A \to M_n(F)$ where U is identified as F^n. Now if $E \supseteq F$ is an extension field then $E \otimes A$ acts on $E \otimes U = E^n$ via the homomorphism $E \otimes A \to E \otimes M_n(F) \cong M_n(E)$, which is also surjective. Since E^n is a simple $M_n(E)$-module it follows that $E \otimes_F U$ is a simple $E \otimes_F A$-module.

(1) $\Rightarrow$ (2): We prove this implication here only in the situation where F is a perfect field, so that all irreducible polynomials with coefficients in F are separable. The result is true in general and is not difficult but requires some technicality that we wish to avoid (see [10, Theorem 3.43]). This implication will not be needed for our application of the result.

Suppose that $\operatorname{End}_A(U)$ is larger than F, so there exists an endomorphism $\phi : U \to U$ that is not scalar multiplication by an element of F. Let α be a root of the characteristic polynomial of ϕ in some field extension $E \supseteq F$: in other words, α is an eigenvalue of ϕ. Then $1_E \otimes \phi : E \otimes_F U \to E \otimes_F U$ is not scalar multiplication by α, because if it were the minimal polynomial of α over F would be a factor of $(X - \alpha)^n$ where $n = \dim_F U$ and by separability of the minimal polynomial we would deduce $\alpha \in F$. Now $1_E \otimes \phi - \alpha \otimes 1_U \in \operatorname{End}_{E\otimes A}(E \otimes_F U)$ is a nonzero endomorphism with nonzero kernel, and since $E \otimes_F U$ is simple this cannot happen, by Schur's Lemma. $\square$

The next theorem is the main result about splitting fields that we will need for the process of reduction modulo p to be described later in this chapter.

Theorem 9.2.6. *Let A be a finite-dimensional algebra over a field F. Then A has a splitting field of finite degree over F. If G is a finite group, it has splitting fields that are finite degree extensions of $\mathbb{Q}$ (in characteristic zero) or of $\mathbb{F}_p$ (in characteristic p).*

Proof. The algebraic closure $\overline{F}$ of F is a splitting field for A, since by Schur's Lemma condition (2) of Proposition 9.2.5 is satisfied for each simple $\overline{F} \otimes_F A$-module. By Proposition 9.2.1, there is a finite extension $E \supseteq F$ so that every simple $\overline{F} \otimes_F A$-module can be written in E. The simple $E \otimes_F A$-modules U that arise like this are absolutely simple, because if $K \supseteq E$ is an extension field for which $K \otimes_E U$ is not simple then $\overline{K} \otimes_E U$ is not simple, where $\overline{K}$ is an algebraic closure of K, and since $\overline{K}$ contains a copy of $\overline{F}$, $\overline{F} \otimes_F U$ cannot be simple since $\overline{F}$ is a splitting field for A, a contradiction.

It is also true that every simple $E \otimes_F A$-module is isomorphic to one of the simple modules U that arise in this way from the algebraic closure $\overline{F}$. For, if V is a simple $E \otimes_F A$-module let $e^2 = e \in E \otimes_F A$ be an idempotent with the property that $eV \neq 0$ but $eV' = 0$ for all simple modules V' not isomorphic to V. Let W be a simple $\overline{F} \otimes_F A$-module that is not annihilated by e. We must have $W \cong \overline{F} \otimes_E V$ since V is the only possible simple module that would give a result not annihilated by e.

Group algebras are defined over the prime field $\mathbb{Q}$ or $\mathbb{F}_p$ (depending on the characteristic), and by what we have just proved $\mathbb{Q}G$ and $\mathbb{F}_pG$ have splitting fields that are finite degree extensions of the prime field. □

We see from the preceding that every simple representation of a finite group may be written over a field that is a finite degree extension of the prime field. In characteristic zero, this means that every representation can be written in such a field, by semisimplicity. It is not always true that every representation of a finite group in positive characteristic can be written in a finite field, and an example of this is given as Exercise 5 of this chapter.

Some other basic facts about splitting fields are left to the exercises at the end of this chapter. Thus, if A is a finite-dimensional algebra over a field F that is a splitting field for A and $E \supset F$ is a field extension, it is the case that every simple $E \otimes_F A$-module can be written in F (Exercises 4 and 8). It is also true for a finite-dimensional algebra that no matter which splitting field we take, after extending scalars we always have the same number of isomorphism classes of simple modules (Exercise 8). Thus, in defining the character table of a finite group, instead of working with complex representations we could have used representations over any splitting field and obtained the same table.

In positive characteristic, the situation is not so straightforward. It is usually not the case that indecomposable modules always remain indecomposable under all field extensions (those that do are termed *absolutely indecomposable*), even when all fields concerned are splitting fields. We can, however, show that if we are working over a splitting field the indecomposable projective modules remain indecomposable under field extension (Exercise 11). The consequence of this is that the Cartan matrix does not change once we have a splitting field, being independent of the choice of the splitting field. Just as when we speak of the character table of group we mean the character table of representations over some splitting field, so in speaking of the Cartan matrix of a group algebra we usually mean the Cartan matrix over some splitting field.

In the case of group algebras, there is a finer result about splitting fields than Theorem 9.2.6. It was first conjectured by Schur and later proved by Brauer as a deduction from Brauer's Induction Theorem. We state the result, but will not use it and do not prove it. The *exponent* of a group G is the least common multiple of the orders of its elements.

Theorem 9.2.7 (Brauer). *Let G be a finite group, F a field, and suppose that F contains a primitive mth root of unity, where m is the exponent of G. Then F is a splitting field for G.*

This theorem tells us that $\mathbb{Q}(e^{\frac{2\pi i}{m}})$ and $\mathbb{F}_p(\zeta)$ are splitting fields for G, where ζ is a primitive mth root of unity in an extension of $\mathbb{F}_p$. Often smaller splitting fields than these can be found, and the determination of minimal splitting fields must be done on a case-by-case basis. For example, we may see as a result of the calculations we have performed earlier in this text that in every characteristic the prime field is a splitting field for S_3—the same is in fact true for all the symmetric groups. However, if we require that a field be a splitting field not only for G but also for all of its subgroups, then $\mathbb{Q}(e^{\frac{2\pi i}{m}})$ and $\mathbb{F}_p(\zeta)$ are the smallest possibilities, because, as we have seen earlier, a cyclic group of order n requires the presence of a primitive nth root of 1 in a splitting field.

Again we will not use it, but it is important to know the following theorem about field extensions. For a proof see [10, Exercise 6.6].

Theorem 9.2.8 (Noether–Deuring). *Let A be a finite-dimensional algebra over a field F and let $E \supseteq F$ be a field extension. Suppose that U and V are A-modules for which $E \otimes_F U \cong E \otimes_F V$ as $E \otimes_F A$-modules. Then $U \cong V$ as A-modules.*

9.3 The Number of Simple Representations in Positive Characteristic

Our next aim is to show that, over a splitting field of characteristic p, the number of non-isomorphic simple representations of a group G equals the number of conjugacy classes of p-regular elements. Several proofs of this result are available, the first appearing in a paper of Brauer from 1932. The proof we shall present is also due to Brauer, coming from 1956. This proof is appealing because it is technically elementary, and it could have appeared earlier in this text once we knew that the radical of a finite-dimensional algebra is nilpotent.

We start with some lemmas. These have to do with a finite-dimensional algebra A over a field k of characteristic p, and we will write

$$S = \text{ linear span in } A \text{ of } \{ab - ba \mid a, b \in A\}$$
$$T = \{r \in A \mid r^{p^n} \in S \text{ for some } n > 0\}.$$

Lemma 9.3.1. *T is a linear subspace of A containing S.*

Proof. We show first that if $a, b \in A$ then $(a+b)^p \equiv a^p + b^p \pmod{S}$. To prove this, we use a modification of the familiar binomial expansion argument, but we must be careful because S need not be an ideal of A, and so it might not be true, for instance, that $aabab \equiv aaabb \pmod{S}$. On expanding $(a+b)^p$, we obtain a sum of terms that are products of length p. Letting a cyclic group of order p permute these products by operating on the positions in the product, two terms are fixed (namely, a^p and b^p) and the remaining terms all occur in orbits of length p, such as $aabab, baaba, abaab, babaa, ababa$ when $p = 5$. The difference of any two of these terms can be expressed as a commutator, such as $aabab - baaba = (aaba)b - b(aaba)$, and so lies in S. It follows that all the terms in an orbit of length p are equal modulo S, and so their sum lies in S, since $p = 0$ in the ground field k.

We next observe that if $a, b \in A$ then

$$(ab - ba)^p \equiv (ab)^p - (ba)^p = a(b(ab)^{p-1}) - (b(ab)^{p-1})a \equiv 0 \pmod{S},$$

so that commutators lie in T. We deduce that $S \subseteq T$ since if $\sum \lambda_i c_i$ is a linear combination of commutators then

$$(\sum \lambda_i c_i)^p \equiv \sum \lambda_i^p c_i^p \equiv 0 \pmod{S}$$

using both formulas we have proved, so that in fact pth powers of elements of S lie in S.

The proof will be completed by showing that T is a linear subspace. Let $a, b \in T$, so that $a^{p^n} \in S$, $b^{p^m} \in S$ with $m \geq n$, say. Then

$$(\lambda a + \mu b)^{p^m} \equiv \lambda^{p^m} a^{p^m} + \mu^{p^m} b^{p^m} \equiv 0 \pmod{S},$$

showing that T is closed under taking linear combinations. Here we used the fact that $a^{p^m} = (a^{p^n})^{p^{m-n}} \in S$ because pth powers of elements of S lie in S. □

Lemma 9.3.2. *If $A = M_n(k)$ is a matrix algebra where k is a field then $S = T =$ matrices of trace zero.*

Proof. Since $\mathrm{tr}(ab - ba) = 0$ we see that S is a subset of the matrices of trace zero. On the other hand, when $i \neq j$ every matrix E_{ij} (zero everywhere except for a 1 in position (i, j)) can be written as a commutator: $E_{ij} = E_{ik}E_{kj} - E_{kj}E_{ik}$, and also $E_{ii} - E_{jj} = E_{ij}E_{ji} - E_{ji}E_{ij}$. Since these matrices span the matrices of trace zero we deduce that S consists exactly of the matrices of trace 0. Now $S \subseteq T \subseteq A$ and S has codimension 1 so either $T = S$ or $T = A$. The matrix E_{11} is idempotent and does not lie in T, so $T = S$. □

Proposition 9.3.3. *Let A be a finite-dimensional algebra over a field of characteristic p that is a splitting field for A. The number of nonisomorphic simple representations of A equals the codimension of T in A.*

Proof. Let us write $T(A)$, $S(A)$, $T(A/\operatorname{Rad}(A))$, $S(A/\operatorname{Rad}(A))$ for the constructions S, T applied to A and $A/\operatorname{Rad}(A)$. Since $\operatorname{Rad}(A)$ is nilpotent it is contained in $T(A)$. Also

$$(S(A)+\operatorname{Rad}(A))/\operatorname{Rad}(A) = S(A/\operatorname{Rad}(A))$$

is easily verified. We claim that $T(A)/\operatorname{Rad}(A) = T(A/\operatorname{Rad}(A))$. For, if $a^{p^n} \in S(A)$ then $(a+\operatorname{Rad}(A))^{p^n} \in (S(A)+\operatorname{Rad}(A))/\operatorname{Rad}(A) = S(A/\operatorname{Rad}(A))$ and this shows that the left-hand side is contained in the right. Conversely, if $(a+\operatorname{Rad}(A))^{p^n} \in S(A/\operatorname{Rad}(A))$ then $a^{p^n} \in S(A)+\operatorname{Rad}(A) \subseteq T(A)$ so $T(A/\operatorname{Rad}(A)) \subseteq T(A)/\operatorname{Rad}(A)$.

Now $A/\operatorname{Rad}(A)$ is a direct sum of matrix algebras. It is apparent that both S and T preserve direct sums, so the codimension of $T(A/\operatorname{Rad}(A))$ in $A/\operatorname{Rad}(A)$ equals the number of simple A-modules, and this equals the codimension of $T(A)$ in A. □

Let p be a prime. An element in a finite group is said to be *p-regular* if it has order prime to p, and *p-singular* if it has order a power of p. The only element that is both p-regular and p-singular is the identity.

Lemma 9.3.4. *Let G be a finite group and p a prime. Each element $x \in G$ can be uniquely written $x = st$ where s is p-regular, t is p-singular, and $st = ts$. If $x_1 = s_1t_1$ is such a decomposition of an element x_1 that is conjugate to x then s is conjugate to s_1, and t is conjugate to t_1.*

Proof. If x has order $n = \alpha\beta$ where α is a power of p and β is prime to p then we may write $1 = \lambda\alpha + \mu\beta$ for integers λ, μ and put $s = x^{\lambda\alpha}$ and $t = x^{\mu\beta}$. If $x = st = s_1t_1$ is a second such decomposition then s_1 commutes with x and hence commutes with s and t which are powers of x. Similarly, t_1 commutes with s and t. Thus, $s_1^{-1}s = t_1t^{-1}$ and now $s_1^{-1}s$ is p-regular and t_1t^{-1} is p-singular, so these products equal 1, and $s_1 = s$, $t_1 = t$. If $x_1 = gxg^{-1}$ then $x_1 = gsg^{-1}gtg^{-1}$ is a decomposition of x_1 as a product of commuting p-regular and p-singular elements. Hence $s_1 = gsg^{-1}$ and $t_1 = gtg^{-1}$ by uniqueness of the decomposition. □

Lemma 9.3.5. *Let k be a field and G a group. Then S is the set of elements of kG with the property that the sum of coefficients from each conjugacy class of G is zero.*

Proof. S is spanned by elements $ab - ba$ with $a, b \in G$. Now $ab - ba = a(ba)a^{-1} - ba$ is the difference of an element and its conjugate. Such elements exactly span the elements of kG that have coefficient sum zero on conjugacy classes. □

We come now to the result that is the goal of these lemmas.

Theorem 9.3.6 (Brauer). *Let k be a splitting field of characteristic p for a finite group G. The number of nonisomorphic simple kG modules equals the number of conjugacy classes of p-regular elements of G.*

Proof. We know that the number of simple kG modules equals the codimension of T in kG. We show this equals the number of p-regular conjugacy classes by showing that if $x_1, \ldots, x_r$ is a set of representatives of the conjugacy classes of p-regular elements of G then $x_1 + T, \ldots, x_r + T$ is a basis of kG/T.

If we write $x = st$ where s is p-regular and t is p-singular, s and t commute, then $(st - s)^{p^n} = s^{p^n}t^{p^n} - s^{p^n} = s^{p^n} - s^{p^n} = 0$ for sufficiently large n, so that $s + T = st + T$. The elements $g + T$, $g \in G$ do span kG/T, and now it follows from the last observation that we may throw out all except the p-regular elements and still have a spanning set. We show that the set that remains is linearly independent.

Suppose that $\sum \lambda_i x_i \in T$ so that for sufficiently large n, $(\sum \lambda_i x_i)^{p^n} \in S$. From the proof of Lemma 9.3.1, we know that $(\sum \lambda_i x_i)^{p^n} \equiv \sum \lambda_i^{p^n} x_i^{p^n} \pmod S$ so that $\sum \lambda_i^{p^n} x_i^{p^n} \in S$. We can find n sufficiently large so that $x_i^{p^n} = x_i$ for all i, since the x_i are p-regular. Now $\sum \lambda_i^{p^n} x_i \in S$. But $x_1, \ldots, x_r$ are independent modulo S by Lemma 9.3.5 so $\lambda_i^{p^n} = 0$ for all i, and hence $\lambda_i = 0$ for all i. This shows that $x_1 + T, \ldots, x_r + T$ are linearly independent. □

Corollary 9.3.7. *Let k be a splitting field of characteristic p for a finite groups G_1 and G_2. The simple $k[G_1 \times G_2]$-modules are precisely the tensor products $S_1 \otimes S_2$ where S_i is a simple kG_i-module, $i = 1, 2$, and the action of $G_1 \times G_2$ is given by $(g_1, g_2)(s_1 \otimes s_2) = g_1 s_1 \otimes g_2 s_2$. Two such tensor products $S_1 \otimes S_2$ and $S_1' \otimes S_2'$ are isomorphic as $k[G_1 \times G_2]$-modules if and only if $S_i \cong S_i'$ as kG_i-modules, $i = 1, 2$.*

We have commented before, after Theorem 4.1.2, that this kind of result is a special case of a more general statement about the simple modules for a tensor product of algebras, and this can be found in [10, Theorem 10.38]. The general argument is not hard, but we can use our expression for the number of simple representations of a group to eliminate half of it.

Proof. We verify that the modules $S_1 \otimes S_2$ are simple. This is so since the image of each kG_i in $\mathrm{End}_k(S_i)$ given by the module action is the full matrix algebra

$\mathrm{End}_k(S_i)$, by the theorem of Burnside that was presented as Exercise 10 in Chapter 2, and which follows from the Artin–Wedderburn Theorem. The image of $k[G_1 \times G_2]$ in $\mathrm{End}_k(S_1 \otimes S_2)$ contains $\mathrm{End}_k(S_1) \otimes \mathrm{End}_k(S_2)$ and so by counting dimensions it is the whole of $\mathrm{End}_k(S_1 \otimes S_2)$. This implies that $S_1 \otimes S_2$ is simple.

On restriction to G_1, $S_1 \otimes S_2$ is a direct sum of copies of S_1, and similarly for G_2, so $S_1 \otimes S_2 \cong S_1' \otimes S_2'$ as $k[G_1 \times G_2]$-modules if and only if $S_1 \cong S_1'$ as kG_1-modules and $S_2 \cong S_2'$ as kG_2-modules.

We conclude by checking that this gives the right number of simple modules for $G_1 \times G_2$. By Brauer's Theorem this is the number of p-regular conjugacy classes of $G_1 \times G_2$. Since (g_1, g_2) is p-regular if and only if both g_1 and g_2 are p-regular, and this element is conjugate in $G_1 \times G_2$ to (g_1', g_2') if and only if $g_1 \sim_{G_1} g_1'$ and $g_2 \sim_{G_2} g_2'$, the number of p-regular classes in $G_1 \times G_2$ is the product of the numbers for G_1 and G_2. □

9.4 Reduction Modulo p and the Decomposition Map

We turn now to the theory of reducing modules from characteristic zero to characteristic p, for some prime p. This is a theory developed principally by Richard Brauer. There is inherent interest in studying the relationships between representations in different characteristics, but aside from this our more specific goals include a remarkable way to compute the Cartan matrix of a group algebra, and a second proof of the symmetry of this matrix. After that, we study the simple characters whose degree is divisible by the order of a Sylow p-subgroup of G (the so-called blocks of defect zero). In the next chapter, the same ideas will be used in a proof that the Cartan matrix is nonsingular.

There will be three rings in the setup for reducing modules to characteristic p. We list them as a triple (F, R, k) where F is a field of characteristic zero equipped with a discrete valuation, R is the valuation ring in F with maximal ideal (π), and $k = R/(\pi)$ is the residue field of R, which is required to have characteristic p. A quick introduction to valuations and valuation rings is given in an appendix. Such a triple is called a *p-modular system*. We may find such systems by taking F to be a finite extension of $\mathbb{Q}$—the complex numbers will not work—and this is one of the reasons that we have studied representations over arbitrary fields.

Given a finite group G, if both F and k are splitting fields for G, we say that the triple is a *splitting p-modular system* for G. If F contains a primitive mth root of unity, where m is the exponent of G, then necessarily R and k also contain primitive mth roots of unity because roots of unity always have valuation 1, and according to Brauer's Theorem 9.2.7 both F and k are then splitting fields. If we do not wish to use Brauer's Theorem we may still deduce from Theorem 9.2.6 the existence of splitting p-modular systems where F is a finite-extension of $\mathbb{Q}$ and k is a finite field.

We start by studying representations of a finite group over a discrete valuation ring R. In Propositions 9.4.3 and 9.4.5, we assume R is complete, but in other results this is not necessary, and sometimes all we need is that R is a principal ideal domain with the field k as a factor ring. We comment also that in the next few results nothing specific about group representations is used, except for the fact that the group algebra of G over a field of characteristic zero is semisimple. Many results apply in the generality of an order in a finite-dimensional semisimple algebra.

Lemma 9.4.1. *Let R be a discrete valuation ring with maximal ideal (π) and residue field $k = R/(\pi)$. Let G be a finite group.*

(1) *If S is a simple RG-module then $\pi S = 0$.*
(2) *The simple RG-modules are exactly the simple kG-modules made into RG-modules via the surjection $RG \to kG$.*
(3) *For each RG-module U, $\pi U \subseteq \mathrm{Rad}(U)$, and in particular, $\pi RG \subseteq \mathrm{Rad}(RG)$.*
(4) *For each RG-module U, we have $(\mathrm{Rad}\, U)/\pi U = \mathrm{Rad}(U/\pi U)$.*

Proof. (1) πS is an RG-submodule of S, so $\pi S = S$ or 0. Since $\mathrm{Rad}\, R = (\pi)$ the R-module homomorphism $S \to S/\pi S$ is essential by Nakayama's Lemma, so that $\pi S \neq S$. Therefore, $\pi S = 0$.

(2) This follows from (1) since $kG = RG/(\pi)G$ and $(\pi)G$ annihilates the simple RG-modules.

(3) This again follows from (1) since if V is a maximal submodule of U then U/V is simple so that $\pi U \subseteq V$, and it follows that πU is contained in all of the maximal submodules of U and hence in their intersection.

(4) $\mathrm{Rad}\, U$ is the intersection of kernels of all the homomorphisms from U to simple modules. These homomorphisms all factor through the quotient homomorphism $U \to U/\pi U$, and so $\mathrm{Rad}\, U$ is the preimage in U of the radical of $U/\pi U$, which is what we have to prove. □

Corollary 9.4.2. *Let R be a discrete valuation ring with maximal ideal (π) and residue field $k = R/(\pi)$. Let G be a finite group. Let P and Q be finitely generated projective RG-modules. Then $P \cong Q$ as RG-modules if and only if $P/\pi P \cong Q/\pi Q$ as kG-modules.*

Proof. If $P/\pi P \cong Q/\pi Q$ as kG-modules then by Lemma 9.4.1 the radical quotients of P and Q are isomorphic, $P/\mathrm{Rad}\, P \cong Q/\mathrm{Rad}\, Q$. Now P and Q are projective covers of their radical quotients, by Nakayama's Lemma 7.3.1, so $P \cong Q$ by uniqueness of projective covers. The converse implication is trivial. □

In the next pair of results, we see that some important properties of idempotents and projective modules, that we have already studied in the case of representations over a field, continue to hold when we work over a complete discrete valuation ring. For our proofs to work, it is important that the discrete valuation ring be complete. The idea of the proofs is the same as for the corresponding results over a field.

Proposition 9.4.3. *Let R be a complete discrete valuation ring with maximal ideal (π) and residue field $k = R/(\pi)$. Let G be a finite group. Every expression $1 = e_1 + \cdots + e_n$ as a sum of orthogonal idempotents in kG can be lifted to an expression $1 = \hat{e}_1 + \cdots + \hat{e}_n$ in RG, where the $\hat{e}_i \in RG$ are orthogonal idempotents with $\hat{e}_i + (\pi) \cdot RG = e_i$. Each idempotent e_i is primitive if and only if its lift $\hat{e}_i$ is primitive.*

Proof. The proof is very like the proofs of Theorem 7.3.5 and Corollaries 7.3.6 and 7.3.7. We start by showing simply that each idempotent $e \in kG$ can be lifted to an idempotent $\hat{e} \in RG$. Consider the surjections of group rings $(R/(\pi^n))G \to (R/(\pi^{n-1}))G$ for each $n \geq 2$. Here $(\pi^{n-1})G/(\pi^n)G$ is a nilpotent ideal in $(R/(\pi^n))G$ and so by Theorem 7.3.5, any idempotent $e_{n-1} + (\pi^{n-1})G \in (R/(\pi^{n-1}))G$ can be lifted to an idempotent $e_n + (\pi^n)G \in (R/(\pi^n))G$. Starting with an element $e_1 \in RG$ for which $e_1 + (\pi)G = e$, we obtain a sequence $e_1, e_2, \ldots$ of elements of RG that successively lift each other modulo increasing powers of (π), and so is a Cauchy sequence in RG. (The metric on RG comes from the valuation on R by taking the distance between two elements to be the maximum of the distances in the coordinate places.) This Cauchy sequence represents an element $\hat{e} \in RG$, since R is complete. Evidently $\hat{e}$ is idempotent, because it is determined by its images modulo the powers of (π) and these are idempotent. It also lifts e.

The argument that sums of orthogonal idempotents can be lifted now proceeds by analogy with the proof of Corollary 7.3.6, and the assertion that e is primitive if and only if $\hat{e}$ is primitive is proved as in Corollary 7.3.7. □

We have seen before in Example 7.3.11, taking $\mathbb{Z}$ as the ground ring, that $\mathbb{Z}G$-modules do not always have projective covers. It is also the case that indecomposable projective $\mathbb{Z}G$-modules do not always have the form $\mathbb{Z}Ge$ where e is idempotent. To give an example of this phenomenon will take us too far afield, but we have seen in Exercise 1 of Chapter 8 that the only nonzero module of the form $\mathbb{Z}Ge$ is $\mathbb{Z}G$ itself, since 0 and 1 are the only idempotents. By using facts from number theory one can show, for example, that when G is cyclic of order 23 there is an indecomposable nonfree projective $\mathbb{Z}G$-module [16, Theorem 6.24]. Such a module cannot be a summand of $\mathbb{Z}G$ because the only non-zero summand is $\mathbb{Z}G$ itself. There is also an example due to Swan [19], when G is the generalized quaternion group of order 32, of a projective $\mathbb{Z}G$-module P for which $\mathbb{Z}G \oplus P \cong \mathbb{Z}G \oplus \mathbb{Z}G$ but $P \not\cong \mathbb{Z}G$. In the next result, we show that, when the ground ring R is a complete discrete valuation ring, such examples can no longer be found.

We will be examining the relationship between RG-modules and their reductions modulo the ideal (π). To facilitate this, we introduce some terminology. Working over a principal ideal domain R, an RG-module L is called an *RG-lattice* if it is finitely generated and free as an R-module. (In more general contexts, an RG-lattice is merely supposed to be projective as an R-module, but since projective modules are free over a principal ideal domain we do not need to phrase the definition that way here. We also sometimes encounter the term *maximal Cohen–Macaulay module* instead of lattice.)

The reason we introduce RG-lattices is that we can reduce them modulo ideals of R. Given an ideal I of R and an RG-lattice L, evidently $V = L/(I \cdot L)$ is an $(R/I)G$-module. We say that V is the *reduction modulo I* of the lattice L, and also that L is a *lift* from R/I to R of V. Not every $(R/I)G$-module need be liftable from R/I to R, as the following example shows.

Example 9.4.4. Let $R = \mathbb{Z}$ and let $I = (p)$ with $p \geq 3$. The group $GL(2, p)$ has a faithful 2-dimensional representation over $\mathbb{F}_p$ that cannot be lifted to $\mathbb{Z}$. Such a lifted representation would have to be faithful also, and on extending the scalars from $\mathbb{Z}$ to $\mathbb{R}$ would provide a faithful 2-dimensional representation over $\mathbb{R}$. It is well known that the only finite subgroups of 2×2 real matrices are cyclic and dihedral, and when $p \geq 3$, $GL(2, p)$ is not one of these, so there is no such faithful representation. On the other hand, the 2-dimensional faithful representation of $GL(2, 2)$ (which is dihedral of order 6) over $\mathbb{F}_2$ does lift to $\mathbb{Z}$.

Proposition 9.4.5. *Let R be a complete discrete valuation ring with maximal ideal (π) and residue field $k = R/(\pi)$. Let G be a finite group.*

(1) For each simple RG-module S, there is a unique indecomposable projective RG-module $\hat{P}_S$ that is the projective cover of S. It has the form $\hat{P}_S = RG\hat{e}_S$ where $\hat{e}_S$ is a primitive idempotent in RG for which $\hat{e}_S \cdot S \neq 0$.

(2) The kG-module $\hat{P}_S/(\pi \cdot \hat{P}_S) \cong P_S$ is the projective cover of S as a kG-module and $\hat{P}_S$ is the projective cover of P_S as an RG-module.

Furthermore, S is the unique simple quotient of $\hat{P}_S$. Thus, for simple kG-modules S and T, $\hat{P}_S \cong \hat{P}_T$ if and only if $S \cong T$.

(3) *Every finitely generated RG-module has a projective cover.*

(4) *Every finitely generated indecomposable projective RG-module is isomorphic to $\hat{P}_S$ for some simple module S.*

(5) *Every finitely generated projective kG-module can be lifted to an RG-lattice. Such a lift is unique up to isomorphism, and necessarily projective. Consequently, an RG-lattice L is a projective RG-module if and only if $L/(\pi \cdot L)$ is a projective kG-module, and projective RG-modules L_1, L_2 are isomorphic if and only if $L_1/(\pi \cdot L_1) \cong L_1/(\pi \cdot L_1)$ as kG-modules.*

Proof. (1) Let $e_S \in kG$ be a primitive idempotent for which $e_S \cdot S \neq 0$ and let $\hat{e}_S \in RG$ be an idempotent that lifts e_S (as in Proposition 9.4.3), so that $\hat{e}_S \cdot S = e_S \cdot S \neq 0$. We define $\hat{P}_S = RG\hat{e}_S$. Then $\hat{P}_S$ is projective, and it is indecomposable since $\hat{e}_S$ is primitive. Furthermore, $\hat{P}_S/(\pi \cdot \hat{P}_S) = kGe_S$, and defining this module to be P_S, it is a projective cover of S as a kG-module by Theorem 7.3.8.

(2) Now $\hat{P}_S/\operatorname{Rad}(\hat{P}_S) \cong P_S/\operatorname{Rad}(P_S)$ by part (4) of Lemma 9.4.1, and this is isomorphic to S. Thus, each of the three morphisms $\hat{P}_S \to P_S$, $\hat{P}_S \to S$ and $P_S \to S$ is essential, by Nakayama's Lemma, and so they are all projective covers. Since S is the radical quotient of $\hat{P}_S$ it is the unique simple quotient of this module. This quotient determines the isomorphism type of $\hat{P}_S$ by the uniqueness of projective covers.

(3) Let U be a finitely generated RG-module. Then $U/\operatorname{Rad} U$ is a kG-module by Lemma 9.4.1, and it is semisimple, so $U/\operatorname{Rad} U \cong S_1 \oplus \cdots S_t$ for various simple modules S_i. Consider the diagram

$$\begin{array}{ccc} & & \hat{P}_{S_1} \oplus \cdots \oplus \hat{P}_{S_t} \\ & & \downarrow \\ U & \longrightarrow & U/\operatorname{Rad} U \end{array}$$

where the vertical arrow is the projective cover of $S_1 \oplus \cdots S_t$ as an RG-module. By projectivity, we obtain a homomorphism $\hat{P}_{S_1} \oplus \cdots \oplus \hat{P}_{S_t} \to U$ that completes the triangle, and it is an essential epimorphism by Proposition 7.3.2. Thus, it is a projective cover.

(4) Let P be a finitely generated projective RG-module. By part (3), it has a projective cover of the form $\alpha : \hat{P}_{S_1} \oplus \cdots \oplus \hat{P}_{S_t} \to P$. Since P is projective α must split, and there is a monomorphism $\beta : P \to \hat{P}_{S_1} \oplus \cdots \oplus \hat{P}_{S_t}$ with $\alpha\beta = 1_P$. Since α is an essential epimorphism β must be an epimorphism also, so

it is an isomorphism. If we suppose that P is indecomposable, then $t = 1$ and $P \cong \hat{P}_{S_1}$.

(5) It is sufficient to prove the assertion for indecomposable projective kG-modules. The indecomposable projective kG-modules all have the form P_S for some simple module S, and we have seen that such a module lifts to $\hat{P}_S$, which is a lattice and is the projective cover of P_S. Suppose that L is any RG-lattice for which $L/\pi L \cong P_S$. Since $\pi L \subseteq \operatorname{Rad} L$ the radical quotient of L is S. The projective cover morphism $\hat{P}_S \to S$ factors as $\hat{P}_S \to L \to S$, giving an isomorphism on radical quotients. It follows that $\hat{P}_S \to L$ is surjective by Nakayama's Lemma, and since the ranks of $\hat{P}_S$ and L are the same, this map is an isomorphism. The remaining deductions are immediate. □

We next examine the relationship between RG-modules and FG-modules where R is a principal ideal domain and F is its field of fractions. Given an RG-lattice L, we may regard L as a subset of $F \otimes_R L$. In this situation, the FG-module $F \otimes_R L$ may be written in R, according to the terminology introduced at the start of this chapter. Conversely, if U is an FG-module, a *full* RG-lattice U_0 in U is defined to be an RG-lattice $U_0 \subseteq U$ that has an R basis which is also an F-basis of U. In this situation, $U \cong F \otimes_R U_0$ and we say that U_0 is an *R-form* of U. Thus, an FG-module that has an R-form can be written in R. We now show that every finitely generated FG-module has an R-form.

Lemma 9.4.6. *Let R be a principal ideal domain with field of fractions F, and let U be a finite-dimensional F-vector space. Any finitely generated R-submodule of U that contains an F-basis of U is a full lattice in U.*

Proof. Let U_0 be a finitely generated R-submodule of U that contains an F-basis of U. Since U_0 is a finitely generated torsion-free R-module, $U_0 \cong R^n$ for some n, and it has an R-basis $x_1, \dots, x_n$. Since U_0 contains an F-basis of U it follows that $x_1, \dots, x_n$ span U over F. We show that $x_1, \dots, x_n$ are independent over F. Suppose that $\lambda_1 x_1 + \cdots \lambda_n x_n = 0$ for certain $\lambda_i \in F$. We may write $\lambda_i = \frac{a_i}{b_i}$ where $a_i, b_i \in R$, since F is the field of fractions of R. Now clearing denominators we have $(\prod b_i)(\lambda_1 x_1 + \cdots \lambda_n x_n) = 0$ which implies that $(\prod b_i)\lambda_i = 0$ for each i since $x_1, \dots, x_n$ is an R-basis. This implies that $\lambda_i = 0$ for all i and hence that $n = \dim U$ and $x_1, \dots, x_n$ is an F-basis of U. □

The kind of phenomenon that the last result is designed to exclude is exemplified by considering subgroups of $\mathbb{R}$ generated by elements that are independent over $\mathbb{Q}$, such as the subgroup $\langle 1, \sqrt{2} \rangle \cong \mathbb{Z}^2$. This is a free abelian group, but its basis is not an $\mathbb{R}$-basis for $\mathbb{R}$. According to the last lemma, such a phenomenon would not occur if $\mathbb{R}$ were the field of fractions of $\mathbb{Z}$; indeed, the finitely generated subgroups of $\mathbb{Q}$ are all cyclic.

Corollary 9.4.7. *Let R be a principal ideal domain with field of fractions F, and let U be a finite-dimensional FG-module. Then there exists an RG-lattice U_0 that is an R-form for U.*

Proof. Let $u_1, \ldots, u_n$ be any basis for U and let U_0 be the R-submodule of U spanned by $\{gu_i \mid i = 1, \ldots, n,\ g \in G\}$. This is a finitely generated R-submodule of U that contains an F-basis of U. Since R is a principal ideal domain with field of fractions F, by the last result U_0 is a full RG-lattice in U, which is what we need to prove. □

We should expect that much of the time when $p \mid |G|$ an FG-module U will contain various nonisomorphic full sublattices. To show how such nonisomorphic sublattices may come about, consider an indecomposable projective RG-module $\hat{P}$. It often happens that $F \otimes_R \hat{P}$ is not simple as an FG-module. Writing $F \otimes_R \hat{P} = S_1 \oplus \cdots \oplus S_t$ as a direct sum of simple FG-modules and taking a full RG-lattice in each S_i, the direct sum of these lattices will not be isomorphic to $\hat{P}$, because $\hat{P}$ is indecomposable. In this manner, we may construct nonisomorphic R-forms of $F \otimes_R \hat{P}$.

For another specific example, consider a cyclic group $G = \langle g \rangle$ of order 2 and let (F, R, k) be a 2-modular system. The regular representation FG contains the full lattice $R \cdot 1 + R \cdot g$ that is indecomposable since its reduction kG is indecomposable. It also contains the full lattice $R(1+g) + R(1-g)$, which is a direct sum of RG-modules and hence is decomposable.

The following result is crucial to the definition of the decomposition map, which will be given afterward.

Theorem 9.4.8 (Brauer–Nesbitt). *Let (F, R, k) be a p-modular system, G a finite group, and U a finitely generated FG-module. Let L_1, L_2 be full RG-lattices in U. Then $L_1/\pi L_1$ and $L_2/\pi L_2$ have the same composition factors with the same multiplicities, as kG-modules.*

Proof. We observe first that $L_1 + L_2$ is also a full RG-lattice in U, by Lemma 9.4.6, so by proving the result first for the pair of lattices L_1 and $L_1 + L_2$ and then for L_2 and $L_1 + L_2$ we see that it suffices to consider the case of a pair of lattices, one contained in the other. We now assume that $L_1 \subseteq L_2$.

As R-modules, L_1 and L_2 are free of the same rank, and so L_2/L_1 is a torsion module. Hence L_2/L_1 has a composition series as an R-module, and hence also as an RG-module, because every series of RG-modules can be extended to give a composition series of R-modules. By working down the terms in a composition series, we see that it suffices to assume that L_1 is a maximal RG-submodule of L_2, and we now make this assumption.

Since L_2/L_1 is a simple RG-module, we have $\pi L_2 \subseteq L_1$, by Lemma 9.4.1, and we consider the chain of sublattices $L_2 \supseteq L_1 \supseteq \pi L_2 \supseteq \pi L_1$. We must show that $L_2/\pi L_2$ and $L_1/\pi L_1$ have the same composition factors. The composition factors of $L_1/\pi L_2$ are common to both $L_2/\pi L_2$ and $L_1/\pi L_1$, and we will complete the proof by showing that $L_2/L_1 \cong \pi L_2/\pi L_1$. In fact, the map

$$\begin{aligned} L_2 &\to \pi L_2/\pi L_1 \\ x &\mapsto \pi x + \pi L_1 \end{aligned}$$

is a surjection with kernel L_1. □

We now define the decomposition matrix for a group G in characteristic p. Suppose that (F, R, k) is a splitting p-modular system for G. The *decomposition matrix D* is the matrix with rows indexed by the simple FG-modules and columns indexed by the simple kG-modules whose entries are the numbers

$$d_{TS} = \text{ multiplicity of } S \text{ as a composition factor of } T_0/\pi T_0,$$

where S is a simple kG-module, T is a simple FG-module and T_0 is a full RG-lattice in T.

By the theorem of Brauer and Nesbitt, these multiplicities are independent of the choice of full lattice T_0 in T. Although it would be possible to define a decomposition matrix without the assumption that the p-modular system should be splitting, this is never done, because apart from the inconvenience of having possibly more than one decomposition matrix, the important relationship that we will see between the Cartan matrix and the decomposition matrix in Corollary 9.5.6 does not hold without the splitting hypothesis.

Example 9.4.9. At the moment, the only technique we have to compute a decomposition matrix is to construct all simple representations in characteristic 0, find lattice forms for them, reduce these modulo the maximal ideal and compute composition factors. A less laborious method, in general, is to use Brauer characters as described in Chapter 10. The approach we have for now does at least allow us to construct the decomposition matrices for S_3 in characteristic 2 and characteristic 3. They are

$$\begin{bmatrix} 1 & 0 \\ 1 & 0 \\ 0 & 1 \end{bmatrix} \quad \text{and} \quad \begin{bmatrix} 1 & 0 \\ 0 & 1 \\ 1 & 1 \end{bmatrix}.$$

In Example 2.1.6, we observed that S_3 has three simple representations over a field of characteristic 0: the trivial representation, the sign representation and a 2-dimensional representation that we constructed in Chapter 1 as a $\mathbb{Z}S_3$-lattice. These three representations index the rows of the decomposition matrices. In characteristic 2, the simple representations are the trivial module and the 2-dimensional irreducible with the same matrices as in characteristic 0, but with the entries interpreted as lying in $\mathbb{F}_2$ (see Example 7.2.2). In characteristic 3, the simple representations are the trivial representation and the sign representation, as seen in Example 6.2.3. The fact that these lists are complete also follows from Theorem 9.3.6. Another approach is to say that all 1-dimensional representations are representations of the abelianization C_2 of S_3, giving one such representation in characteristic 2 and two in characteristic 3. Now use a count of dimensions of the simple modules together with Theorem 7.3.9 to show that the lists are correct.

It is now a question of calculating composition factors of the reductions from characteristic 0. The 1-dimenional representations remain simple on reduction. The 2-dimensional representation has been given as a $\mathbb{Z}$-form, and it suffices to compute composition factors when the matrices are interpreted as have entries in $\mathbb{F}_2$ and $\mathbb{F}_3$. Over $\mathbb{F}_2$ this representation is simple, as has been observed.

Example 9.4.10. When G is a p-group and k has characteristic p, the decomposition matrix has a single column and an entry for each ordinary simple character, that entry being the degree of the character. This is a consequence of Proposition 6.2.1.

The third example is sufficiently important that we state it as a separate result. It describes the situation when $|G|$ is invertible in k, so that both FG and kG are semisimple.

Theorem 9.4.11. *Let (F, R, k) be a splitting p-modular system for the group G and suppose that $|G|$ is relatively prime to p. Then each simple FG-module reduces to a simple kG-module of the same dimension. This process establishes a bijection between the simple FG-modules and the simple kG-modules, so that when the two sets of simple modules are taken in corresponding order, the decomposition matrix is the identity matrix. The bijection preserves the decomposition into simples of tensor products, Hom spaces as well as induction and restriction of modules.*

Proof. If necessary, we may extend (F, R, k) to a larger p-modular system in which R is complete by completing R with respect to its maximal ideal and replacing F by the field of fractions of the completion (k remains unchanged in this process). Since F and k are splitting fields, distinct simple modules remain distinct simple modules under field extension, and their properties under taking tensor products, Hom groups, induction, and restriction do not change. We thus assume R is complete.

Let T be a simple FG-module with full RG-lattice T_0. Then $T_0/\pi T_0 \cong S_1 \oplus \cdots \oplus S_n$ for various simple kG-modules S_i, since kG is semisimple. For the same reason these modules are projective, so by Proposition 9.4.5, they lift to projective RG-lattices $\hat{S}_1, \ldots, \hat{S}_n$ that are the projective covers of $S_1, \ldots S_n$. Thus, the projective cover of T_0 is a homomorphism $\hat{S}_1 \oplus \cdots \oplus \hat{S}_n \to T_0$, and this is an isomorphism since the R-ranks of the two modules are the same. We deduce that $T \cong (F \otimes_R \hat{S}_1) \oplus \cdots \oplus (F \otimes_R \hat{S}_n)$, and so $n = 1$ since T is simple. Thus, every reduction of a simple module is simple. Equally, every simple kG-module is a composition factor of the reduction of some simple FG-module, since it is a composition factor of the reduction of FG, and so every simple kG-module does appear as the reduction of a simple FG-module.

For any FG-modules U and V with R-forms U_0 and V_0, $U_0 \otimes_R V_0$ is an R-form of $U \otimes_F V$ and

$$(U_0 \otimes_R V_0)/\pi(U_0 \otimes_R V_0) \cong (U_0/\pi U_0) \otimes_k (V_0/\pi V_0).$$

From this, we can see that if $U \otimes_F V$ decomposes in a certain way as a direct sum of simple modules, on reduction modulo π it gives rise to a corresponding decomposition of $(U_0/\pi U_0) \otimes_k (V_0/\pi V_0)$ as a direct sum of simple modules. In a similar way, we see that $\mathrm{Hom}_F(U, V)$ and $\mathrm{Hom}_k(U_0/\pi U_0, V_0/\pi V_0)$ decompose in a corresponding fashion, as do the induction and restriction of corresponding modules. □

Coming out of the proof of the last result we see that in the situation where $|G|$ is relatively prime to p, an RG-module L is projective if and only if it is projective as an R-module, and furthermore that for each FG-module U, all R-forms of U are isomorphic as RG-modules. We leave the details of this as Exercise 14.

When $|G|$ is divisible by p the decomposition matrix cannot be the identity, because as a consequence of Theorem 9.3.6 it is not even square. We state without proof a theorem which says that when G is p-solvable the decomposition matrix does at least contain the identity matrix as a submatrix of the maximum possible size. A group G is said to be *p-solvable* if it has a chain of subgroups

$$1 = G_n \lhd \cdots \lhd G_1 \lhd G_0 = G$$

so that each factor G_i/G_{i+1} is either a p-group or a group of order prime to p. A proof can be found in [10, Theorem 22.1].

Theorem 9.4.12 (Fong–Swan–Rukolaine). *Let (F, R, k) be a splitting p-modular system for a p-solvable group G. Then every simple kG-module is the reduction modulo (π) of an RG-lattice.*

9.5 The cde Triangle

It is conceptually helpful to express the decomposition matrix as the matrix of a linear map, and to this end we introduce three groups, which are instances of Grothendieck groups. These groups should properly be defined in an abstract fashion after which we would prove that they have a certain structure under the hypotheses in force. For our purposes, it is more direct to skip the abstract step and define the Grothendieck groups in terms of this structure.

Let (F, R, k) be a splitting p-modular system for a finite group G, and suppose that F and R are complete with respect to the valuation. We define

$$\begin{aligned} G_0(FG) &= \text{the free abelian group with the isomorphism types} \\ &\quad\ \text{of simple } FG\text{-modules as a basis,} \\ G_0(kG) &= \text{the free abelian group with the isomorphism types} \\ &\quad\ \text{of simple } kG\text{-modules as a basis,} \\ K_0(kG) &= \text{the free abelian group with the isomorphism types} \\ &\quad\ \text{of indecomposable projective } kG\text{-modules as a basis.} \end{aligned}$$

Thus, $G_0(FG)$ has rank equal to the number of conjugacy classes of G, and both $G_0(kG)$ and $K_0(kG)$ have rank equal to the number of p-regular conjugacy classes of G. If T is a simple FG-module, we write $[T]$ for the corresponding basis element of $G_0(FG)$. Similarly, if S is a simple kG-module, we write $[S]$ for the corrsponding basis element of $G_0(kG)$, and if P is an indecomposable projective kG-module, we write $[P]$ for the corresponding basis element of $K_0(kG)$. Extending this notation, if U is any kG-module with composition factors $S_1, \dots, S_r$ occurring with multiplicities $n_1, \dots, n_r$ in some composition series of U, we write

$$[U] = n_1[S_1] + \cdots + n_r[S_r] \in G_0(kG).$$

The fact that this is well defined depends on the Jordan–Hölder theorem. There is a similar interpretation of $[V] \in G_0(FG)$ if V happens to be an FG-module. This time because V is semisimple it is the direct sum of its composition factors, and so if $V = T_1^{n_1} \oplus \cdots \oplus T_r^{n_r}$, we put $[V] = n_1[T_1] + \cdots + n_r[T_r] \in G_0(FG)$. In the same way, if $P = P_1^{n_1} \oplus \cdots \oplus P_r^{n_r}$ where the P_i are indecomposable projective kG-modules, we put

$$[P] = n_1[P_1] + \cdots + n_r[P_r] \in K_0(kG).$$

Since simple FG-modules biject with their characters, we may identify $G_0(FG)$ with the subset of the space of class functions $\mathbb{C}^{cc(G)}$ consisting of the $\mathbb{Z}$-linear combinations of the characters of the simple modules as considered in

Chapter 3. Such $\mathbb{Z}$-linear combinations of characters are termed *virtual characters* of G, so $G_0(FG)$ is the group of virtual characters of FG.

We now define the homomorphisms of the cde triangle, which is as follows:

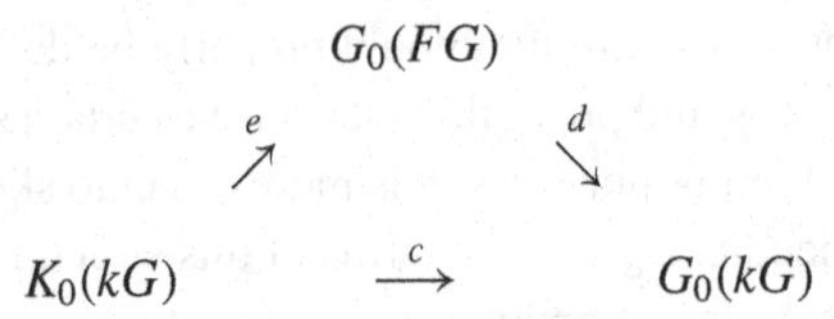

$$\begin{array}{ccc} & G_0(FG) & \\ {}^{e}\nearrow & & \searrow^{d} \\ K_0(kG) & \xrightarrow{c} & G_0(kG) \end{array}$$

The definition of the homomorphism e on the basis elements of $K_0(kG)$ is that if P_S is an indecomposable projective kG-module then $e([P_S]) = [F \otimes_R \hat{P}_S]$. As observed in Proposition 9.4.5, the lift $\hat{P}_S$ is unique up to isomorphism, and so this map is well defined. The *decomposition map* d is defined thus on basis elements: if V is a simple FG-module containing a full RG-lattice V_0, we put $d([V]) = [V_0/\pi V_0]$. By Theorem 9.4.8, this is well defined, and in fact the formula works for arbitrary finite-dimensional FG-modules V, not just the simple ones. This definition means that the matrix of d is the transpose of the decomposition matrix. The homomorphism c is called the *Cartan map* and is defined by $c([P_S]) = [P_S]$, where on the left the symbol $[P_S]$ means the basis element of $K_0(kG)$ corresponding to the indecomposable projective P_S, and on the right $[P_S]$ is an element of $G_0(kG)$. From the definitions, we see that the matrix of c is the Cartan matrix: $[P_T] = \sum_{\text{simple } S} c_{ST}[S]$ for each simple kG-module T.

Proposition 9.5.1. *$c = de$.*

Proof. It is simply a question of following through the definitions of these homomorphisms. If P_S is an indecomposable projective kG-module then $e[P_S] = [F \otimes_R \hat{P}_S]$. To compute $d[F \otimes_R \hat{P}_S]$, we choose any full RG-sublattice of $F \otimes_R \hat{P}_S$ and reduce it modulo (π). Taking $\hat{P}_S$ to be that lattice, its reduction is P_S and so $de([P_S]) = [P_S] = c([P_S])$. □

To investigate the properties of the cde triangle, we study the relationship between homomorphisms between lattices and between their reductions modulo (π).

Proposition 9.5.2. *Let (F, R, k) be a p-modular system. Let U, V be FG-modules containing full RG-lattices U_0 and V_0.*

(1) $\mathrm{Hom}_{RG}(U_0, V_0)$ *is a full R-lattice in* $\mathrm{Hom}_{FG}(U, V)$.
(2) $\pi\,\mathrm{Hom}_{RG}(U_0, V_0) = \mathrm{Hom}_{RG}(U_0, \pi V_0)$ *as a subset of* $\mathrm{Hom}_{RG}(U_0, V_0)$.
(3) Suppose that U_0 is a projective RG-lattice. Then

$$\begin{aligned} \mathrm{Hom}_{RG}(U_0, V_0)/\pi\,\mathrm{Hom}_{RG}(U_0, V_0) &\cong \mathrm{Hom}_{RG}(U_0, V_0/\pi V_0) \\ &\cong \mathrm{Hom}_{kG}(U_0/\pi U_0, V_0/\pi V_0). \end{aligned}$$

Proof. (1) We should explain how it is that $\mathrm{Hom}_{RG}(U_0, V_0)$ may be regarded as a subset of $\mathrm{Hom}_{FG}(U, V)$. The most elementary approach is to take R-bases $u_1, \dots, u_r$ for U_0 and $v_1, \dots, v_s$ for V_0. These are also F-bases for U and V. Any RG-homomorphism $U_0 \to V_0$ can be represented with respect to these bases by a matrix with entries in R. Regarding it as a matrix with entries in F, it represents an FG-module homomorphism $U \to V$.

To see that $\mathrm{Hom}_{RG}(U_0, V_0)$ is in fact a sublattice of $\mathrm{Hom}_{FG}(U, V)$, we observe that $\mathrm{Hom}_{RG}(U_0, V_0) \subseteq \mathrm{Hom}_R(U_0, V_0) \cong R^{rs}$ where $r = \dim U$, $s = \dim V$. The latter is a free R-module, so $\mathrm{Hom}_{RG}(U_0, V_0)$ is an R-lattice since R is a principal ideal domain. We show that it is full in $\mathrm{Hom}_{FG}(U, V)$. Using the bases for U_0, V_0, let $\phi : U \to V$ be an FG-module homomorphism. Then $\phi(u_i) = \sum \lambda_{ji} v_j$ with $\lambda_{ji} \in F$. Choose $a \in R$ so that $a\lambda_{ji} \in R$ for all i, j. Then $a\phi : U_0 \to V_0$, showing that ϕ belongs to $F \cdot \mathrm{Hom}_{RG}(U_0, V_0)$. Therefore, $\mathrm{Hom}_{RG}(U_0, V_0)$ spans $\mathrm{Hom}_{FG}(U, V)$ over F.

(2) The map $V_0 \to \pi V_0$ given by $x \mapsto \pi x$ is an RG-isomorphism so the morphisms $U_0 \to \pi V_0$ are precisely those that arise as composites $U_0 \to V_0 \xrightarrow{\pi} \pi V_0$, which are in turn the elements of $\pi \,\mathrm{Hom}_{RG}(U_0, V_0)$.

(3) Consider $\mathrm{Hom}(U_0, V_0) \to \mathrm{Hom}(U_0, V_0/\pi V_0)$. Its kernel is $\mathrm{Hom}(U_0, \pi V_0)$, which equals $\pi \,\mathrm{Hom}(U_0, V_0)$. Since U_0 is projective, the map is surjective, and it gives rise to the first isomorphism. For the second, all homomorphisms $\alpha : U_0 \to V_0/\pi V_0$ contain πU_0 in the kernel, and so factor as $U_0 \to U_0/\pi U_0 \xrightarrow{\beta} V_0/\pi V_0$. The correspondence of α and β provides the isomorphism. □

Corollary 9.5.3. *Suppose U_0 and V_0 are full RG-lattices in U and V, and U_0 is projective. Then*

$$\dim_F \mathrm{Hom}_{FG}(U, V) = \dim_k \mathrm{Hom}_{kG}(U_0/\pi U_0, V_0/\pi V_0).$$

Proof. Both sides equal $\mathrm{rank}_R \mathrm{Hom}_{RG}(U_0, V_0)$ by parts (1) and (3) of the last result. □

We now identify the map e as the transpose of d.

Theorem 9.5.4. *Let (F, R, k) be a splitting p-modular system for G and suppose that R is complete with respect to its valuation.*

(1) *Let S be a simple kG-module and let T be a simple FG-module containing a full RG-lattice T_0. The multiplicity of T in $F \otimes_R \hat{P}_S$ equals the multiplicity of S as a composition factor of $T_0/\pi T_0$.*
(2) *With respect to the given bases of $G_0(FG)$, $G_0(kG)$, and $K_0(kG)$ the matrix of e is D and the matrix of d is D^T, where D is the decomposition matrix.*

The given bases of the Grothendieck groups are the bases whose elements are the symbols $[T]$, $[S]$, and $[P_S]$ where T is a simple FG-module, and S is a simple kG-module.

Proof. (1) Applying the last corollary to the full RG-lattice $\hat{P}_S$ of $F \otimes_R \hat{P}_S$, we obtain $\dim_F \operatorname{Hom}_{FG}(F \otimes_R \hat{P}_S, T) = \dim_k \operatorname{Hom}_{kG}(P_S, T_0/\pi T_0)$. The left side equals $\dim_F \operatorname{End}_{FG}(T)$ times the multiplicity of T in $F \otimes_R \hat{P}_S$, and the right side equals $\dim_k \operatorname{End}_{kG}(S)$ times the multiplicity of S as a composition factor of $T_0/\pi T_0$. The splitting hypothesis implies that the endomorphism rings both have dimension 1, and the result follows.

(2) We have already observed when defining d that its matrix is D^T. The entries e_{TS} in the matrix of e are defined by

$$e([P_S]) = [F \otimes_R \hat{P}_S] = \sum_T e_{TS} T,$$

so that e_{TS} is the multiplicity of T in $F \otimes_R \hat{P}_S$. By part (1), $e_{TS} = d_{TS}$. □

We comment that part (1) of the preceding theorem gives a second proof of the Brauer–Nesbitt Theorem 9.4.8 that the decomposition numbers ($d_{TS} =$ the multiplicity of S as a composition factor of $T_0/\pi T_0$) are defined independently of the choice of lattice T_0, since they have just been shown to be equal to quantities that do not depend on this choice. It also shows that the decomposition numbers are independent of the choice of p-modular system (F, R, k), provided it is splitting.

Example 9.5.5. This result allows us to compute the characters of the indecomposable projective RG-modules $\hat{P}_S$ (or more properly the characters of the FG-modules $F \otimes_R \hat{P}_S$). Using the decomposition matrices for S_3 that were previously computed, we see that in characteristic 2,

$$\chi_{\hat{P}_1} = \chi_1 + \chi_\epsilon$$
$$\chi_{\hat{P}_2} = \chi_2,$$

and in characteristic 3,

$$\chi_{\hat{P}_1} = \chi_1 + \chi_2$$
$$\chi_{\hat{P}_\epsilon} = \chi_\epsilon + \chi_2.$$

We now have a second proof of the symmetry of the Cartan matrix, but perhaps more importantly an extremely good way to calculate it. The effectiveness of this approach will be increased once we know about Brauer characters, which are treated in Chapter 10. We will also prove in Corollary 10.2.4 that the Cartan matrix is invertible.

Corollary 9.5.6. *Let (F, R, k) be a splitting p-modular system for G. Then the Cartan matrix $C = D^T D$. Thus, C is symmetric.*

Example 9.5.7. When $G = S_3$ the Cartan matrices in characteristic 2 and in characteristic 3, expressed as a product $D^T D$, are

$$\begin{bmatrix} 1 & 1 & 0 \\ 0 & 0 & 1 \end{bmatrix} \begin{bmatrix} 1 & 0 \\ 1 & 0 \\ 0 & 1 \end{bmatrix} = \begin{bmatrix} 2 & 0 \\ 0 & 1 \end{bmatrix} \quad \text{and} \quad \begin{bmatrix} 1 & 0 & 1 \\ 0 & 1 & 1 \end{bmatrix} \begin{bmatrix} 1 & 0 \\ 0 & 1 \\ 1 & 1 \end{bmatrix} = \begin{bmatrix} 2 & 1 \\ 1 & 2 \end{bmatrix}.$$

The decomposition matrices were calculated in Example 9.4.9. In characteristic 2, the Cartan matrix follows from Example 7.2.2, and in characteristic 3, it follows from Proposition 8.3.3. The decomposition matrix factorization provides a new way to compute the Cartan matrices.

Example 9.5.8. Theorem 9.5.4 and Corollary 9.5.6 fail without the hypothesis that the p-modular system is splitting. An example of this is provided by the alternating group A_4 with the 2-modular system $(\mathbb{Q}_2, \mathbb{Z}_2, \mathbb{F}_2)$. The Cartan matrix of $\mathbb{F}_2 A_4$ was computed in Exercise 2 of Chapter 8, and we leave it as a further exercise to compute the matrices of d and e here.

The equality of dimensions that played the key role in the proof of Theorem 9.5.4 can be nicely expressed in terms of certain bilinear pairings between the various Grothendieck groups. On the vector space of class functions on G, we already have defined a Hermitian form and on the subgroup $G_0(FG)$ it restricts to give a bilinear form

$$\langle \ , \ \rangle : G_0(FG) \times G_0(FG) \to \mathbb{Z}$$

specified by $\langle [U], [V] \rangle = \dim \operatorname{Hom}_{FG}(U, V)$ when U and V are FG-modules. We also have a pairing

$$\langle \ , \ \rangle : K_0(kG) \times G_0(kG) \to \mathbb{Z}$$

specified by $\langle [P], [V] \rangle = \dim \operatorname{Hom}_{kG}(P, V)$ when P is a projective kG-module and V is a kG-module. By Proposition 7.4.1, this quantity depends only on the composition factors of V, not on the actual module V, and so this pairing is well defined. We claim that each of these bilinear pairings is nondegenerate, since in each case the free abelian groups have bases that are dual to each other. Thus, if U and V are simple FG-modules, we have $\langle [U], [V] \rangle = \delta_{[U],[V]}$, and if S and T are simple kG-modules, we have $\langle [P_S], [T] \rangle = \delta_{[S],[T]}$. The equality of dimensions that appeared in the proof of Theorem 9.5.4 can now be expressed as follows. If $x \in K_0(kG)$ and $y \in G_0(FG)$ then $\langle e(x), y \rangle = \langle x, d(y) \rangle$. This formalism is an expression of the fact that e and d are the transpose of each other.

9.6 Blocks of Defect Zero

Blocks of a group algebra were introduced in Chapter 3 as the ring direct summands of that algebra, or equivalently the primitive central idempotents to which the summands correspond. They will be studied in a much more complete fashion in Chapter 12. Before that, we study the properties of a special kind of block, known as a block of defect zero. Taking a splitting p-modular system (F, R, k) for a group G, a *block of defect zero* may be defined as a ring summand of kG having a projective simple module. We will see that such a block can have only one simple module, it is a matrix algebra, and there is a unique ordinary simple character that reduces to it. These simple representations, over k and over F, are also called blocks of defect zero, by abuse of terminology. The notion of the defect of a block will be explained in Chapter 12, and for now we must accept this term as nothing more than a name. The theory to be explained generalizes the situation of Theorem 9.4.11, which described the case where all blocks have defect zero.

We will find that blocks of defect zero have a useful consequence in predicting zeros in the character table of G: such a simple complex character (of degree is divisible by the p-part of $|G|$) is zero on all elements of G of order divisible by p. It is easy to identify from the character table, and it happens quite often that the zeros in a character table arise from blocks of defect zero. This will be explained in Corollary 9.6.3. Blocks of defect zero are also useful because they remain simple on reduction mod p, which is helpful when calculating decomposition matrices. They arise naturally in the context of representations of groups of Lie type in defining characteristic in the form of the Steinberg representation. Blocks of defect zero are one of the main ingredients in one of the most significant conjectures in the representation theory of finite groups: Alperin's weight conjecture.

We will make the hypothesis in Theorem 9.6.1 that R should be complete, but this is just a convenience and the assertions are still true without this hypothesis. The result can be proved without this hypothesis by first completing an arbitrary p-modular system.

Theorem 9.6.1. *Let (F, R, k) be a splitting p-modular system in which R is complete, and let G be a group of order $p^d q$ where q is prime to p. Let T be an FG-module of dimension n, containing a full RG-sublattice T_0. The following are equivalent.*

(1) *$p^d \mid n$ and T is a simple FG-module.*
(2) *The homomorphism $RG \to \mathrm{End}_R(T_0)$ that gives the action of RG on T_0 identifies $\mathrm{End}_R(T_0) \cong M_n(R)$ with a ring direct summand of RG.*
(3) *T is a simple FG-module and T_0 is a projective RG-module.*

(4) *The homomorphism* $kG \to \mathrm{End}_k(T_0/\pi T_0)$ *identifies* $\mathrm{End}_k(T_0/\pi T_0) \cong M_n(k)$ *with a ring direct summand of kG.*
(5) *As a kG-module,* $T_0/\pi T_0$ *is simple and projective.*

Proof. (1) $\Rightarrow$ (2) Suppose that (1) holds. We will use the formula obtained in Theorem 3.6.2 for the primitive central idempotent e associated to T, namely,

$$e = \frac{n}{|G|} \sum_{g \in G} \chi_T(g^{-1})g$$

where χ_T is the character of T. Observe that $n/|G| \in R$ because $p^d \mid n$, and also $\chi_T(g^{-1})$ lies in R since it is a sum of roots of unity, and roots of unity (in some extension ring if necessary) have valuation 1, so lie in R. Thus, $e \in RG$. It follows that $RG = eRG \oplus (1-e)RG$ as a direct sum of rings.

The homomorphism $\rho : FG \to \mathrm{End}_F(T)$ that expresses the action of G on T identifies eFG with the matrix algebra $\mathrm{End}_F(T)$, and has kernel $(1-e)FG$. The restriction of ρ to RG takes values in $\mathrm{End}_R(T_0)$ because T_0 is a full RG-sublattice of T. The kernel of this restriction is $(1-e)FG \cap RG = (1-e)RG$ which is a direct summand of RG. We will show that this restricted homomorphism is surjective to $\mathrm{End}_R(T_0)$, and from this, it will follow that $eRG \cong \mathrm{End}_R(T_0) \cong M_n(R)$, a direct summand of RG.

As an extension of the formula in Theorem 3.6.2 for the primitive central idempotent corresponding to T, we claim that if $\phi \in \mathrm{End}_F(T)$ then

$$\phi = \frac{n}{|G|} \sum_{g \in G} \mathrm{tr}(\rho(g^{-1})\phi)\rho(g).$$

To demonstrate this, it suffices to consider the case $\phi = \rho(h)$ where $h \in G$, since these elements span $\mathrm{End}_F(T)$. In this case,

$$\begin{aligned}\frac{n}{|G|} \sum_{g \in G} \mathrm{tr}(\rho(g^{-1})\rho(h))\rho(g) &= \rho(h)\frac{n}{|G|} \sum_{g \in G} \mathrm{tr}(\rho(g^{-1}h))\rho(h^{-1}g)\\ &= \rho(h)\rho(e)\\ &= \rho(h)\end{aligned}$$

using the previously obtained formula for e. This shows that the claimed formula holds when $\phi = \rho(h)$, and hence holds in general.

Finally, we may see that the restriction of ρ is a surjective homomorphism $RG \to \mathrm{End}_R(T_0)$, since any $\phi \in \mathrm{End}_R(T_0)$ is the image under ρ of

$$\frac{n}{|G|} \sum_{g \in G} \mathrm{tr}(\rho(g^{-1})\phi)g \in RG.$$

This completes the proof of this implication.

(2) $\Rightarrow$ (3) Certainly T_0 is projective as a module for $\mathrm{End}_R(T_0)$ since it identifies with the module of column vectors for this matrix algebra. Assuming (2), we have that T_0 is a projective RG-module, since RG acts via its summand eRG which identifies with the matrix algebra. Furthermore, FG acts on $T \cong F \otimes_R T_0$ as column vectors for a matrix algebra over F, so T is a simple FG-module.

(2) $\Rightarrow$ (4) The decomposition $RG = eRG \oplus (1-e)RG$ with $eRG \cong M_n(R)$ is preserved on reducing modulo (π), and we obtain $kG = \bar{e}kG \oplus (1-\bar{e})kG$ where $\bar{e}$ is the image of e in kG. Furthermore, $\bar{e}kG \cong M_n(k)$ because it is the reduction module (π) of $M_n(R)$, and the action of kG on $T_0/\pi T_0$ is via projection onto $\bar{e}kG$.

(3) $\Rightarrow$ (5) Since T_0 is a direct summand of a free RG-module it follows that $T_0/\pi T_0$ is a direct summand of a free kG-module, and hence is projective. Furthermore, we claim that $T_0/\pi T_0$ is indecomposable. To see this, write $T_0/\pi T_0 = P_{S_1} \oplus \cdots \oplus P_{S_t}$ where the P_{S_i} are indecomposable projectives, so that $T_0/\pi T_0$ is the projective cover of $S_1 \oplus \cdots \oplus S_t$ as an RG-module. It follows that $T_0 \cong \hat{P}_{S_1} \oplus \cdots \oplus \hat{P}_{S_t}$ and $T \cong F \otimes_R \hat{P}_{S_1} \oplus \cdots \oplus F \otimes_R \hat{P}_{S_t}$. Hence $t = 1$ since T is simple. Theorem 9.5.4 the column of the decomposition matrix corresponding to S_1 consists of zeros except for an entry 1 in the row of T. Since $C = D^T D$, the multiplicity of S_1 as a composition factor of P_{S_1} is 1. We know from Theorem 8.5.5 that $\mathrm{Soc}\, P_{S_1} \cong S_1$ and also $P_{S_1}/\mathrm{Rad}\, P_{S_1} \cong S_1$, so that S_1 occurs as a composition factor of P_{S_1} with multiplicity at least 2 unless $P_{S_1} = S_1$. This shows that $T_0/\pi T_0$ is simple.

(4) $\Rightarrow$ (5) This is analogous to the proof (2) $\Rightarrow$ (3).

(5) $\Rightarrow$ (1) Since $T_0/\pi T_0$ is projective its dimension is divisible by p^d by Corollary 8.1.3, and this dimension equals $\mathrm{rank}_R T_0 = \dim T$. If T were not simple as an FG-module, we would be able to write $T = U \oplus V$, and taking full RG-lattices U_0, V_0 the composition factors of $T_0/\pi T_0$ would be the same as those of $U_0/\pi U_0 \oplus V_0/\pi V_0$. Since $T_0/\pi T_0$ is in fact simple, this situation cannot occur, and T is simple. □

Part (5) of Theorem 9.6.1 appears to depend on the FG-module T, but this is not really the case. If P is any simple projective kG-module, it can be lifted to an RG-module $\hat{P}$ and taking $T = F \otimes_R \hat{P}$ we have a module with a full RG-lattice T_0 for which $T_0/\pi T_0 \cong P$. Thus, every simple projective kG-module is acted on by kG via projection onto a matrix algebra direct summand of kG, and the module T just constructed is always simple.

Notice in Theorem 9.6.1 that since the full RG-sublattice T_0 is arbitrary, every full RG-sublattice of T is projective. Since such a full lattice T_0 is the projective cover of $T_0/\pi T_0$, which is a simple module defined independently of the choice of T_0, all such lattices T_0 are isomorphic as RG-modules.

Looking at the various equivalent parts of Theorem 9.6.1, one might be led to suspect that if k is a field of characteristic p and S is a simple kG-module of dimension divisible by the largest power of p that divides $|G|$ then S is necessarily projective, but in fact this conclusion does not always hold. It is known from [20] that McLaughlin's simple group has an absolutely simple module in characteristic 2 of dimension $2^9 * 7$. The module is not projective and the 2-part of the group order is 2^7.

We present the consequence of Theorem 9.6.1 for character tables in Corollary 9.6.3, with a preliminary lemma about character values of projective modules before that.

Lemma 9.6.2. *Let (F, R, k) be a p-modular system and G a finite group. Let $\hat{P}$ be a projective RG-module and χ the character of $F \otimes_R \hat{P}$. Then $\chi(1)$ is divisible by the order of a Sylow p-subgroup of G, and if $g \in G$ has order divisible by p then $\chi(g) = 0$.*

Proof. Since $\hat{P}/\pi\hat{P}$ is a projective kG-module it has dimension divisible by the order of a Sylow p-subgroup of G (Corollary 8.1.3), and this dimension equals the rank of $\hat{P}$, that in turn equals $\chi(1) = \dim F \otimes_R \hat{P}$.

Consider now an element $g \in G$ of order divisible by p. To show that $\chi(g) = 0$, it suffices to consider $\hat{P}$ as an $R\langle g\rangle$-module, and as such it is still projective. We may suppose that $\hat{P}$ is an indecomposable projective $R\langle g\rangle$-module.

Let us write $g = st$ where s is p-regular, t is p-singular, and $st = ts$, as in Lemma 9.3.4, so $\langle g\rangle = \langle s\rangle \times \langle t\rangle$. As in Example 8.2.1, we can write $\hat{P}/\pi\hat{P} = S \otimes k\langle t\rangle$ where S is a simple $k\langle s\rangle$-module. Since $k\langle s\rangle$ is semisimple and S is projective as a $k\langle s\rangle$-module, we can lift S to a projective $R\langle s\rangle$-module $\hat{S}$ for which $\hat{S}/\pi\hat{S} = S$. Now $\hat{P}/\pi\hat{P} = S \otimes k\langle t\rangle = S \otimes k \uparrow_{\langle s\rangle}^{\langle g\rangle} = S \uparrow_{\langle s\rangle}^{\langle g\rangle}$. This lifts to $\hat{S} \uparrow_{\langle s\rangle}^{\langle g\rangle}$, which is a projective $R\langle g\rangle$-module. It is the projective cover of $\hat{P}/\pi\hat{P}$, so by uniqueness of projective covers $\hat{P} \cong \hat{S} \uparrow_{\langle s\rangle}^{\langle g\rangle}$. We deduce that $\chi = \chi_{\hat{S}} \uparrow_{\langle s\rangle}^{\langle g\rangle}$. It follows that $\chi(g) = 0$ from the formula for an induced character, since no conjugate of g lies in $\langle s\rangle$. □

Corollary 9.6.3. *Let T be a simple FG-module, where F is a splitting field for G of characteristic 0, and let χ_T be the character of T. Let p be a prime, and suppose that the highest power of p that divides $|G|$ also divides the degree $\chi_T(1)$. Then if g is any element of G of order divisible by p, we have $\chi_T(g) = 0$.*

Proof. This combines Lemma 9.6.2 with Theorem 9.6.1. If F does not initially appear as part of a p-modular system, we may replace F by a subfield that is a splitting field and which is a finite extension of $\mathbb{Q}$, since $\mathbb{Q}G$ has a splitting field of this form and by the argument of Theorem 9.2.6, it may be chosen to be a subfield of F. We may write T in this subfield without changing its character χ_T and the hypothesis about the power of p that divides $\chi_T(1)$ remains the same. Take the valuation on F determined by a maximal ideal $\mathfrak{p}$ of the ring of integers for which $\mathfrak{p} \cap \mathbb{Z} = (p)$, and complete F with respect to this valuation to get a splitting, complete, p-modular system. We may now apply Theorem 9.6.1. □

Example 9.6.4. The character table information given by the last corollary can be observed in many examples. Thus, the simple character of degree 2 of the symmetric group S_3 is zero except on the 2-regular elements, and the simple characters of S_4 of degree 3 are zero except on the 3-regular elements. All of the zeros in the character table of A_5 may be accounted for in this way, and all except one of the zeros in the character table of $GL(3, 2)$. It is notable that to prove this result about representations in characteristic zero we have used technical machinery from characteristic p.

9.7 Summary of Chapter 9

- Every finite-dimensional algebra over a field has a splitting field of finite degree.
- If two modules are isomorphic after extending the ground field, they were originally isomorphic.
- If k is a splitting field of characteristic p, the number of simple kG-modules equals the number of p-regular conjugacy classes in G.
- Idempotents lift from the group ring over the residue field to the group ring over a complete discrete valuation ring.
- Projective covers exist over RG when R is a complete discrete valuation ring. The indecomposable projective RG-modules are the projective covers $\hat{P}_S$, where S is simple.
- When R is a principal ideal domain with field of fractions F, every FG-module can be written in R.
- The decomposition map is well defined and $C = D^T D$. The decomposition matrix also computes the ordinary characters of the projectives $\hat{P}_S$. We obtain a second proof that the Cartan matrix is symmetric.
- If k is a splitting field of characteristic p where $p \nmid |G|$, the representations of G over $\mathbb{C}$ and over k are "the same" in a certain sense.
- Blocks of defect zero are identified by the fact that the degree of their ordinary character is divisible by the p-part of $|G|$. Such characters vanish on elements of order divisible by p. They remain irreducible on reduction mod p, where they give a projective module and a matrix ring summand of kG.

9.8 Exercises for Chapter 9

1. Let $E = \mathbb{F}_p(t)$ be a transcendental extension of the field with p elements and let F be the subfield $\mathbb{F}_p(t^p)$. Write $\alpha = t^p \in F$, so that $t^p - \alpha = 0$. Let $A = E$, regarded as an F-algebra.

(a) Show that A has a simple module that is not absolutely simple.

(b) Show that E is a splitting field for A, and that the regular representation of $E \otimes_F A$ is a uniserial module. Show that $\mathrm{Rad}(E \otimes_F A) \neq E \otimes_F \mathrm{Rad}(A)$.
[Notice that $E \otimes_F \mathrm{Rad}(A)$ is always contained in the radical of $E \otimes_F A$ when A is a finite dimensional algebra, being a nilpotent ideal.]

(c) Show that A is not isomorphic to FG for any group G.

2. Let R be a complete discrete valuation ring with residue field k of characteristic p. Let $g \in GL(n, k)$ be an $n \times n$-matrix with entries in k. Suppose g has finite order s and suppose that both R and k contain primitive sth roots of unity. Show that there is an $n \times n$-matrix $\hat{g} \in GL(n, R)$ of order s whose reduction to k is g.

3. Let G be a cyclic group, F a field, and S a simple FG-module. Show that $E = \mathrm{End}_{FG}(S)$ is a field with the property that $E \otimes_F S$ is a direct sum of modules that are all absolutely simple.

4. Let A be a finite-dimensional algebra over a field F and suppose that F is a splitting field for A. Let $E \supseteq F$ be a field extension. Prove that $\mathrm{Rad}(E \otimes_F A) \cong E \otimes_F \mathrm{Rad}(A)$.
[The observation at the end of Exercise 1(b) may help here.]

5. Let $G = C_2 \times C_2$ be generated by elements a and b, and let E be a field of characteristic 2. Let $t \in E$ be any element, which may be algebraic or transcendental over $\mathbb{F}_2$. Let $\rho : G \to GL_2(E)$ be the representation with

$$\rho(a) = \begin{bmatrix} 1 & 1 \\ 0 & 1 \end{bmatrix}, \quad \rho(b) = \begin{bmatrix} 1 & t \\ 0 & 1 \end{bmatrix}.$$

Show that this representation is *absolutely indecomposable*, meaning that it remains indecomposable under all field extensions. Show also that this representation cannot be written in any proper subfield of $\mathbb{F}_2(t)$.

6. Let (F, R, k) be a p-modular system and G a finite group. Show that if $U = U_1 \oplus U_2$ is a finite-dimensional FG-module and L is a full RG-lattice in U then $L \cap U_1, L \cap U_2$ are full RG-lattices in U_1 and U_2, but that it need not be true that $L = (L \cap U_1) \oplus (L \cap U_2)$.
[Consider the regular representation when $G = C_2$.]

7. Show that, over a splitting 2-modular system (F, R, k), the dihedral group D_{30} has seven 2-blocks of defect zero and two further 1-dimensional ordinary characters. Hence find the degrees of the simple kD_{30}-modules. Find the decomposition matrix and Cartan matrix in characteristic 2. Show that $kD_{30} \cong kC_2 \oplus M_2(k)^7$ as rings.

8. Let A be a finite-dimensional F-algebra.

(a) Suppose that F is a splitting field for A, and let $E \supseteq F$ be a field extension. Show that every simple $E \otimes_F A$-module can be written in F. Deduce that A and $E \otimes_F A$ have the same number of isomorphism classes of simple modules.
[Bear in mind the result of Exercise 4.]

(b) Show that if E_1 and E_2 are different splitting fields for A, then $E_1 \otimes_F A$ and $E_2 \otimes_F A$ have the same number of isomorphism classes of simple modules.
[Assume that E_1 and E_2 are subfields of some larger field.]

9. Let A be a finite-dimensional F-algebra and let $E \supseteq F$ be a field extension.

(a) Suppose that S is an A module with the property that $E \otimes_F S$ is an absolutely simple $E \otimes_F A$-module. Show that S is absolutely simple as an A-module.

(b) Suppose now that E is a splitting field for A and that every simple A-module remains simple on extending scalars to E. Show that F is a splitting field for A.

10. Let A be a finite dimensional F-algebra and $E \supseteq F$ a field extension.

(a) Show that if $U \to V$ is an essential epimorphism of A-modules then $E \otimes_F U \to E \otimes_F V$ is an essential epimorphism of $E \otimes_F A$-modules.

(b) Show that if $P \to U$ is a projective cover then so is $E \otimes_F P \to E \otimes_F U$.

11. Let A be a finite dimensional F-algebra where F is a splitting field for A. Let P be an indecomposable projective A-module. Show that if $E \supseteq F$ is any field extension then $E \otimes_F P$ is indecomposable and projective as an $E \otimes_F A$-module. Show further that every indecomposable projective $E \otimes_F A$-module can be written in F.

12. Let U be the 2-dimensional representation of S_3 over $\mathbb{Q}$ that is defined by requiring that with respect to a basis u_1, u_2 the elements $(1, 2, 3)$ and $(1, 2)$ act by matrices

$$\begin{bmatrix} 0 & -1 \\ 1 & -1 \end{bmatrix} \quad \text{and} \quad \begin{bmatrix} 1 & -1 \\ 0 & -1 \end{bmatrix}.$$

Let U_0 be the $\mathbb{Z}S_3$-lattice that is the $\mathbb{Z}$-span of u_1 and u_2 in U.

(a) Show that $U_0/3U_0$ has just three submodules as a module for $(\mathbb{Z}/3\mathbb{Z})S_3$, namely, 0, the whole space, and a 1-dimensional submodule. Deduce that $U_0/3U_0$ is not semisimple.
(b) Now let U_1 be the $\mathbb{Z}$-span of the vectors $2u_1 + u_2$ and $-u_1 + u_2$ in U. Show (for example, by drawing a picture in which the angle between u_1 and u_2 is 120°, or else algebraically) that U_1 is a $\mathbb{Z}S_3$-lattice in U, and that it has index 3 in U_0. Write down matrices that give the action of $(1, 2, 3)$ and $(1, 2)$ on U_1 with respect to the new basis. Show that $U_1/3U_1$ also has just three submodules as a $(\mathbb{Z}/3\mathbb{Z})S_3$-module, but that it is not isomorphic to $U_0/3U_0$. Identify U_0/U_1 as a $(\mathbb{Z}/3\mathbb{Z})S_3$-module.
(c) Prove that U_1 is the unique $\mathbb{Z}S_3$-sublattice of U_0 of index 3.
(d) Let $\mathbb{Z}_3$ be the ring of 3-adic integers. Show that the $\mathbb{Z}_3S_3$-sublattices of $\mathbb{Z}_3 \otimes_{\mathbb{Z}} U_0$ are totally ordered by inclusion.

13. Let (F, R, k) be a splitting p-modular system and G a finite group. Let T be an FG-module with the property that every full RG-sublattice of T is indecomposable and projective. Show that T is simple of dimension divisible by p^n, where $p^n \mid |G|$, $p^{n+1} \nmid |G|$.

14. Let (F, R, k) be a splitting p-modular system for the group G and suppose that $|G|$ is relatively prime to p.

(a) Let

$$FG = \bigoplus_{i=1}^{m} M_{a_i}(F), \quad kG = \bigoplus_{i=1}^{n} M_{b_i}(k)$$

be the decompositions into direct sums of matrix algebras given by Wedderburn's Theorem 2.1.3. Show that $m = n$ and that the list of numbers $a_1, \ldots, a_m$ is the same as the list of numbers $b_1, \ldots, b_n$ after reordering them.
(b) Let L be a finitely generated RG-module. Show that L is projective if and only if it is projective as an R-module. Show further that for each finite-dimensional FG-module U, all R-forms of U are isomorphic as RG-modules.

15. Consider the cde triangle for A_4 with the 2-modular system $(\mathbb{Q}_2, \mathbb{Z}_2, \mathbb{F}_2)$, where $\mathbb{Q}_2$ and $\mathbb{Z}_2$ denote the 2-adic rationals and integers. Compute the matrices D^T and E of the maps d and e with respect to the bases for the Grothendieck groups described in this chapter. Verify that $E \neq D$, but that the Cartan matrix does satisfy $C = D^T E$ and is not symmetric. [Compare Chapter 8, Exercise 2. You may assume that $\mathbb{Q}_2$ contains no primitive third root of 1. This fact follows from the discussion of roots of unity at the start of the next chapter, together with the fact that $\mathbb{F}_2$ does not contain a primitive third root of 1.]

16. Give a proof of the following result by following the suggested steps.

Theorem. *Let $E \supset F$ be a field extension of finite degree and let A be an F-algebra. Let U and V be A-modules. Then*

$$E \otimes_F \mathrm{Hom}_A(U, V) \cong \mathrm{Hom}_{E\otimes_F A}(E \otimes_F U, E \otimes_F V)$$

via an isomorphism $\lambda \otimes_F f \mapsto (\mu \otimes_F u \mapsto \lambda\mu \otimes_F f(u))$.

(a) Verify that there is indeed a homomorphism as indicated.
(b) Let $x_1, \dots, x_n$ be a basis for E as an F-vector space. Show that for any F-vector space M, each element of $E \otimes_F M$ can be written uniquely in the form $\sum_{i=1}^n x_i \otimes_F m_i$ with $m_i \in M$.
(c) Show that if an element $\sum_{i=1}^n x_i \otimes f_i \in E \otimes_F \mathrm{Hom}_A(U, V)$ maps to 0 then $\sum_{i=1}^n x_i \otimes f_i(u) = 0$ for all $u \in U$. Deduce that the homomorphism is injective.
(d) Show that the homomorphism is surjective as follows: given an $E \otimes_F A$-module homomorphism $g : E \otimes_F U \to E \otimes_F V$, write $g(1 \otimes_F u) = \sum_{i=1}^n x_i \otimes f_i(u)$ for some elements $f_i(u) \in V$. Show that this defines A-module homomorphisms $f_i : U \to V$. Show that g is the image of $\sum_{i=1}^n x_i \otimes f_i$.

17. Let $E \supset F$ be a field extension of finite degree and let A be an F-algebra.

(a) Let $0 \to U \to V \to W \to 0$ be a short exact sequence of A-modules. Show that the sequence is split if and only if the short exact sequence $0 \to E \otimes_F U \to E \otimes_F V \to E \otimes_F W \to 0$ is split.
(b) Let U be an A-module. Show that U is projective if and only if $E \otimes_F U$ is projective as an $E \otimes_F A$-module.
(c) Let U be an A-module, S an absolutely simple A-module, and let $L_S(U)$ denote the largest submodule of U that is a direct sum of copies of S, as in Corollary 1.2.6. Show that $L_{E\otimes_F S}(E \otimes_F U) \cong E \otimes_F L_S(U)$. Deduce for fixed points that $(E \otimes_F U)^G \cong E \otimes_F (U^G)$. Prove a similar result for the largest quotient of U that is a direct sum of copies of S and hence a similar result for the fixed quotient of U.

(d) Show that if F is assumed to be a splitting field for A and U is an A-module then $\text{Rad}(E \otimes_F U) \cong E \otimes_F \text{Rad}(U)$ and $\text{Soc}(E \otimes_F U) \cong E \otimes_F \text{Soc}(U)$. Explain why this does not contradict the conclusion of Exercise 1.

[Hint: use the most promising of the equivalent conditions of Propositions 7.1.2 and 7.1.3 in combination with the result of Exercise 16.]

10
Brauer Characters

After the success of ordinary character theory in computing with representations over fields of characteristic zero we wish for a corresponding theory of characters of representations in positive characteristic. Brauer characters provide such a theory and we describe it in this chapter, starting with the definition and continuing with their main properties, which are similar in many respects to those of ordinary characters. We arrange the Brauer characters in tables that satisfy orthogonality relations, although of a more complicated kind than what we saw with ordinary characters. We will find that the information carried by Brauer characters is exactly that of the composition factors of a representation, and not more. They are very effective if that is the information we require, but they do not tell us anything more about the range of complicated possibilities that we find with representations in positive characteristic. Using Brauer characters is, however, usually the best way to compute a decomposition matrix, and hence a Cartan matrix in view of Corollary 9.5.6. We obtain in Corollary 10.2.4 the deductions that the Cartan matrix is invertible and the decomposition matrix has maximum rank.

10.1 The Definition of Brauer Characters

Whenever we work with Brauer characters for a finite group G in characteristic p we will implicitly assume that we have a p-modular system (F, R, k) where both F and k are splitting fields for G and all of its subgroups. This has the implication that F and k both contain a primitive ath root of unity, where a is the l.c.m. of the orders of the p-regular elements of G (the elements of order prime to p). If we wish to define the Brauer character of a kG-module U where k or F do not contain a primitive ath root of unity, we first extend the scalars so that the p-modular system does have this property.

We now examine the relationship between the roots of unity in F and in k. Let us put

$$\mu_F = \{a\text{th roots of 1 in } F\}$$
$$\mu_k = \{a\text{th roots of 1 in } k\}.$$

Lemma 10.1.1. *With this notation,*

(1) $\mu_F \subseteq R$, *and*
(2) the quotient homomorphism $R \to R/(\pi) = k$ *gives an isomorphism* $\mu_F \to \mu_k$.

We will write the bijection between ath roots of unity in F and in k as $\hat{\lambda} \mapsto \lambda$, so that if λ is an ath root of unity in k then $\hat{\lambda}$ is the root of unity in R which maps onto it.

Proof. We have $\mu_F \subseteq R$ since roots of unity have value 1 under the valuation. The polynomial $X^a - 1$ is separable both in $F[X]$ and $k[X]$ since its formal derivative $\frac{d}{dX}(X^a - 1) = aX^{a-1}$ is not zero and has no factors in common with $X^a - 1$, so both μ_F and μ_k are cyclic groups of order a. We claim that the quotient homomorphism $R \to R/(\pi) = k$ gives an isomorphism $\mu_F \to \mu_k$. This is because the polynomial $X^a - 1$ over F reduces to the polynomial that is written the same way over k, and so its linear factors over F must reduce to the complete set of linear factors over k. This gives a bijection between the sets of linear factors. We obtain a bijection between the two groups of roots of unity, in such a way that $X - \hat{\lambda}$ reduces to $X - \lambda$. □

Let $g \in G$ be a p-regular element, and let $\rho : G \to GL(U)$ be a representation over k. Then $\rho(g)$ is diagonalizable, since $k\langle g\rangle$ is semisimple and all eigenvalues of $\rho(g)$ lie in k, being ath roots of unity. If the eigenvalues of $\rho(g)$ are $\lambda_1, \ldots, \lambda_n$ we put

$$\phi_U(g) = \hat{\lambda}_1 + \cdots + \hat{\lambda}_n,$$

and this is the *Brauer character* of U. It is a function that is only defined on the p-regular elements of G, and takes values in a field of characteristic zero, which we may always take to be $\mathbb{C}$.

Example 10.1.2. Working over $\mathbb{F}_2$, the specification $\rho(g) = \left[\begin{smallmatrix} 0 & 1 \\ 1 & 1 \end{smallmatrix}\right]$ provides a 2-dimensional representation U of the cyclic group $\langle g\rangle$ of order 3. The characteristic polynomial of this matrix is $t^2 + t + 1$ and its eigenvalues are the primitive cube roots of unity in $\mathbb{F}_4$. These lift to primitive cube roots of unity in

$\mathbb{C}$, and so $\phi_U(g) = e^{2\pi i/3} + e^{4\pi i/3} = -1$. It is very tempting in this situation to observe that the trace of $\rho(g)$ is 1, which can be lifted to $1 \in \mathbb{C}$, and thus to suppose that $\phi_U(g) = 1$; however, this supposition would be incorrect. A correct statement along these lines appears as part of Exercise 6.

We list the immediate properties of Brauer characters.

Proposition 10.1.3. *Let (F, R, k) be a p-modular system, let G be a finite group, and let U, V, S be finite-dimensional kG-modules.*

(1) $\phi_U(1) = \dim_k U$.
(2) ϕ_U is a class function on p-regular conjugacy classes.
(3) $\phi_U(g^{-1}) = \overline{\phi_U(g)} = \phi_{U^*}(g)$.
(4) $\phi_{U\otimes V} = \phi_U \cdot \phi_V$.
(5) If $0 \to U \to V \to W \to 0$ is a short exact sequence of kG-modules then $\phi_V = \phi_U + \phi_W$. In particular, ϕ_U depends only on the isomorphism type of U. Furthermore, if U has composition factors S, each occurring with multiplicity n_S, then $\phi_U = \sum_S n_S\phi_S$.
(6) If U is liftable to an RG-lattice $\hat{U}$ (so $U = \hat{U}/\pi\hat{U}$) and the ordinary character of $\hat{U}$ is $\chi_{\hat{U}}$, then $\phi_U(g) = \chi_{\hat{U}}(g)$ on p-regular elements $g \in G$.

Proof. (1) In its action on U, the identity has $\dim_k U$ eigenvalues all equal to 1. They all lift to 1 and the sum of the lifts is $\dim_k U$.

(2) This follows because g and xgx^{-1} have the same eigenvalues.

(3) The eigenvalues of g^{-1} on U are the inverses of the eigenvalues of g on U, as are the eigenvalues of g on U^* (since here g acts by the inverse transpose matrix). The lifting of roots of unity is a group homomorphism, so the result follows since if $\hat{\lambda}$ lifts λ then $\hat{\lambda}^{-1}$ lifts λ^{-1}.

(4) If g is a p-regular element then U and V have bases $u_1, \dots, u_r$ and $v_1, \dots, v_s$ consisting of eigenvectors of g with eigenvalues $\lambda_1, \dots, \lambda_r$ and $\mu_r, \dots, \mu_s$, respectively. Now the tensors $u_i \otimes u_j$ form a basis of eigenvectors of $U \otimes V$ with eigenvalues $\lambda_i\mu_j$. Their lifts are $\widehat{\lambda_i\mu_j} = \hat{\lambda}_i\hat{\mu}_j$ since lifting is a group homomorphism, and $\sum_{i,j}\hat{\lambda}_i\hat{\mu}_j = (\sum_i\hat{\lambda}_i)(\sum_j\hat{\mu}_j)$ so that $\phi_{U\otimes V}(g) = \phi_U(g)\phi_V(g)$.

(5) If g is a p-regular element then $k\langle g\rangle$ is semisimple so that $V \cong U \oplus W$ as $k\langle g\rangle$-modules. It follows that the eigenvalues of g on V are the union of the eigenvalues on V and on W (taken with multiplicity), and from this $\phi_V(g) = \phi_U(g) + \phi_W(g)$ follows. If $U \cong U_1$ we may consider the sequence $0 \to U_1 \to U \to 0 \to 0$ to see that $\phi_U = \phi_{U_1}$. The final sentence follows by an inductive argument.

(6) If g is p-regular and acts with eigenvalues $\mu_1, \ldots, \mu_n$ on $\hat{U}$ then g acts on $U = \hat{U}/\pi\hat{U}$ with eigenvalues $\mu_1 + (\pi), \ldots, \mu_n + (\pi)$. Since $\phi_U(g)$ is the sum of the lifts of these last quantities we have $\phi(g) = \mu_1 + \cdots \mu_n = \chi_{\hat{U}}(g)$. □

We arrange the Brauer characters in tables but, unlike the case of ordinary characters, there are now two significant tables that we construct: the table of values of Brauer characters of simple modules, and the table of values of Brauer characters of indecomposable projective modules. By Theorems 9.3.6 and 7.3.9, if F and k are splitting fields, both these tables are square. We will eventually establish that they satisfy orthogonality relations that generalize those for ordinary characters, but we first present some examples.

Example 10.1.4. Let $G = S_3$. We have seen that in both characteristic 2 and characteristic 3 the simple representations of S_3 lift to characteristic zero (Example 9.4.9), and so the Brauer characters of the simple modules form tables that are part of the ordinary character table of S_3. The indecomposable projective modules for a group always lift to characteristic zero, but if we do not have some information such as the Cartan matrix or the decomposition matrix it is hard to know a priori what their characters might be. In the case of S_3, we have already computed the decomposition matrices in Example 9.4.9, and the Cartan matrices in Example 9.5.7. The Brauer characters of indecomposable projective modules are obtained by forming the linear combinations of simple Brauer characters given by the columns of the Cartan matrix. We now present the tables of Brauer characters of the simple and indecomposable projective modules.

S_3
Ordinary characters

g $\|C_G(g)\|$	() 6	(12) 2	(123) 3
χ_1	1	1	1
χ_{sign}	1	-1	1
χ_2	2	0	-1

S_3
Brauer simple $p = 2$

g $\|C_G(g)\|$	() 6	(123) 3
ϕ_1	1	1
ϕ_2	2	-1

S_3
Brauer projective $p = 2$

g $\|C_G(g)\|$	() 6	(123) 3
η_1	2	2
η_2	2	-1

S_3
Brauer simple $p = 3$

g $\|C_G(g)\|$	() 6	(12) 2
ϕ_1	1	1
ϕ_{sign}	1	-1

S_3
Brauer projective $p = 3$

g $\|C_G(g)\|$	() 6	(12) 2
η_1	3	1
η_{sign}	3	-1

Example 10.1.5. When $G = S_4$, the ordinary character table was constructed in Example 3.3.5, and it is as follows.

S_4
Ordinary characters

g	()	(12)	(12)(34)	(1234)	(123)
$\lvert C_G(g)\rvert$	24	4	8	4	3
χ_1	1	1	1	1	1
χ_{sign}	1	-1	1	-1	1
χ_2	2	0	2	0	-1
χ_{3a}	3	-1	-1	1	0
χ_{3b}	3	1	-1	-1	0

In characterstic 2, S_4 has two simple modules, namely, the trivial module and the 2-dimensional module of S_3, made into a module for S_4 via the quotient homomorphism $S_4 \to S_3$. Both of these lift to characteristic zero, and so the Brauer characters of the simple 2-modular simple representations are

S_4
Brauer simple $p = 2$

g	()	(123)
$\lvert C_G(g)\rvert$	24	3
ϕ_1	1	1
ϕ_2	2	-1

which is the same as for S_3. The Brauer characters of the reductions modulo 2 of the ordinary characters of S_4 are

S_4
Reductions from characteristic 0 to 2

g	()	(123)
$\lvert C_G(g)\rvert$	24	3
χ_1	1	1
χ_{sign}	1	1
χ_2	2	-1
χ_{3a}	3	0
χ_{3b}	3	0

We see from this that the sign representation reduces modulo 2 to the trivial representation, and reductions of the two 3-dimensional representations each have the 2-dimensional representation and the trivial representation as composition factors with multiplicity 1. This is because the corresponding Brauer characters are expressible as sums of the simple Brauer characters in this way, and since the simple Brauer characters are visibly independent there is a unique such expression. This expression has to be the expression given by the composition factor multiplicities in the manner of Proposition 10.1.3(5). It follows that the decomposition and Cartan matrices for S_4 at the prime 2 are

$$D = \begin{bmatrix} 1 & 0 \\ 1 & 0 \\ 0 & 1 \\ 1 & 1 \\ 1 & 1 \end{bmatrix} \quad \text{and} \quad C = \begin{bmatrix} 1 & 1 & 0 & 1 & 1 \\ 0 & 0 & 1 & 1 & 1 \end{bmatrix} \begin{bmatrix} 1 & 0 \\ 1 & 0 \\ 0 & 1 \\ 1 & 1 \\ 1 & 1 \end{bmatrix} = \begin{bmatrix} 4 & 2 \\ 2 & 3 \end{bmatrix}.$$

Knowing the Cartan matrix and the simple Brauer characters we may now compute the Brauer characters of the indecomposable projective representations by taking the linear combinations of the simple Brauer characters given by the columns of the Cartan matrix.

In characteristic 3, the trivial representation and the sign representation are distinct 1-dimensional representations, and we also have two nonisomorphic 3-dimensional representations that are the reductions modulo 3 of the two 3-dimensional ordinary representations. This is because these 3-dimensional representations are blocks of defect zero, and by Theorem 9.6.1, they remain simple on reduction modulo 3. This constructs four simple representations in characteristic 3, and this is the complete list by Theorem 9.3.6 because S_4 has four 3 regular conjugacy classes. Thus, the table of Brauer characters of simple modules in characteristic 3 is

S_4
Brauer simple $p = 3$

g $\lvert C_G(g)\rvert$	() 24	(12) 4	(12)(34) 8	(1234) 4
ϕ_1	1	1	1	1
ϕ_{sign}	1	-1	1	-1
ϕ_{3a}	3	-1	-1	1
ϕ_{3b}	3	1	-1	-1

and each is the reduction of a simple module from characteristic zero. The remaining 2-dimensional ordinary representation has Brauer character values $2, 0, 2, 0$, and since this Brauer character is uniquely expressible as a linear combination of simple characters, namely, the trivial Brauer character plus the sign Brauer character, these two 1-dimensional modules are the composition factors of any reduction modulo 3 of the 2-dimensional representation. We see that the decomposition and Cartan matrices for S_4 in characteristic 3 are

$$D = \begin{bmatrix} 1 & 0 & 0 & 0 \\ 0 & 1 & 0 & 0 \\ 1 & 1 & 0 & 0 \\ 0 & 0 & 1 & 0 \\ 0 & 0 & 0 & 1 \end{bmatrix} \quad \text{and} \quad C = D^T D = \begin{bmatrix} 2 & 1 & 0 & 0 \\ 1 & 2 & 0 & 0 \\ 0 & 0 & 1 & 0 \\ 0 & 0 & 0 & 1 \end{bmatrix}.$$

This information now allows us to compute the Brauer characters of the projective representations.

10.2 Orthogonality Relations and Grothendieck Groups

In the last examples, we exploited the fact that the Brauer characters of the simple representations turned out to be independent. In fact they always are, and we prove this as a consequence of orthogonality relations for Brauer characters. The development is similar to what we did with ordinary characters, the extra ingredient being that some of the modules we work with must be projective, since these can be lifted from characteristic p to characteristic 0. We start with an expression for dimensions of homomorphism spaces in terms of character values.

Proposition 10.2.1. *Let (F, R, k) be a p-modular system and let G be a finite group. Suppose that P and U are finite-dimensional kG-modules and that P is projective. Then,*

$$\dim \operatorname{Hom}_{kG}(P, U) = \frac{1}{|G|} \sum_{p\text{-regular } g \in G} \phi_P(g^{-1})\phi_U(g).$$

Proof. We may assume without loss of generality that R is complete. For, if necessary, replace R by its completion at (π), and let F be the field of fractions of R. Making this change does not alter the residue field k or the Brauer characters of representations, so the equation we have to establish is unaltered. Assuming that R is complete we may now lift projective modules from kG to RG.

We make use of the isomorphism $\operatorname{Hom}_{kG}(U, V) \cong \operatorname{Hom}_{kG}(U \otimes_k V^*, k)$ whenever U and V are finite-dimensional kG-modules, which holds since both sides are isomorphic to $(U^* \otimes_k V)^G$, using Proposition 3.1.3 and Lemma 3.2.1. Now $\operatorname{Hom}_{kG}(P, U) \cong \operatorname{Hom}_{kG}(P \otimes U^*, k)$ and $P \otimes U^*$ is a projective kG-module by Proposition 8.1.4. Thus, it lifts to a projective RG-lattice $\widehat{P \otimes_k U^*}$ and we have

$$\begin{aligned}
\dim \operatorname{Hom}_{kG}(P, U) &= \dim \operatorname{Hom}_{kG}(P \otimes U^*, k) \\
&= \operatorname{rank} \operatorname{Hom}_{RG}(\widehat{P \otimes_k U^*}, R) \\
&= \dim \operatorname{Hom}_{FG}(F \otimes_R (\widehat{P \otimes_k U^*}), F) \\
&= \frac{1}{|G|} \sum_{g \in G} \chi_{F \otimes_R (\widehat{P \otimes_k U^*})}(g^{-1})\chi_k(g). \\
&= \frac{1}{|G|} \sum_{g \in G} \chi_{F \otimes_R (\widehat{P \otimes_k U^*})}(g^{-1}).
\end{aligned}$$

We claim that

$$\chi_{F\otimes_R(\widehat{P\otimes_k U^*})}(g^{-1}) = \begin{cases} \phi_P(g^{-1})\phi_U(g) & \text{if } g \text{ is } p\text{-regular} \\ 0 & \text{otherwise,} \end{cases}$$

and from this the result follows. If g is not p-regular, it has order divisible by p and the character value is zero by Proposition 9.6.2, since $\widehat{P\otimes_k U^*}$ is projective. When g is p-regular, we calculate the character value by using the fact that it depends only on the structure of P and U as $k\langle g\rangle$-modules. Since $k\langle g\rangle$ is a semisimple algebra both P and U^* are now projective, and they lift to $R\langle g\rangle$-lattices whose tensor product $\hat{P}\otimes_R \widehat{U^*}$ is isomorphic to $\widehat{P\otimes_k U^*}$ as $R\langle g\rangle$-modules, since both of these are projective covers as $R\langle g\rangle$-modules of $P\otimes_k U^*$. From this, we see that

$$\begin{aligned} \chi_{F\otimes_R\widehat{P\otimes_k U^*}}(g^{-1}) &= \chi_{(F\otimes\hat{P})\otimes(F\otimes\widehat{U^*})}(g^{-1}) \\ &= \chi_{F\otimes\hat{P}}(g^{-1})\chi_{F\otimes\widehat{U^*}}(g^{-1}) \\ &= \phi_P(g^{-1})\phi_U(g) \end{aligned}$$

as required. □

It is convenient to interpret the formula of the last proposition in terms of an inner product on a space of functions, in a similar way to what we did with ordinary characters. Let $p\text{-reg}(G)$ denote the set of conjugacy classes of p-regular elements of G, so that $p\text{-reg}(G) \subseteq \mathrm{cc}(G)$ where the latter denotes the set of all conjugacy classes of G. Since Brauer characters are constant on the conjugacy classes of p-regular elements, we may regard them as elements of the vector space $\mathbb{C}^{p-\mathrm{reg}(G)}$ of functions

$$p\text{-reg}(G) \to \mathbb{C}.$$

We define a Hermitian form on this vector space by

$$\langle \phi, \psi\rangle = \frac{1}{|G|}\sum_{p\text{-regular } g\in G} \overline{\phi(g)}\psi(g)$$

and just as with the similarly defined bilinear form on $\mathbb{C}^{\mathrm{cc}(G)}$, we note that

$$\langle \phi\theta, \psi\rangle = \langle \phi, \theta^*\psi\rangle$$

where $\theta^*(g) = \overline{\theta(g)}$ is the complex conjugate of $\theta(g)$. If ϕ and ψ are the Brauer characters of representations we have $\phi^*(g) = \phi(g^{-1})$ so that

$\langle \phi, \psi \rangle = \langle \psi, \phi \rangle = \langle \phi^*, \psi^* \rangle = \langle \psi^*, \phi^* \rangle$. With the notation of this bilinear form the last result now says that if P and U are finite-dimensional kG-modules with P projective then

$$\dim \operatorname{Hom}_{kG}(P, U) = \langle \phi_P, \phi_U \rangle.$$

Theorem 10.2.2 (Row orthogonality relations for Brauer characters). *Let G be a finite group and k a splitting field for G of characteristic p. Let $S_1, \dots, S_n$ be a complete list of nonisomorphic simple kG-modules, with projective covers $P_{S_1}, \dots, P_{S_n}$. Then the Brauer characters $\phi_{S_1}, \dots, \phi_{S_n}$ of the simple modules form a basis for $\mathbb{C}^{p\text{-reg}(G)}$, as do also the Brauer characters $\phi_{P_{S_1}}, \dots, \phi_{P_{S_n}}$ of the indecomposable projective modules. These two bases are dual to each other with respect to the bilinear form, in that*

$$\langle \phi_{P_{S_i}}, \phi_{S_j} \rangle = \delta_{i,j}.$$

The bilinear form on $\mathbb{C}^{p\text{-reg}(G)}$ is nondegenerate.

Proof. Since P_{S_i} has S_i as its unique simple quotient we have $\operatorname{Hom}_{kG}(P_{S_i}, S_j) = 0$ unless $i = j$, in which case $\operatorname{Hom}_{kG}(P_{S_i}, S_j) \cong \operatorname{End}_{kG}(S_i)$. Because k is a splitting field, this endomorphism ring is k and so $\langle \phi_{P_{S_i}}, \phi_{S_j} \rangle = \delta_{i,j}$. Everything follows from this and the fact that the number of nonisomorphic simple modules equals the number of p-regular conjugacy classes of G. Thus, if $\sum_{i=1}^n \lambda_j \phi_{S_j} = 0$ we have $\lambda_j = \langle \phi_{P_{S_i}}, \sum_{i=1}^n \lambda_j \phi_{S_j} \rangle = 0$, which shows that the ϕ_{S_j} are independent, and hence form a basis. By a similar argument, the $\phi_{P_{S_i}}$ also form a basis. The matrix of the bilinear form with respect to these bases is the identity matrix and it is nondegenerate. □

This result says that the rows of the matrices of Brauer characters of simples and projectives are orthogonal to each other, provided that entries from each column are weighted by the reciprocal of the centralizer order of the element that parametrizes the column, this being the number of conjugates of the element divided by $|G|$. We will give examples of this later. The result implies, of course, that the Brauer characters of the simple kG-modules are linearly independent as functions on the set of p-regular conjugacy classes of G, a fact we observed and used in the earlier examples. It has the consequence that the information contained in a Brauer character is exactly that of composition factor multiplicities and we see this as part (3) of the next result. Since there is the tacit assumption with Brauer characters that we extend the field of definition to include roots of unity, we describe the behavior of composition factors under field extension.

Corollary 10.2.3. *Let $E \supseteq k$ be a field extension and let U and V be finite-dimensional kG-modules.*

(1) *If S and T are nonisomorphic simple kG-modules then the EG-modules $E \otimes_k S$ and $E \otimes_k T$ have no composition factors in common.*
(2) *U and V have the same composition factors as kG-modules if and only if $E \otimes_k U$ and $E \otimes_k V$ have the same composition factors as EG-modules.*
(3) *U and V have the same composition factors if and only if their Brauer characters ϕ_U and ϕ_V are equal.*

Proof. (1) As in Theorem 7.3.8, let $f_S \in kG$ be an idempotent such that $f_S S = S$ and $f_S T = 0$. Now $f_S(E \otimes_k S) = E \otimes_k S$ and $f_S(E \otimes_k T) = 0$, so f_S acts as the identity on all composition factors of $E \otimes_k S$, but as zero on all composition factors of $E \otimes_k T$. Thus, $E \otimes_k S$ and $E \otimes_k T$ can have no composition factors in common.

(2) The implication from left to right is immediate since extending scalars is exact. Conversely, the composition factors of $E \otimes_k U$ are precisely the composition factors of the $E \otimes_k S$ where S is a composition factor of U. Since nonisomorphic S and T give disjoint sets of composition factors in $E \otimes_k S$ and $E \otimes_k T$, if the composition factors of $E \otimes_k U$ and $E \otimes_k V$ are the same, the composition factors of U and V that give rise to them must be the same.

(3) By part (2), we may extend k if necessary to assume that it is a splitting field. We know from part (5) of Proposition 10.1.3 that ϕ_U is the sum of the Brauer characters of the composition factors of U, and similarly for ϕ_V, so if U and V have the same composition factors then $\phi_U = \phi_V$. Conversely, the composition factors of U are determined by ϕ_U since by Theorem 10.2.2, the Brauer characters of simple modules form a basis for $\mathbb{C}^{p-\text{reg}(G)}$. Hence if $\phi_U = \phi_V$ then U and V must have the same composition factors. □

Computing the Brauer character of a module and expressing it as a linear combination of simple characters is a very good way of finding the composition factors of the module, and some examples of this appear in the exercises. As another application, we may deduce that if S is a simple kG-module then $S \cong S^*$ if and only if the Brauer character ϕ_S takes real values, since this is the condition that $\phi_S = \overline{\phi_S} = \phi_{S^*}$.

It also follows from Theorem 10.2.2 that the Brauer characters of the indecomposable projective modules are independent functions on $p-\text{reg}(G)$, and so Brauer characters enable us to distinguish between projective modules. Since the Brauer characters of a module are determined by its composition factors this has the following consequence.

Corollary 10.2.4. *Let P and Q be finitely generated projective kG-modules, where k is a splitting field of characteristic p. Then, P and Q are isomorphic if and only if they have the same composition factors. The Cartan matrix is invertible. The decomposition matrix has maximum rank. The mapping e in the cde triangle is injective.*

By invertibility of the Cartan matrix we mean invertibility as a matrix with entries in $\mathbb{Q}$ or, equivalently, that its determinant is nonzero.

Proof. Write $P = P_{S_1}^{a_1} \oplus \cdots \oplus P_{S_n}^{a_n}$ and $Q = P_{S_1}^{b_1} \oplus \cdots \oplus P_{S_n}^{b_n}$ so that $\phi_P = \sum_{i=1}^{n} a_i \phi_{P_{S_i}}$ and $\phi_Q = \sum_{i=1}^{n} b_i \phi_{P_{S_i}}$. Then $P \cong Q$ if and only if $a_i = b_i$ for all i, if and only if $\sum_{i=1}^{n} a_i \phi_{P_{S_i}} = \sum_{i=1}^{n} b_i \phi_{P_{S_i}}$ since the $\phi_{P_{S_i}}$ are linearly independent, if and only if $\phi_P = \phi_Q$. By the last corollary, this happens if and only if P and Q have the same composition factors.

We claim that the kernel of the Cartan homomorphism $c : K_0(kG) \to G_0(kG)$ is zero. Any element of $K_0(kG)$ can be written $[P] - [Q]$ where P and Q are projective modules, and such an element lies in the kernel if and only if P and Q have the same composition factors. This forces $P \cong Q$, so that the kernel is zero. It now follows that the Cartan homomorphism is an isomorphism.

For the assertion about the decomposition matrix we use the fact (Corollary 9.5.6) that $C = D^T D$. Thus, $\operatorname{rank} C \leq \operatorname{rank} D$. The number of columns of D equals the number of columns of C, and this number must be the rank of D. By Theorem 9.5.4, the matrix of e is D, so that e is injective. □

A finer result than this is true. The determinant of the Cartan matrix over a splitting field of characteristic p is known to be a power of p, and the decomposition map $d : G_0(FG) \to G_0(kG)$ is, in fact, a surjective homomorphism of abelian groups. The assertion in Corollary 10.2.4 that its matrix has maximum rank implies only that the cokernel of the decomposition map is finite. These stronger results may be proved by a line of argument that originates with the induction theorem of Brauer, a result that we have omitted. The surjectivity of of the decomposition map is also implied by the Fong–Swan–Rukolaine Theorem 9.4.12 in the case of p-solvable groups.

We may write the row orthogonality relations in various ways. The most rudimentary way, for simple kG-modules S and T, is the equation

$$\frac{1}{|G|} \sum_{p\text{-regular } g \in G} \phi_{P_T}(g^{-1}) \phi_S(g) = \begin{cases} 1 & \text{if } S \cong T \\ 0 & \text{otherwise.} \end{cases}$$

We can also express this as a matrix product. Let Φ be the table of Brauer character values of simple kG-modules, Π the table of Brauer character values of indecomposable projective modules, and let B be the diagonal matrix whose entries are $\frac{1}{|C_G(g)|}$ as g ranges through the p-regular classes. The row orthogonality relations may now be written as

$$\overline{\Pi} B \Phi^T = I.$$

Note that the independence of the Brauer characters of simple modules and of projective modules is equivalent to the property that Φ and Π are invertible matrices, and also that $\Pi = C\Phi$ where C is the Cartan matrix. From this, we now deduce the column orthogonality relations for Brauer characters, which say that each column of Π is orthogonal to the remaining columns of $\overline{\Phi}$, and has product with the corresponding column of $\overline{\Phi}$ equal to the order of the centralizer of the group element that indexes the column.

Proposition 10.2.5 (Column orthogonality relations for Brauer characters). *With the preceding notation,*

$$\overline{\Phi}^T \Pi = \begin{bmatrix} |C_G(x_1)| & 0 & \cdots & 0 \\ 0 & |C_G(x_2)| & & \\ \vdots & & \ddots & \vdots \\ 0 & & \cdots & |C_G(x_n)| \end{bmatrix}$$

where $x_1, \dots, x_n$ are representatives of the p-regular conjugacy classes of elements of G. Thus

$$\sum_{\text{simple } S} \phi_S(g^{-1})\phi_{P_S}(h) = \begin{cases} |C_G(g)| & \text{if } g \text{ and } h \text{ are conjugate} \\ 0 & \text{otherwise.} \end{cases}$$

Proof. We take the equation $\overline{\Pi} B \Phi^T = I$ and multiply on the left by $\overline{\Pi}^{-1}$ and on the right by $(\Phi^T)^{-1}$ to get $B = \overline{\Pi}^{-1}(\Phi^T)^{-1}$. Inverting both sides gives $B^{-1} = \Phi^T \overline{\Pi}$. We finally take the complex conjugate of both sides, observing that B is a real matrix. □

The orthogonality relations can be used to determine the composition factors of a representation in a similar way to the procedure with ordinary characters. The idea is that we obtain the composition factor multiplicity of a simple kG-module S in another module U as $\langle \phi_{P_S}, \phi_U \rangle$ (assuming k is a splitting field). However, this possibility is less useful than in characteristic zero because we need to know the Brauer characters of the indecomposable projective P_S to make it work. Usually, we would only have this information once we already have fairly complete information about the simple modules, so that we know the decomposition matrix, the Cartan matrix and hence the table Π of Brauer characters of projectives. Such an approach is less useful in constructing the table of simple Brauer characters. By contrast, in characteristic zero we can test for simplicity of a character and subtract known character summands from a character of interest without complete character table information.

The orthogonality relations can also be used to decompose a module that is known to be projective into its projective summands, and here they are perhaps more useful. Working over a splitting field, the idea is that if P is a projective kG-module and S is a simple kG-module, then the multiplicity of P_S as a direct summand of P equals $\langle \phi_P, \phi_S \rangle$ (assuming k is a splitting field).

Example 10.2.6. We present an example in which the orthogonality relations for Brauer characters are used to find the composition factors of a module, and also the decomposition of a projective module into indecomposable projective summands. We have already seen that the table Φ of simple Brauer characters of S_4 in characteristic 3 is

S_4
Brauer simple $p = 3$

g $\lvert C_G(g)\rvert$	() 24	(12) 4	(12)(34) 8	(1234) 4
ϕ_1	1	1	1	1
ϕ_{sign}	1	−1	1	−1
ϕ_{3a}	3	−1	−1	1
ϕ_{3b}	3	1	−1	−1

and the Cartan matrix $C = D^T D$ is

S_4
Cartan matrix $p = 3$

	η_1	η_{sign}	η_{3a}	η_{3b}
ϕ_1	2	1	0	0
ϕ_{sign}	1	2	0	0
ϕ_{3a}	0	0	1	0
ϕ_{3b}	0	0	0	1

Thus the table Π of Brauer characters of projectives is

S_4
Brauer projective $p = 3$

g $\lvert C_G(g)\rvert$	() 24	(12) 4	(12)(34) 8	(1234) 4
η_1	3	1	3	1
η_{sign}	3	−1	3	−1
η_{3a}	3	−1	−1	1
η_{3b}	3	1	−1	−1

In these tables of Brauer characters of projective modules, we are writing $\eta_S = \phi_{P_S}$ for the Brauer character of the projective cover of S. Consider the first of the 3-dimensional simple modules, which we denote $3a$. It is a block of defect zero, being the reduction mod 3 of a 3-dimensional simple module in characteristic zero, and hence is projective. The Brauer character of its tensor square is

g $\lvert C_G(g)\rvert$	$()$ 24	(12) 4	$(12)(34)$ 8	(1234) 4
$\phi_{3a\otimes 3a} = \phi_{3a}\phi_{3a}$	9	1	1	1

Taking first the inner products $\langle \phi_{P_S}, \phi_{3a\otimes 3a}\rangle$ with the Brauer characters of projectives, we obtain the numbers

$$\frac{27}{24} + \frac{1}{4} + \frac{3}{8} + \frac{1}{4} = 2$$
$$\frac{27}{24} - \frac{1}{4} + \frac{3}{8} - \frac{1}{4} = 1$$
$$\frac{27}{24} - \frac{1}{4} - \frac{1}{8} + \frac{1}{4} = 1$$
$$\frac{27}{24} + \frac{1}{4} - \frac{1}{8} - \frac{1}{4} = 1$$

which means that $3a \otimes 3a$ has composition factors 1 (with multiplicity 2), ϵ, $3a$, $3b$ where ϵ is the sign representation. On the other hand, $3a \otimes 3a$ is itself projective, and we get more information if we take the inner products $\langle \phi_{3a\otimes 3a}, \phi_S\rangle$ with the simple Brauer characters, giving numbers

$$\frac{9}{24} + \frac{1}{4} + \frac{1}{8} + \frac{1}{4} = 1$$
$$\frac{9}{24} - \frac{1}{4} + \frac{1}{8} - \frac{1}{4} = 0$$
$$\frac{27}{24} - \frac{1}{4} - \frac{1}{8} + \frac{1}{4} = 1$$
$$\frac{27}{24} + \frac{1}{4} - \frac{1}{8} - \frac{1}{4} = 1.$$

This shows that $3_a \otimes 3_a \cong P_1 \oplus 3a \oplus 3b$ as kG-modules.

10.3 The cde Triangle in Terms of Brauer Characters

The spaces of functions defined on conjugacy classes of G that we have been considering are very closely related to the Grothendieck groups defined in Section 9.5. As always, (F, R, k) is a p-modular system containing primitive ath roots of unity, where a is the l.c.m. of the orders of the p-regular elements of G.

We know that as S ranges through the isomorphism classes of simple kG-modules the Grothendieck group $G_0(kG)$ has the elements $[S]$ as a basis, the group $K_0(kG)$ has the elements $[P_S]$ as a basis, and the space $\mathbb{C}^{p-\text{reg}(G)}$ has the Brauer characters ϕ_S and also the Brauer characters ϕ_{P_S} as bases. Similarly, when T ranges through the isomorphism classes of simple FG-modules the Grothendieck group $G_0(FG)$ has the elements $[T]$ as a basis, and the space $\mathbb{C}^{\text{cc}(G)}$ has the ordinary characters χ_T as a basis. There are thus group homomorphisms

$$K_0(kG) \to \mathbb{C}^{p-\text{reg}(G)},$$
$$G_0(kG) \to \mathbb{C}^{p-\text{reg}(G)},$$
$$G_0(FG) \to \mathbb{C}^{\text{cc}(G)}$$

defined on the basis elements $[P] \in K_0(kG)$, $[S] \in G_0(kG)$, and $[T] \in G_0(FG)$, by

$$[P] \mapsto \phi_P, \quad [S] \mapsto \phi_S, \quad \text{and} \quad [T] \mapsto \chi_T$$

and in fact these same formulas hold whenever P is an arbitrary finitely generated projective module and S, T are arbitrary finitely generated modules. Extending scalars from $\mathbb{Z}$ to $\mathbb{C}$ they define isomorphisms of vector spaces

$$\mathbb{C} \otimes_{\mathbb{Z}} K_0(kG) \cong \mathbb{C}^{p-\text{reg}(G)},$$
$$\mathbb{C} \otimes_{\mathbb{Z}} G_0(kG) \cong \mathbb{C}^{p-\text{reg}(G)},$$
$$\mathbb{C} \otimes_{\mathbb{Z}} G_0(FG) \cong \mathbb{C}^{\text{cc}(G)}.$$

There is further structure that we have not mentioned yet, which is that there is a multiplication defined on the three Grothendieck groups by $[U] \cdot [V] := [U \otimes V]$, the same formula working for all three groups but with the modules U and V interpreted suitably over FG or kG and projective or not as appropriate. Since tensor product over the ground field preserves exact sequences, this definition makes sense for arbitrary (finitely generated) modules. This multiplication makes $G_0(FG)$ and $G_0(kG)$ into rings with identity, the identity element being the class of the trivial module. We see that the isomorphisms just given preserve the product structure.

Proposition 10.3.1. *After extending scalars to $\mathbb{C}$, the Grothendieck groups $\mathbb{C} \otimes_{\mathbb{Z}} G_0(FG)$ and $\mathbb{C} \otimes_{\mathbb{Z}} G_0(kG)$ are semisimple rings.*

Proof. The isomorphisms to $\mathbb{C}^{\text{cc}(G)}$ and $\mathbb{C}^{p-\text{reg}(G)}$ are ring isomorphisms, and the latter algebras are direct sums of copies of $\mathbb{C}$, which are semisimple. □

This means that after extending scalars to $\mathbb{C}$ we may identify the cde triangle of Section 9.5 with the following diagram of vector spaces:

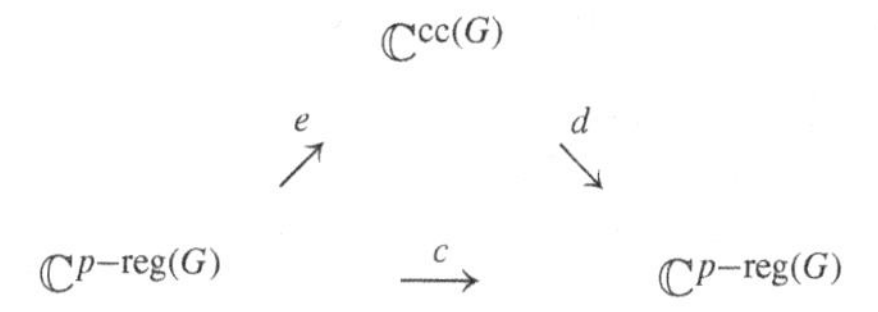

where, as before, the maps c, d, e have matrices C, D^T, and D, respectively.

Proposition 10.3.2. *Let (F, R, k) be a splitting p-modular system for G and consider the cde triangle as a diagram of complex vector spaces, as just indicated. The mappings d and e are described as follows: if $\phi \in \mathbb{C}^{p-\text{reg}(G)}$ then $e(\phi)$ is the function that is the same as ϕ on p-regular conjugacy classes and is zero on the other conjugacy classes. If $\chi \in \mathbb{C}^{\text{cc}(G)}$ then $d(\chi)$ is the restriction of χ to the p-regular conjugacy classes. The Cartan homomorphism c is an isomorphism, the decomposition map d is surjective, and e is injective. The image of e is the space of class functions that are nonzero only on p-regular classes.*

Proof. The description of $e(\phi)$ follows from Proposition 9.6.2 and the definition of e, whereas the description of d comes from part (6) of Proposition 10.1.3. The Cartan homomorphism is an isomorphism because the Cartan matrix is nonsingular, and because the decomposition matrix D has maximal rank, d is surjective, and e is injective. The image of e is a space whose dimension is the number of p-regular conjugacy classes of G, and it is contained in the space of maps whose support is the set of p-regular conjugacy classes, so we must have equality. □

Corollary 10.3.3. *Let (F, R, k) be a splitting p-modular system for G. Two finitely generated projective RG-modules P and Q are isomorphic if and only if $F \otimes_R P$ and $F \otimes_R Q$ are isomorphic as FG-modules, which happens if and only if the ordinary characters of P and Q are equal.*

Proof. By Corollary 10.2.4, the elements $[P_S]$ form a basis of $G_0(kG)$. They are sent by e to the $[\hat{P}_S]$, and since e is injective the latter elements are linearly independent in $G_0(FG)$. From the identification of $\mathbb{C} \otimes_{\mathbb{Z}} G_0(FG)$ with the class functions on G, we see that the ordinary characters of the $\hat{P}_S$ are linearly independent. Since P and Q are direct sums of the $\hat{P}_S$ by Proposition 9.4.5, the multiplicities of the direct summands and hence the isomorphism types of P and Q are determined by their ordinary characters. □

We conclude by explaining how the decomposition matrix is computed using Brauer characters. In fact, we have already seen this done in Examples 10.1.4 and 10.1.5, but the missing ingredient there was that we did not know that the Brauer characters of the simple representations are linearly independent. In those examples this fact was observed directly by inspection of the values of the characters. The entries of the decomposition matrix are the composition factor multiplicities of the simple kG-modules S when the simple FG-modules T are reduced to k. Given the Brauer characters ϕ_S and an ordinary character χ_T the Brauer character of a reduction of T is $d(\chi_T)$, which is the truncation of χ_T to the p-regular elements of G. Expressing $d(\chi_T) = \sum_S d_{TS}\phi_S$ as a linear combination of the ϕ_S gives the decomposition numbers d_{TS}. This linear combination may be computed by the usual techniques of linear algebra since the Brauer characters of the indecomposable projective modules are not usually known at this point, making the orthogonality relations unvailable. This approach is generally the most effective way to compute the decomposition matrix, and hence the Cartan matrix.

10.4 Summary of Chapter 10

- Brauer characters take values in $\mathbb{C}$. They satisfy similar properties to ordinary characters, except that they are only defined on p-regular elements of G.
- Two Brauer characters ϕ_U and ϕ_V are equal if and only if U and V have the same composition factors.
- The tables of Brauer characters of simple modules and of projective modules satisfy orthogonality relations.
- Brauer characters are a very useful tool in computing the decomposition and Cartan matrices.
- The Cartan matrix is invertible. The decomposition matrix has maximal rank and the mapping e in the cde triangle is injective.

10.5 Exercises for Chapter 10

1. The simple group $GL(3,2)$ has order $168 = 8 \cdot 3 \cdot 7$. The following is part of its ordinary character table (the numbers that label the conjugacy classes of elements in the top row indicate the order of the elements):

$GL(3,2)$
Ordinary characters

g	1	2	4	3	$7a$	$7b$
$\lvert C_G(g)\rvert$	168	8	4	3	7	7
χ_1	1	1	1	1	1	1
χ_3	3	-1	1		α	$\overline{\alpha}$
$\chi_?$						
χ_6	6	2	0		-1	-1
χ_7	7	-1	-1	1		
χ_8	8			-1	1	1

Here $\alpha = \zeta_7 + \zeta_7^2 + \zeta_7^4$ where $\zeta_7 = e^{2\pi i/7}$. Note that $\alpha^2 = \overline{\alpha} - 1$ and $\alpha\overline{\alpha} = 2$.

(a) Obtain the complete character table of $GL(3,2)$.
(b) Compute the table of Brauer characters of simple $\mathbb{F}_2[GL(3,2)]$-modules.
(c) Find the decomposition matrix and Cartan matrix of $GL(3,2)$ at the prime 2.
(d) Write down the table of Brauer characters of projective $\mathbb{F}_2[GL(3,2)]$-modules.
(e) Determine the direct sum decomposition of the module $8 \otimes 3$ (where 8 and 3 denote the simple $\mathbb{F}_2[GL(3,2)]$-modules of those dimensions shown in the table), as a direct sum of indecomposable modules.
(f) Determine the composition factors of $3 \otimes 3$ and $3 \otimes 3^*$.
[Note that 3 can be taken to be the natural 3-dimensional $\mathbb{F}_2[GL(3,2)]$-module. One approach is to use the orthogonality relations.]

2. It so happens that $GL(3,2) \cong PSL(2,7)$. In this exercise, we regard this group as $PSL(2,7)$ and assume the construction of the simple modules over $\mathbb{F}_7$ given in Exercise 25 of Chapter 6.

(a) Construct the table of Brauer characters of simple $\mathbb{F}_7[PSL(2,7)]$-modules.

[It will help to observe that all elements of order prime to 7 in $SL(2,7)$ act semisimply on $\mathbb{F}_7^2$ by Maschke's Theorem. From this, it follows that there is a unique element of order 2 in $SL(2,7)$, namely, $-I$, since its eigenvalues must both be -1, and it is semisimple. We may deduce that an element of order 8 in $SL(2,7)$ represents an element of order 4 in $PSL(2,7)$, and its square represents an element of order 2, since $-I$ represents 1 in $PSL(2,7)$. We may now determine the eigenvalues of 7-regular elements in their action on the symmetric powers of the 2-dimensional module.]

(b) Compute the decomposition and Cartan matrices for $PSL(2,7)$ in characteristic 7. Show that the projective cover of the trivial module, P_1, has just four submodules, namely, 0, P_1 and two others.

3. Let k be a splitting field for D_{30} in characteristic 3. Using the fact that $D_{30} \cong C_3 \rtimes D_{10}$ has a normal Sylow 3-subgroup, show that kD_{30} has four simple modules of dimensions 1, 1, 2, and 2 and that the values of the Brauer characters on the simple modules are the same as the ordinary character table of D_{10}. We will label the simple modules k_1, k_ϵ, Ua, and Ub, respectively. Using the method of Proposition 8.3.3, show that the indecomposable projectives have the form

$$P_{k_1} = \begin{matrix} \bullet k_1 \\ | \\ \bullet k_\epsilon \\ | \\ \bullet k_1 \end{matrix} \quad P_{k_\epsilon} = \begin{matrix} \bullet k_\epsilon \\ | \\ \bullet k_1 \\ | \\ \bullet k_\epsilon \end{matrix} \quad P_{Ua} = \begin{matrix} \bullet Ua \\ | \\ \bullet Ua \\ | \\ \bullet Ua \end{matrix} \quad P_{Ub} = \begin{matrix} \bullet Ub \\ | \\ \bullet Ub \\ | \\ \bullet Ub \end{matrix}$$

Deduce that the Cartan matrix is

$$C = \begin{bmatrix} 2 & 1 & 0 & 0 \\ 1 & 2 & 0 & 0 \\ 0 & 0 & 3 & 0 \\ 0 & 0 & 0 & 3 \end{bmatrix}$$

and that $kD_{20} \cong kD_6 \oplus M_2(kC_3) \oplus M_2(kC_3)$ as rings. Compute the decomposition matrix and calculate a second time the Cartan matrix as $D^T D$.

4. Let H be a subgroup of G. If $\theta \in \mathbb{C}^{p-\text{reg}(H)}$ is a function defined on the set of p-regular conjugacy classes of elements of H we may define an *induced* function $\theta \uparrow_H^G$ on the p-regular conjugacy classes of elements of G by

$$\begin{aligned} \theta \uparrow_H^G (g) &= \frac{1}{|H|} \sum_{\substack{t \in G \\ t^{-1}gt \in H}} \theta(t^{-1}gt) \\ &= \sum_{\substack{t \in [G/H] \\ t^{-1}gt \in H}} \theta(t^{-1}gt). \end{aligned}$$

This is the same as the formula in Proposition 4.3.5 that was used to define induction of ordinary class functions.

(a) Let U be a finite-dimensional kH-module with Brauer character ϕ_U, where k is a field of characteristic p. Prove that $\phi_{U\uparrow_H^G} = \phi_U \uparrow_H^G$, and that $e(\theta \uparrow_H^G) = e(\theta) \uparrow_H^G$ and $d(\chi \uparrow_H^G) = d(\chi) \uparrow_H^G$ if χ is a class function.

(b) Similarly, define the restriction $\psi \downarrow_H^G$ where $\psi \in \mathbb{C}^{p-\text{reg}(G)}$ and show that similar formulas hold.

(c) Using the Hermitian forms defined on these spaces of functions, show that $\langle \theta \uparrow_H^G, \psi \rangle = \langle \theta, \psi \downarrow_H^G \rangle$ always holds.

5. Let (F, R, k) be a splitting p-modular system for G. Let P and Q be finitely-generated projective RG-modules such that $F \otimes_R P \cong F \otimes_R Q$ as FG-modules. Show that $P \cong Q$.

6. Let (F, R, k) be a p-modular system for some prime p. For the following two statements, show by example that the first is false in general, and that the second is true.

(a) Suppose that $\hat{A}$ is an invertible matrix with entries in R and let A be the matrix with entries in k obtained by reducing the entries of $\hat{A}$ modulo (π). Then the eigenvalues of $\hat{A}$ are the lifts of the eigenvalues of A.

(b) Suppose that $\hat{A}$ is an invertible matrix with entries in R and let A be the matrix with entries in k obtained by reducing the entries of $\hat{A}$ module (π). Suppose further that $\hat{A}$ has finite order, and that this order is prime to p. Then the eigenvalues of $\hat{A}$ are the lifts of the eigenvalues of A. Furthermore, the order of A is the same as the order of $\hat{A}$.

7. Show that when k is an arbitrary field of characteristic p, the number of isomorphism classes of simple kG-modules is at most the number of p-regular conjugacy classes in G.

8. Let (F, R, k) be a splitting p-modular system for G. Show that the primitive central idempotents in kG are given by a formula similar to that of Theorem 3.6.2:

$$e_i = \sum_{g \in G} \left(\frac{d_i}{|G|} \chi_i(g^{-1}) + (\pi) \right) g \in kG,$$

where $\chi_1, \ldots, \chi_r$ are the simple characters of G in characteristic zero.

9. (Modular version of Molien's Theorem. Copy the approach of Chapter 4 Exercise 19 and assume Exercise 6.) Let G be a finite group and let (F, R, k) be a splitting p-modular system for G, for some prime p.

(a) Let $\rho : G \to GL(V)$ be a representation of G over k, and for each n let $\phi_{S^n(V)}$ be the Brauer character of the nth symmetric power of V. For each p-regular $g \in G$, let $\widehat{\rho(g)}$ be a matrix with entries in R of the same order as $\rho(g)$ that reduces modulo (π) to $\rho(g)$. Show that for each p-regular $g \in G$ there is an equality of formal power series

$$\sum_{n=0}^{\infty} \phi_{S^n(V)}(g) t^n = \frac{1}{\det(1 - t\widehat{\rho(g)})}.$$

Here t is an indeterminate, and the determinant that appears in this expression is of a matrix with entries in the polynomial ring $\mathbb{C}[t]$, so that the determinant is a polynomial in t. On expanding the rational function on the right, we obtain a formal power series that is asserted to be equal to the formal power series on the left.

[Choose a basis for V so that g acts diagonally, with eigenvalues $\xi_1, \ldots, \xi_d$. Show that on both sides of the equation the coefficient of t^n is equal to $\sum_{i_1+\cdots+i_d=n} \xi_1^{i_1} \cdots \xi_d^{i_d}$.]

(b) If W is a simple kG-module, we may write the multiplicity of W as a composition factor of $S^n(V)$ as $\langle \phi_{P_W}, \phi_{S^n(V)} \rangle$ and consider the formal power series

$$M_V(W) = \sum_{i=0}^{\infty} \langle \phi_{P_W}, \phi_{S^n(V)} \rangle t^n.$$

Show that

$$M_V(W) = \frac{1}{|G|} \sum_{g \in p\text{-reg}(G)} \frac{\phi_{P_W}(g^{-1})}{\det(1 - t\widehat{\rho(g)})}.$$

(c) When $G = S_3$ and V is the 2-dimensional simple $\mathbb{F}_2 S_3$-module, show that

$$\begin{aligned} M_V(k) &= \frac{1+t^3}{(1-t^2)(1-t^3)} \\ &= 1 + t^2 + 2t^3 + t^4 + 2t^5 + 3t^6 + 2t^7 + 3t^8 + 4t^9 + 3t^{10} + \cdots \\ M_V(V) &= \frac{t(1+t)}{(1-t^2)(1-t^3)} \\ &= t + t^2 + t^3 + 2t^4 + 2t^5 + 2t^6 + 3t^7 + 3t^8 + 3t^9 + 4t^{10} + \cdots . \end{aligned}$$

Show that $S^8(V)$ is the direct sum as a $\mathbb{F}_2 S_3$-module of three copies of V and a 3-dimensional module whose composition factors are all trivial. [Note that the ring of invariants $(S^\bullet(V))^G$ has series $\frac{1}{(1-t^2)(1-t^3)}$ in this case, that is not the same as $M_V(k)$.]

(d) When $G = GL(3,2)$ and V is the natural 3-dimensional $\mathbb{F}_2 GL(3,2)$-module, use the information from Exercise 1 to show that

$$\begin{aligned} M_V(k) &= \frac{1 - t + t^4 - t^7 + t^8}{(1-t)(1-t^3)(1-t^7)}, \\ M_V(V) &= \frac{t(1 - t^3 + t^4 + t^5)}{(1-t)(1-t^3)(1-t^7)}, \\ M_V(V^*) &= \frac{t^2(1 + t - t^2 + t^5)}{(1-t)(1-t^3)(1-t^7)}, \\ M_V(8) &= \frac{t^4}{(1-t)(1-t^3)(1-t^7)}. \end{aligned}$$

11

Indecomposable Modules

We recall from Chapter 6 that a module is said to be *indecomposable* if it cannot be expressed as the direct sum of two nonzero submodules. We have already considered many indecomposable modules, but mainly only those that are projective or simple. An exception to this in Chapter 6 is that we classified in their entirety the indecomposable modules for a cyclic p-group over a field of characteristic p. To understand indecomposable modules in general might be considered one of the goals of representation theory, because such understanding would enable us to say something about all finite-dimensional representations, as they are direct sums of indecomposables. We will explain in this chapter that this goal is not realistic, but that nevertheless there is much that can be said. We try to give an overview of the theory, much of which has as its natural context the abstract representation theory of algebras. We can only go into detail with a small part of this general theory, focusing on what it has to say for group representations, and the particular techniques that are available in this case.

Our first task is to understand the structure of the endomorphism rings of indecomposable modules. This leads to the Krull–Schmidt Theorem, which specifies the extent to which decompositions as direct sums of indecomposable modules are unique. We go on to describe in detail the indecomposable modules in a particular situation: the case of a group with a cyclic normal Sylow p-subgroup. We next describe the theory of relative projectivity and discuss representation type, from which the conclusion is that most of the time we cannot hope to get a description of all indecomposable modules. Vertices, sources, Green correspondence, and the Heller operator complete the topics covered.

11.1 Indecomposable Modules, Local Rings, and the Krull–Schmidt Theorem

We start with an extension of Proposition 7.2.1.

Proposition 11.1.1. *Let U be a module for a ring A with a 1. Expressions*

$$U = U_1 \oplus \cdots \oplus U_n$$

as a direct sum of submodules biject with expressions $1_U = e_1 + \cdots + e_n$ for the identity $1_U \in \mathrm{End}_A(U)$ as a sum of orthogonal idempotents. Here e_i is obtained from U_i as the composite of projection and inclusion $U \to U_i \to U$, and U_i is obtained from e_i as $U_i = e_i(U)$. The summand U_i is indecomposable if and only if e_i is primitive.

Proof. We must check several things. Two constructions are indicated in the statement of the proposition: given a direct sum decomposition of U, we obtain an idempotent decomposition of 1_U, and vice-versa. It is clear that the idempotents constructed from a module decomposition are orthogonal and sum to 1_U. Conversely, given an expression $1_U = e_1 + \cdots + e_n$ as a sum of orthogonal idempotents, every element $u \in U$ can be written $u = e_1u + \cdots + e_nu$ where $e_iu \in e_iU = U_i$. In any expression $u = u_1 + \cdots u_n$ with $u_i \in e_iU$, we have $e_ju_i \in e_je_iU = 0$ if $i \neq j$ so $e_iu = e_iu_i = u_i$, and this expression is uniquely determined. Thus, the expression $1_U = e_1 + \cdots + e_n$ gives rise to a direct sum decomposition.

We see that U_i decomposes as $U_i = V \oplus W$ if and only if $e_i = e_V + e_W$ can be written as a sum of orthogonal idempotents, and so U_i is indecomposable if and only if e_i is primitive. □

Corollary 11.1.2. *An A-module U is indecomposable if and only if the only nonzero idempotent in* $\mathrm{End}_A(U)$ *is* 1_U.

Proof. From the proposition, U is indecomposable if and only if 1_U is primitive, and this happens if and only if 1_U and 0 are the only idempotents in $\mathrm{End}_A(U)$. This last implication in the forward direction follows since any idempotent e gives rise to an expression $1_U = e + (1_U - e)$ as a sum of orthogonal idempotents, and in the opposite direction there simply are no nontrivial idempotents to allow us to write $1_U = e_1 + e_2$. □

The equivalent conditions of the next result are satisfied by the endomorphism ring of an indecomposable module, but we first present them in abstract. The connection with indecomposable modules will be presented in Corollary 11.1.5.

Proposition 11.1.3. *Let B be a ring with 1. The following are equivalent:*

(1) *B has a unique maximal left ideal.*
(2) *B has a unique maximal right ideal.*
(3) *$B/\operatorname{Rad}(B)$ is a division ring.*
(4) *The set of elements in B that are not invertible forms a left ideal.*
(5) *The set of elements in B that are not invertible forms a right ideal.*
(6) *The set of elements in B that are not invertible forms a 2-sided ideal.*

Proof. (1) $\Rightarrow$ (3) Let I be the unique maximal left ideal of B. Since $\operatorname{Rad}(B)$ is the intersection of the maximal left ideals, it follows that $I = \operatorname{Rad}(B)$. If $a \in B - I$ then Ba is a left ideal not contained in I, so $Ba = B$. Thus, there exists $x \in B$ with $xa = 1$. Furthermore, $x \notin I$, so $Bx = B$ also and there exists $y \in B$ with $yx = 1$. Now $yxa = a = y$ so a and x are 2-sided inverses of one another. This implies that B/I is a division ring.

(1) $\Rightarrow$ (6) The argument just presented shows that the unique maximal left ideal I is in fact a 2-sided ideal, and every element not in I is invertible. This implies that every noninvertible element is contained in I. Equally, no element of I can be invertible, so I consists of the noninvertible elements, and they form a 2-sided ideal.

(3) $\Rightarrow$ (1) If I is a maximal left ideal of B then $I \supseteq \operatorname{Rad}(B)$ and so corresponds to a left ideal of $B/\operatorname{Rad}(B)$, which is a division ring. It follows that either $I = \operatorname{Rad}(B)$ or $I = B$, and so $\operatorname{Rad}(B)$ is the unique maximal left ideal of B.

(4) $\Rightarrow$ (1) Let J be the set of noninvertible elements of B and I a maximal left ideal. Then no element of I is invertible, so $I \subseteq J$. Since J is an ideal, we have equality, and I is unique.

(6) $\Rightarrow$ (4) This implication is immediate, and so we have established the equivalence of conditions (1), (3), (4), and (6).

Since conditions (3) and (6) are left-right symmetric, it follows that they are also equivalent to conditions (2) and (5), by analogy with the equivalence with (1) and (4). □

We will call a ring B satisfying any of the equivalent conditions of the last proposition a *local ring*. Any commutative ring that is local in the usual sense (i.e., it has a unique maximal ideal) is evidently local in this noncommutative sense. As for noncommutative examples of local rings, we see from Proposition 6.3.3 part (3) that if G is a p-group and k is a field of characteristic p then the group algebra kG is a local ring. This is because its radical is the augmentation ideal and the quotient by the radical is k, which is a division ring, thus verifying condition (3) of Proposition 11.1.3.

We have seen in Corollary 11.1.2 a characterization of indecomposable modules as modules whose endomorphism ring only has idempotents 0 and 1. We now make the connection with local rings.

Proposition 11.1.4. *(1) In a local ring, the only idempotents are 0 and 1.*

(2) Suppose that B is an R-algebra that is finitely generated as an R-module, where R is a complete discrete valuation ring or a field. If the only idempotents in B are 0 and 1 then B is a local ring.

Proof. (1) In a local ring B, any idempotent e other than 0 and 1 would give a nontrivial direct sum decomposition of $B = Be \oplus B(1 - e)$ as left B-modules, and so B would have more than one maximal left ideal, a contradiction.

(2) Suppose that 0 and 1 are the only idempotents in B, and let (π) be the maximal ideal of R. Just as in the proof of part (1) of Proposition 9.4.1, we see that π annihilates every simple B-module, and so $\pi B \subseteq \mathrm{Rad}(B)$. This implies that $B/\mathrm{Rad}(B)$ is a finite-dimensional $R/(\pi)$-algebra. If $e \in B/\mathrm{Rad}(B)$ is idempotent then by the argument of Proposition 9.4.3 it lifts to an idempotent of B, which must be 0 or 1. Since e is the image of this lifting, it must also be 0 or 1. Now $B/\mathrm{Rad}(B) \cong M_{n_1}(\Delta_1) \oplus \cdots \oplus M_{n_t}(\Delta_t)$ for certain division rings Δ_i, since this is a semisimple algebra, and the only way this algebra would have just one nonzero idempotent is if $t = 1$ and $n_1 = 1$. This shows that condition (3) of the last proposition is satisfied. □

We put these pieces together:

Corollary 11.1.5. *Let U be a module for a ring A.*

(1) If $\mathrm{End}_A(U)$ is a local ring then U is indecomposable.

(2) Suppose that R is a complete discrete valuation ring or a field, A is an R-algebra, and U is finitely generated as an R-module. Then U is indecomposable if and only if $\mathrm{End}_A(U)$ is a local ring. In particular, this holds if $A = RG$ where G is a finite group.

Proof. (1) This follows from Corollary 11.1.2 and Proposition 11.1.4.

(2) From Corollary 11.1.2 and Proposition 11.1.4, again all we need to do is to show that $\mathrm{End}_A(U)$ is finitely generated as an R-module. Let $R^m \to U$ be a surjection of R-modules. Composition with this surjection gives a homomorphism $\mathrm{End}_A(U) \to \mathrm{Hom}_R(R^m, U)$, and it is an injection since $R^m \to U$ is surjective (using the property of Hom from homological algebra that it is "left exact" and the fact that A-module homomorphisms are a subset of R-module homomorphisms). Thus, $\mathrm{End}_A(U)$ is realized as an R-submodule of $\mathrm{Hom}_R(R^m, U) \cong U^m$, which is a finitely generated R-module. Since R is Noetherian, the submodule is also finitely generated. □

The next result is a version of the Krull–Schmidt Theorem. We first present it in greater generality than for group representations.

Theorem 11.1.6 (Krull–Schmidt). *Let A be a ring with a 1, and suppose that U is an A-module that has two A-module decompositions*

$$U = U_1 \oplus \cdots \oplus U_r = V_1 \oplus \cdots \oplus V_s$$

where, for each i, $\mathrm{End}_A(U_i)$ is a local ring and V_i is an indecomposable A-module. Then $r = s$ and the summands U_i and V_j are isomorphic in pairs when taken in a suitable order.

Proof. The proof is by induction on $\max\{r, s\}$. When this number is 1, we have $U = U_1 = V_1$, and this starts the induction.

Now suppose $\max\{r, s\} > 1$ and the result is true for smaller values of $\max\{r, s\}$. For each j, let $\pi_j : U \to V_j$ be projection onto the jth summand with respect to the decomposition $U = V_1 \oplus \cdots \oplus V_s$, and let $\iota_j : V_j \hookrightarrow U$ be inclusion. Then $\sum_{j=1}^{s} \iota_j \pi_j = 1_U$. Now let $\beta : U \to U_1$ be projection with respect to the decomposition $U = U_1 \oplus \cdots \oplus U_r$ and $\alpha : U_1 \hookrightarrow U$ be inclusion so that $\beta\alpha = 1_{U_1}$. We have

$$1_{U_1} = \beta \left(\sum_{j=1}^{s} \iota_j \pi_j \right) \alpha = \sum_{j=1}^{s} \beta \iota_j \pi_j \alpha$$

and since $\mathrm{End}_A(U_1)$ is a local ring it follows that at least one term $\beta\iota_j\pi_j\alpha$ must be invertible. By renumbering the V_j if necessary we may suppose that $j = 1$, and we write $\phi = \beta\iota_1\pi_1\alpha$. Now $(\phi^{-1}\beta\iota_1)(\pi_1\alpha) = 1_{U_1}$ and so $\pi_1\alpha : U_1 \to V_1$ is split mono and $\phi^{-1}\beta\iota_1 : V_1 \to U_1$ is split epi. It follows that $\pi_1\alpha(U_1)$ is a direct summand of V_1. Since V_1 is indecomposable, we have $\pi_1\alpha(U_1) = V_1$ and $\pi_1\alpha : U_1 \to V_1$ must be an isomorphism.

We now show that $U = U_1 \oplus V_2 \oplus \cdots \oplus V_s$. Because $\pi_1\alpha$ is an isomorphism, π_1 is one-to-one on the elements of U_1. Also π_1 is zero on $V_2 \oplus \cdots \oplus V_s$ and it follows that $U_1 \cap (V_2 \oplus \cdots \oplus V_s) = 0$, since any element of the intersection is detected by its image under π_1, and this must be zero. The submodule $U_1 + V_2 + \cdots + V_s$ contains $V_2 + \cdots + V_s = \mathrm{Ker}\,\pi_1$ and so corresponds via the first isomorphism theorem for modules to a submodule of $\pi_1(U) = V_1$. In fact, π_1 is surjective and so $U_1 + V_2 + \cdots + V_s = U$. It follows that $U = U_1 \oplus V_2 \oplus \cdots \oplus V_s$.

We now deduce that $U/U_1 \cong U_2 \oplus \cdots \oplus U_r \cong V_2 \oplus \cdots \oplus V_s$. It follows by induction that $r = s$ and the summands are isomorphic in pairs, which completes the proof. □

Note that the proof of Theorem 11.1.6 shows that an "exchange lemma" property holds for the indecomposable summands in the situation of the theorem. After the abstraction of general rings, we state the Krull–Schmidt Theorem in the context of finite group representations, just to make things clear.

Corollary 11.1.7. *Let R be a complete discrete valuation ring or a field and G a finite group. Suppose that U is a finitely generated RG-module that has two decompositions*

$$U = U_1 \oplus \cdots \oplus U_r = V_1 \oplus \cdots \oplus V_s$$

where the U_i and V_j are indecomposable RG-modules. Then $r = s$ and the summands U_i and V_j are isomorphic in pairs when taken in a suitable order.

Proof. We have seen in Corollary 11.1.5 that the rings $\mathrm{End}_{RG}(U_i)$ are local, so that Theorem 11.1.6 applies. □

11.2 Groups with a Normal Cyclic Sylow p-Subgroup

Before developing any more abstract theory, we present an example where we can describe explicitly all the indecomposable modules. The example will help to inform our later discussion. Let k be a field of characteristic p and suppose that G has a normal cyclic Sylow p-subgroup H. Thus, by the Schur–Zassenhaus Theorem, $G = H \rtimes K$ where $H = \langle x \rangle \cong C_{p^n}$ is cyclic of order p^n and K is a group of order relatively prime to p. In this situation, we have seen in Proposition 8.3.3 that all the indecomposable projective modules are uniserial, and we also gave a description of their composition factors. Since projective and injective modules are the same thing for a group algebra over a field, it follows that all the indecomposable injective modules are uniserial too. Such an algebra, where the indecomposable projective and injective modules are all uniserial, is called a *Nakayama algebra*.

Proposition 11.2.1. *Let A be a finite-dimensional Nakayama algebra over a field. Then every indecomposable module is uniserial and is a homomorphic image of an indecomposable projective module. Thus, the homomorphic images of indecomposable projectives form a complete list of indecomposable modules.*

Proof. Let M be an indecomposable A-module and let ℓ be its Loewy length. Consider any surjective homomorphism $A^d \to M$ where A^d is a free module of some rank d. At least one indecomposable summand P of A^d must have image with Loewy length ℓ in M (otherwise M would have shorter Loewy length than ℓ). Let U be the image of this $P \to M$, so U is a uniserial submodule of M

of Loewy length ℓ. Consider now the injective hull $U \to I$ of U. Since U has a simple socle the injective module I is indecomposable. By the property of injectivity, we obtain a factorization $U \to M \to I$. The image of M in I can have Loewy length no larger than ℓ, and since this is the Loewy length of U the image must have Loewy length equal to ℓ. Because I is uniserial and has a unique submodule of each Loewy length, it follows that U and M have the same image in U. Composing with the inverse of the isomorphism from U to its image in I, we get a homomorphism $M \to U$ that splits $U \to M$, so that U is a direct summand of M. Since M is indecomposable, $U = M$.

We have just established that M is a homomorphic image of an indecomposable projective. Equally, any such module is indecomposable since it has a unique simple quotient, so we have a complete list of indecomposables. □

Corollary 11.2.2. *Let k be a field of characteristic p and suppose that G has a normal cyclic Sylow p-subgroup H, so that $G = H \rtimes K$ where $H = \langle x \rangle \cong C_{p^n}$ is cyclic of order p^n and K is a group of order relatively prime to p. The number of isomorphism classes of indecomposable kG-modules equals $p^n \cdot l_k(K)$ where $l_k(K)$ is the number of isomorphism classes of simple kK-modules.*

Proof. We have seen in Proposition 8.3.3 that each indecomposable projective is uniserial of Loewy length p^n, so has p^n nonisomorphic homomorphic images. Since there are $l_k(K)$ simple kG-modules and hence indecomposable projectives, the result follows. □

Example 11.2.3. From these results, we can immediately describe all the indecomposable modules for cyclic groups or, indeed, abelian groups with a cyclic Sylow p-subgroup, the dihedral group D_{2p} in characteristic p when p is odd, or for groups such as the non-abelian group of order 21 in characteristic 7 (as well as other more complicated groups). For the groups just mentioned, the subgroup K is abelian of order prime to p, so $l_k(K) = |K|$ and the number of isomorphism classes of indecomposable modules happens to equal $|G|$. Thus, when p is odd, D_{2p} has $2p$ indecomposable modules and when p is 7, $C_7 \rtimes C_3$ has 21 indecomposable modules: they are the homomorphic images of the three projectives shown in Example 8.3.4.

11.3 Relative Projectivity

The relationship between the representations of a group and those of its subgroups are one of the most important tools in representation theory. In the context of modular representations of groups, this relationship shows itself in a refinement of the notion of projectivity, namely, relative projectivity. We will use it to give a criterion for actual projectivity and to determine the group rings of finite representation type. Finally, we will describe the theory of vertices and sources of indecomposable modules, which appear as part of Green correspondence.

Let H be a subgroup of G and R a commutative ring with 1. An RG-module is said to be *H-free* if it has the form $V \uparrow_H^G$ for some RH-module V. It is *H-projective*, or *projective relative to H*, if it is a direct summand of a module of the form $V \uparrow_H^G$ for some RH-module V.

For example, the regular representation $RG \cong R \uparrow_1^G$ is 1-free, and projective modules are 1-projective. If R is a field then every 1-projective module is projective, but if R is not a field and V is an R-module that is not projective as an R-module, then $V \uparrow_1^G$ is 1-free but not projective as an RG-module. Every RG-module is G-projective.

To investigate relative projectivity, we first deal with some technicalities. We have seen the pervasive importance of the group ring element $\sum_{g \in G} g$ at every stage of the development of representation theory. As an operator on any representation of G, it has image contained in the G-fixed points. We now consider something more general: for a subgroup H of G and an RG-module U, we define the *relative trace map* $\mathrm{tr}_H^G : U^H \to U^G$. To define this, we choose a set of representatives $g_1, \ldots, g_n$ of the left cosets of H in G, so $G = g_1 H \cup \cdots \cup g_n H$. If $u \in U^H$, we define $\mathrm{tr}_H^G(u) = \sum_{i=1}^n g_i u$. To complete the picture, we mention that when $H \leq G$ there is an inclusion of fixed points that we denote $\mathrm{res}_H^G : U^G \hookrightarrow U^H$. When $H \leq G$ and $g \in G$, there is also a map $c_g : U^H \to U^{({}^gH)}$ given by $u \mapsto gu$. These operations behave like induction, restriction, and conjugation of modules.

Lemma 11.3.1. *Let U be an RG-module and let $K \leq H \leq G$ and L be subgroups of G.*

(1) *The homomorphism $\mathrm{tr}_H^G : U^H \to U^G$ is well defined.*
(2) $\mathrm{tr}_H^G \mathrm{tr}_K^H = \mathrm{tr}_K^G$,
$\mathrm{res}_K^H \mathrm{res}_H^G = \mathrm{res}_K^G$,
$c_g \mathrm{tr}_K^H = \mathrm{tr}_{{}^gK}^{{}^gH} c_g$,
$c_g \mathrm{res}_K^H = \mathrm{res}_{{}^gK}^{{}^gH} c_g$, *and*
$\mathrm{tr}_H^H = 1 = \mathrm{res}_H^H$.
(3) $\mathrm{tr}_K^H \mathrm{res}_K^H$ *is multiplication by the index $|H : K|$.*
(4) *(Mackey formula)* $\mathrm{res}_L^G \mathrm{tr}_H^G = \sum_{g \in [L \backslash G/H]} \mathrm{tr}_{L \cap {}^gH}^L \mathrm{res}_{L \cap {}^gH}^{{}^gH} c_g$.

Proof. (1) We could have chosen different coset representatives of H in G, in which case the different set we might have picked would have the form $g_1 h_1, \ldots, g_n h_n$ for certain elements $h_1, \ldots, h_t$ in H. Now the definition of tr_H^G

would be $\mathrm{tr}_H^G(u) = \sum_{i=1}^t g_i h_i u$, but this equals $\sum_{i=1}^t g_i u$ as before, since u is fixed by all of the h_i.

(2) Many of these formulas are quite straightforward, and we only prove the first and the third. For the first, if we take left coset representatives $h_1, \dots, h_m$ of K in H and left coset representatives $g_1, \dots, g_n$ of H in G then the set of all elements $g_i h_j$ is a set of left coset representatives for K in G and their sum is $(g_1 + \cdots g_n)(h_1 + \cdots + h_m)$. Now

$$\mathrm{tr}_K^G(u) = (g_1 + \cdots g_n)(h_1 + \cdots + h_m)u = (g_1 + \cdots g_n)\mathrm{tr}_K^H(u) = \mathrm{tr}_H^G \mathrm{tr}_K^H(u).$$

For the third formula, the elements ${}^g h_1, \dots, {}^g h_m$ are a set of left coset representatives for ${}^gK$ in ${}^gH$ and so $\mathrm{tr}_{{}^gK}^{{}^gH} c_g(u) = \sum_{i=1}^m {}^g h_i g u = \sum_{i=1}^m g h_i u = c_g \mathrm{tr}_K^H(u)$.

(3) Continuing with the notation in use, if u is fixed by H then

$$\mathrm{tr}_K^H \mathrm{res}_K^H(u) = \sum_{i=1}^m h_i u = \sum_{i=1}^m u = mu$$

where $m = |H : K|$.

(4) Considering the action from the left of L on the cosets G/H, for any $g \in G$ the L-orbit containing the coset gH consists of the cosets agH where a lies in L and ranges through representatives of the cosets $L/(L \cap {}^gH)$. This is because the cosets in the L-orbit are the agH with $a \in L$, and $a_1 gH = a_2 gH$ if and only if $a_2^{-1} a_1 \in \mathrm{Stab}_L(gH) = L \cap {}^gH$, which happens if and only if a_1 and a_2 lie in the same coset of $L \cap {}^gH$ in L. Thus, on partitioning the cosets G/H into L-orbits, we see that $\sqcup_{g \in [L\backslash G/H]} \{ag \mid a \in [L/(L \cap {}^gH)]\}$ is a set of left coset representatives for H in G. Hence if $u \in U^H$,

$$\mathrm{tr}_H^G u = \sum_{g \in L\backslash G/H]} \sum_{a \in [L/(L \cap {}^gH)]} agu = \sum_{g \in [L\backslash G/H]} \mathrm{tr}_{L \cap {}^gH}^L c_g u. \qquad \square$$

The most important situation where we will use the relative trace map is when the RG-module U is a space of homomorphisms $\mathrm{Hom}_R(X, Y)$ between RG-modules X and Y. In this situation, $\mathrm{Hom}_R(X, Y)^H = \mathrm{Hom}_{RH}(X, Y)$ for any subgroup H, so that the relative trace map from H to G is a homomorphism of R-modules $\mathrm{tr}_H^G : \mathrm{Hom}_{RH}(X, Y) \to \mathrm{Hom}_{RG}(X, Y)$.

In the following result it would have been technically correct to insert res_H^G in several places, but since this operation is simply the inclusion of fixed points it seems more transparent to leave it out.

Lemma 11.3.2. *Suppose that $\alpha : U \to V$ and $\gamma : W \to X$ are homomorphisms of RG-modules and that $\beta : V \downarrow_H^G \to W \downarrow_H^G$ is an RH-module homomorphism. Then $(\mathrm{tr}_H^G \beta) \circ \alpha = \mathrm{tr}_H^G(\beta \circ \alpha)$ and $\gamma \circ (\mathrm{tr}_H^G \beta) = \mathrm{tr}_H^G(\gamma \circ \beta)$.*

Proof. Let $g_1, \dots, g_n$ be a set of left coset representatives for H in G and $u \in U$. Then

$$(\mathrm{tr}_H^G \beta)\alpha(u) = \sum_{i=1}^{n} g_i\beta(g_i^{-1}\alpha u) = \sum_{i=1}^{n} g_i\beta\alpha(g_i^{-1}u) = \mathrm{tr}_H^G(\beta\alpha)(u).$$

Similarly

$$\gamma(\mathrm{tr}_H^G \beta)(u) = \gamma \sum_{i=1}^{n} g_i\beta(g_i^{-1}u) = \sum_{i=1}^{n} g_i\gamma\beta(g_i^{-1}u) = \mathrm{tr}_H^G(\gamma\beta)(u).$$

□

Corollary 11.3.3. *Let U be an RG-module and let H be a subgroup of G.*

(1) The image of $\mathrm{tr}_H^G : \mathrm{End}_{RH}(U) \to \mathrm{End}_{RG}(U)$ is an ideal.
(2) The map $\mathrm{res}_H^G : \mathrm{End}_{RG}(U) \to \mathrm{End}_{RH}(U)$ is a ring homomorphism.

Proof. The first property is immediate from Lemma 11.3.2, which implies that the image of tr_H^G is closed under composition with elements of $\mathrm{End}_{RG}(U)$. The second statement is not a corollary, but it is immediate because res_H^G is the inclusion map. □

It will help us to consider the adjoint properties of induction and restriction of modules in detail. We have seen in Corollary 4.3.8 that when $H \le G$, U is an RH-module and V is an RG-module, we have

$$\mathrm{Hom}_{RG}(U \uparrow_H^G, V) \cong \mathrm{Hom}_{RH}(U, V \downarrow_H^G).$$

There may be many such isomorphisms, but there is a choice that is *natural* in U and V. This means that whenever $U_1 \to U_2$ is an RG-module homomorphism the resulting square

$$\begin{array}{ccc}
\mathrm{Hom}_{RG}(U_1 \uparrow_H^G, V) & \xrightarrow{\cong} & \mathrm{Hom}_{RH}(U_1, V \downarrow_H^G) \\
\uparrow & & \uparrow \\
\mathrm{Hom}_{RG}(U_2 \uparrow_H^G, V) & \xrightarrow{\cong} & \mathrm{Hom}_{RH}(U_2, V \downarrow_H^G)
\end{array}$$

commutes, as does the square

$$\begin{array}{ccc}
\mathrm{Hom}_{RG}(U \uparrow_H^G, V_1) & \xrightarrow{\cong} & \mathrm{Hom}_{RH}(U, V_1 \downarrow_H^G) \\
\downarrow & & \downarrow \\
\mathrm{Hom}_{RG}(U \uparrow_H^G, V_2) & \xrightarrow{\cong} & \mathrm{Hom}_{RH}(U, V_2 \downarrow_H^G)
\end{array}$$

whenever $V_1 \to V_2$ is a homomorphism of RG-modules. In this situation, we say that the operation $\uparrow_H^G$: RH-modules $\to$ RG-modules is *left adjoint* to $\downarrow_H^G$:

RG-modules $\to$ RH-modules. (These "operations" are in fact *functors*.) We say also that $\downarrow_H^G$ is *right adjoint* to $\uparrow_H^G$, and this relationship is called an *adjunction*.

We have also seen in Corollary 4.3.8 that

$$\mathrm{Hom}_{RG}(V, U\uparrow_H^G) \cong \mathrm{Hom}_{RH}(V\downarrow_H^G, U)$$

in the same circumstances. Again, if this isomorphism can be given naturally in both U and V (meaning that the corresponding squares commute) then induction $\uparrow_H^G$ is right adjoint to restriction $\downarrow_H^G$. It is, in fact, the case for representations of finite groups that induction is both the left and right adjoint of restriction.

In Proposition 11.3.4, we will need to know in detail about the natural isomorphisms that arise in these adjunctions, and we now describe them. They depend on certain distinguished homomorphisms called the *unit* and *counit* of the adjuntions. For each RH-module U, we will define RH-module homomorphisms

$$\mu : U \to U\uparrow_H^G\downarrow_H^G$$
$$\nu : U\uparrow_H^G\downarrow_H^G \to U,$$

and for each RG-module V, we will define RG-module homomorphisms

$$\eta : V \to V\downarrow_H^G\uparrow_H^G$$
$$\epsilon : V\downarrow_H^G\uparrow_H^G \to V.$$

In the language of category theory, μ and ϵ are the unit and counit of the adjuntion that shows that $\uparrow_H^G$ is left adjoint to $\downarrow_H^G$, and η, ν are the unit and counit of the adjuntion that shows that $\uparrow_H^G$ is right adjoint to $\downarrow_H^G$.

To define these homomorphisms, choose a set of left coset representatives $\{g_1, \ldots, g_n\}$ of H in G with $g_1 = 1$. For each RH-module U, let

$$\mu : U \to U\uparrow_H^G\downarrow_H^G = \bigoplus_{i=1}^{n} g_i \otimes U$$

be the inclusion into the summand $1 \otimes U$, so $\mu(u) = 1 \otimes u$, and let $\nu : U\uparrow_H^G\downarrow_H^G \to U$ be projection onto this summand. We see that μ is a monomorphism, ν is an epimorphism, and their composite is the identity. If V is an RG-module, we define the RG-module homomorphisms $\eta : V \to V\downarrow_H^G\uparrow_H^G$ and $\epsilon : V\downarrow_H^G\uparrow_H^G \to V$ by $\eta(v) = \sum_{i=1}^{n} g_i \otimes g_i^{-1}v$ and $\epsilon(\sum \lambda_x x \otimes u) = \sum \lambda_x xu$. In fact, regarding $\mathrm{tr}_H^G : \mathrm{Hom}_{RH}(V, V\downarrow_H^G\uparrow_H^G) \to \mathrm{Hom}_{RG}(V, V\downarrow_H^G\uparrow_H^G)$ we have $\eta = \mathrm{tr}_H^G(\mu)$ where μ has domain V regarded as an RH-module by restriction. Similarly, $\epsilon = \mathrm{tr}_H^G(\nu)$, and this shows that η and ϵ are defined independently of the choice of coset representatives. We see directly that η is a monomorphism, ϵ is an epimorphism and their composite is multiplication by $|G : H|$.

We now construct mutually inverse natural isomorphisms

$$\mathrm{Hom}_{RG}(U\uparrow_H^G, V) \cong \mathrm{Hom}_{RH}(U, V\downarrow_H^G).$$

Given an RG-module homomorphism $\alpha : U \uparrow_H^G \to V$, we obtain an RH-module homomorphsim

$$U \xrightarrow{\mu} U \uparrow_H^G \downarrow_H^G \xrightarrow{\alpha \downarrow_H^G} V \downarrow_H^G$$

and given an RH-module homomorphism $\beta : U \to V \downarrow_H^G$, we obtain an RG-module homomorphism

$$U \uparrow_H^G \xrightarrow{\beta \uparrow_H^G} V \downarrow_H^G \uparrow_H^G \xrightarrow{\epsilon} V.$$

We check that these two constructions are mutually inverse, and are natural in U and V. These are the desired isomorphisms.

We next construct mutually inverse natural isomorphisms

$$\mathrm{Hom}_{RG}(V, U \uparrow_H^G) \cong \mathrm{Hom}_{RH}(V \downarrow_H^G, U)$$

whenever U is an RH-module and V is an RG-module. Given an RG-module homomorphism $\gamma : V \to U \uparrow_H^G$ we obtain an RH-module homomorphism

$$V \downarrow_H^G \xrightarrow{\gamma} U \uparrow_H^G \downarrow_H^G \xrightarrow{\nu} U$$

and given an RH-module homomorphism $\delta : V \downarrow_H^G \to U$, we obtain an RG-module homomorphism

$$V \xrightarrow{\eta} V \downarrow_H^G \uparrow_H^G \xrightarrow{\delta \uparrow_H^G} U \uparrow_H^G .$$

Again we check that these operations are natural and mutually inverse.

The conditions in the next result characterize relative projectivity in a way that extends familiar characterizations of projectivity.

Proposition 11.3.4. *Let G be a finite group with a subgroup H. The following are equivalent for an RG-module U.*

(1) U is H-projective.
(2) Whenever we have homomorphisms

$$\begin{array}{ccc} & & U \\ & & \downarrow \psi \\ V & \xrightarrow{\phi} & W \end{array}$$

where ϕ is an epimorphism and for which there exists an RH-module homomorphism $U \downarrow_H^G \to V \downarrow_H^G$ making the diagram commute, then there exists an RG-module homomorphism $U \to V$ making the diagram commute.

(3) *Whenever* $\phi : V \to U$ *is a homomorphism of RG-modules such that* $\phi \downarrow_H^G : V \downarrow_H^G \to U \downarrow_H^G$ *is a split epimorphism of RH-modules, then* ϕ *is a split epimorphism of RG-modules.*

(4) *The surjective homomorphism of RG-modules*

$$\begin{aligned} U \downarrow_H^G \uparrow_H^G = RG \otimes_{RH} U &\to U \\ x \otimes u &\mapsto xu \end{aligned}$$

is split.

(5) U *is a direct summand of* $U \downarrow_H^G \uparrow_H^G$.

(6) *(Higman's criterion)* 1_U *lies in the image of*

$$\mathrm{tr}_H^G : \mathrm{End}_{RH}(U) \to \mathrm{End}_{RG}(U).$$

Proof. (1) $\Rightarrow$ (2) We first prove this implication in the special case when U is an induced module $T \uparrow_H^G$. Suppose we have a diagram of RG-modules

$$\begin{array}{ccc} & & T \uparrow_H^G \\ & & \downarrow \psi \\ V & \xrightarrow{\phi} & W \end{array}$$

and a homomorphism of RH-modules $\alpha : T \uparrow_H^G \downarrow_H^G \to V \downarrow_H^G$ so that $\psi = \phi\alpha$. Under the adjoint correspondence ψ corresponds to the composite

$$T \xrightarrow{\mu} T \uparrow_H^G \downarrow_H^G \xrightarrow{\psi} W \downarrow_H^G$$

and we have a commutative triangle of RH-module homomorphisms

$$\begin{array}{ccc} & & T \\ & \swarrow^{\alpha\mu} & \downarrow \psi\mu \\ V \downarrow_H^G & \xrightarrow{\phi \downarrow_H^G} & W \downarrow_H^G \end{array}$$

By the adjoint correspondence (and its naturality) this corresponds to a commutative triangle of RG-module homomorphisms

$$\begin{array}{ccc} & & T \uparrow_H^G \\ & \swarrow^{\epsilon(\alpha\mu)\uparrow_H^G} & \downarrow \psi \\ V & \xrightarrow{\phi} & W \end{array}$$

which proves this implication for the module $T \uparrow_H^G$.

Now consider a module U that is a summand of $T \uparrow_H^G$, and let

$$U \xrightarrow{\iota} T \uparrow_H^G \xrightarrow{\pi} U$$

be inclusion and projection. We suppose there is a homomorphism

$$\alpha : U \downarrow_H^G \to V \downarrow_H^G$$

so that $\phi\alpha = \psi$. The homomorphism $\psi\pi : T \uparrow_H^G \to W$ has the property that $\psi\pi = \phi(\alpha\pi)$ and so by what we proved there is a homomorphism of RG-modules $\beta : T \uparrow_H^G \to V$ so that $\phi\beta = \psi\pi$. Now $\phi\beta\iota = \psi\pi\iota = \psi$ so that $\beta\iota : U \to V$ is an RG-module homomorphism that makes the triangle commute.

(2) $\Rightarrow$ (3) This follows immediately on applying (2) to the diagram

$$\begin{array}{ccc} & & U \\ & & \downarrow{\scriptstyle 1_U} \\ V & \longrightarrow & U \end{array}$$

(3) $\Rightarrow$ (4) We know that $\epsilon : U \downarrow_H^G \uparrow_H^G \to U$ is split as an RH-module homomorphism by $\mu : U \to U \downarrow_H^G \uparrow_H^G$. Applying condition (3) it splits as an RG-module homomorphism.

(4) $\Rightarrow$ (5) and (5) $\Rightarrow$ (1) are immediate.

(5) $\Rightarrow$ (6) We let μ, ν denote the maps defined prior to this proposition. We will prove that $1_{U \downarrow_H^G \uparrow_H^G} = \mathrm{tr}_H^G(\mu\nu)$ by direct computation. Writing $U \downarrow_H^G \uparrow_H^G = V_1 \oplus V_2$ where $V_1 = U$, we can represent $\mu\nu$ as a matrix

$$\mu\nu = \begin{bmatrix} f_{11} & f_{21} \\ f_{12} & f_{22} \end{bmatrix}$$

where $f_{ij} : V_i \to V_j$. Then

$$\mathrm{tr}_H^G(\mu\nu) = \begin{bmatrix} \mathrm{tr}_H^G f_{11} & \mathrm{tr}_H^G f_{21} \\ \mathrm{tr}_H^G f_{12} & \mathrm{tr}_H^G f_{22} \end{bmatrix} = \begin{bmatrix} 1 & 0 \\ 0 & 1 \end{bmatrix}$$

and from this, we see that for every summand of $U \downarrow_H^G \uparrow_H^G$ (and in particular for U) the identity map on that summand is in the image of tr_H^G.

(6) $\Rightarrow$ (5) Write $1_U = \mathrm{tr}_H^G \alpha$ for some morphism $\alpha : U \downarrow_H^G \to U \downarrow_H^G$. Now α corresponds by the adjoint correspondence to the composite homomorphism

$$U \xrightarrow{\eta} U \downarrow_H^G \uparrow_H^G \xrightarrow{\alpha \uparrow_H^G} U \downarrow_H^G \uparrow_H^G .$$

We claim that $\alpha \uparrow_H^G \eta$ splits ϵ: for

$$\epsilon\alpha \uparrow_H^G \eta = \mathrm{tr}_H^G(\nu)\alpha \uparrow_H^G \eta = \mathrm{tr}_H^G(\nu\alpha \uparrow_H^G \eta) = \mathrm{tr}_H^G(\alpha) = 1_U. \qquad \square$$

Condition (6) of Proposition 11.3.4 is in fact equivalent to the surjectivity of $\mathrm{tr}_H^G : \mathrm{End}_{RH}(U) \to \mathrm{End}_{RG}(U)$ This is because the image of tr_H^G is an ideal in $\mathrm{End}_{RG}(U)$, and so it equals $\mathrm{End}_{RG}(U)$ if and only if it contains 1_U. Conditions

(2), (3), and (4) are modeled on conditions associated with the notion of projectivity. There are dual conditions associated with the notion of an injective module, obtained by reversing the arrows and interchanging the words "epimorphism" and "monomorphism." These conditions are also equivalent to the ones listed in this result. In fact, the notion of relative projectivity in the context of group algebras of finite groups is the same as that of relative injectivity. We leave the proof to the reader, bearing in mind Corollary 4.3.8(5), which says that induction and coinduction are the same.

Proposition 11.3.5. *Suppose that H is a subgroup of G and that $|G : H|$ is invertible in the ring R. Then every RG-module is H-projective.*

Proof. For any RG-module U, we may write $1_U = \frac{1}{|G:H|}\mathrm{tr}_H^G 1_U$, thus verifying condition (6) of the last result. □

Corollary 11.3.6. *Suppose that H is a subgroup of G for which $|G : H|$ is invertible in the ring R, and let U be an RG-module. Then U is projective as an RG-module if and only if $U \downarrow_H^G$ is projective as an RG-module.*

Proof. We already know that if U is projective then $U \downarrow_H^G$ is projective, no matter what subgroup H is. Conversely, if $U \downarrow_H^G$ is projective it is a summand of a free module RH^n. Since U is H-projective, it is a summand of $U \downarrow_H^G \uparrow_H^G$, which is a summand of $RH^n \uparrow_H^G \cong RG^n$. Therefore, U is projective. □

Example 11.3.7. This criterion for projectivity would have simplified matters when we were considering the projective modules for groups of the form $G = H \rtimes K$ in Chapter 8. In this situation, we saw that RH becomes an RG-module where H acts by left multiplication and K acts by conjugation. If K has order prime to p and R is a field of characteristic p (or a discrete valuation ring with residue field of characteristic p) it follows from the corollary that RH is projective as an RG-module, because on restriction to H it is projective and the index of H in G is invertible. On the other hand, we may also regard RK as an RG-module via the homomorphism $G \to K$, and if now H has order prime to p then RK is a projective RG-module, because it is projective on restriction to RK and the index of K in G is prime to p.

Example 11.3.8. In the exercises to Chapter 6, the simple $\mathbb{F}_pSL(2, p)$-modules were considered. The goal of Exercise 25 of Chapter 6 was to show that the symmetric powers $S^r(U_2)$ are all simple $\mathbb{F}_pSL(2, p)$-modules when $0 \le r \le p - 1$, where U_2 is the 2-dimensional space on which $SL(2, p)$ acts as invertible transformations of determinant 1. The order of $SL(2, p)$ is $p(p^2 - 1)$ and so a Sylow p-subgroup of this group is cyclic of order p. From Exercise 23 of Chapter 6, it follows that, on restriction to the Sylow p-subgroup of upper uni-triangular matrices, $S^r(U_2)$ is indecomposable of dimension $r+1$ when $0 \le r \le p-1$. From the classification of indecomposable modules for a cyclic group of order p, we deduce that $S^{p-1}(U_2)$ is projective as a module for the Sylow p-subgroup. It follows from the last corollary that $S^{p-1}(U_2)$ is projective as an $\mathbb{F}_p SL(2,p)$-module. This module is thus a simple projective $\mathbb{F}_p SL(2,p)$-module or, in other words, a block of defect zero, which therefore lifts to characteristic zero. Starting from information in characteristic p, we have thus deduced the existence of an ordinary simple character of $SL(2,p)$ of degree p.

11.4 Finite Representation Type

In trying to understand the representation theory of a ring, we might hope to be able to describe all the indecomposable modules, since this would allow us to construct all modules up to isomorphism by taking direct sums. The possibility of such a description depends on there being sufficiently few indecomposable modules for a classification to be a reasonable goal, and its utility would depend on there being a description of them that can be understood in some way. Unfortunately, for the majority of group rings, we encounter in positive characteristic, a classification of the indecomposable modules that is understandable seems to be an unreasonable expectation. On the other hand, there are some special cases where the indecomposable modules can indeed be classified, and we will indicate in this section and the next what these are.

We say that a ring A has *finite representation type* if and only if there are only finitely many isomorphism classes of indecomposable A-modules; otherwise, we say that A has *infinite representation type*. We have seen in Theorem 6.1.2 that if G is a cyclic p-group and k is a field of characteristic p, then kG has finite representation type. Our immediate goal in this section is to characterize the groups for which kG has finite representation type. We first reduce it to a question about p-groups.

Proposition 11.4.1. *Let R be a discrete valuation ring with residue field of characteristic p or a field of characteristic p, and let P be a Sylow p-subgroup of a finite group G. Then RG has finite representation type if and only if RP has finite representation type.*

Proof. Since $|G:P|$ is invertible in R, by Proposition 11.3.5, every indecomposable RG-module is a summand of some module $T\uparrow_P^G$, and we may assume that T is indecomposable, since if $T = T_1 \oplus T_2$ then $T\uparrow_P^G = T_1\uparrow_P^G \oplus T_2\uparrow_P^G$, and by the Krull–Schmidt Theorem the indecomposable summands of $T\uparrow_P^G$ are the indecomposable summands of $T_1\uparrow_P^G$ together with the indecomposable summands of $T_2\uparrow_P^G$. If RP has finite representation type then there are only finitely many modules $T \uparrow_P^G$ with T indecomposable, and these have only finitely many isomorphism types of summands by the Krull–Schmidt Theorem. Hence RG has finite representation type.

Conversely, every RP-module U is a direct summand of $U \uparrow_P^G \downarrow_P^G$ and hence is a direct summand of some module $V \downarrow_P^G$. If U is indecomposable, we may assume V is indecomposable. Now if RG has finite representation type there are only finitely many isomorphism types of summands of modules $V \downarrow_P^G$, by the Krull–Schmidt Theorem, and hence RP has finite representation type. □

We have already seen in Theorem 6.1.2 that cyclic p-groups have finite representation type over a field of characteristic p, so by Proposition 11.4.1, groups with cyclic Sylow p-subgroups have finite representation type. We will show the converse in Theorem 11.4.3, and as preparation for this, we now show that $k[C_p \times C_p]$ has infinitely many nonisomorphic indecomposable modules, where k is a field of characteristic p. Apart from the use of this in establishing infinite representation type, it is useful to see how indecomposable modules may be constructed, and in the case of $C_2 \times C_2$, it will lead to a classification of the indecomposable modules.

We will first describe infinitely many modules of different dimensions for $k[C_p \times C_p]$, and after that we will prove that they are indecomposable. Let $G = C_p \times C_p = \langle a \rangle \times \langle b \rangle$. For each $n \geq 1$, we define a module M_{2n+1} of dimension $2n+1$ with basis $u_1, \ldots, u_n, v_0, \ldots, v_n$ and an action of G given as follows:

$$\begin{array}{lll} a(u_i) = u_i + v_{i-1}, & b(u_i) = u_i + v_i & \text{where } 1 \leq i \leq n \\ a(v_i) = v_i, & b(v_i) = v_i & \text{where } 0 \leq i \leq n. \end{array}$$

It is easier to see what is going on if we write this as

$$\begin{array}{lll} (a-1)u_i = v_{i-1}, & (b-1)u_i = v_i & \text{where } 1 \leq i \leq n \\ (a-1)v_i = 0, & (b-1)v_i = 0 & \text{where } 0 \leq i \leq n \end{array}$$

and describe M_{2n+1} diagrammatically:

$$\begin{array}{ccccccccc} & u_1 & & u_2 & & & & u_n & \\ & \bullet & & \bullet & & & & \bullet & \\ {\scriptstyle a-1} \swarrow & & \searrow {\scriptstyle b-1} \quad {\scriptstyle a-1} \swarrow & & \searrow {\scriptstyle b-1} & \cdots & {\scriptstyle a-1} \swarrow & & \searrow {\scriptstyle b-1} \\ \bullet & & \bullet & & & & & & \bullet \\ v_0 & & v_1 & & & & & & v_n \end{array}$$

We may check that this is indeed a representation of G by verifying that

$$(a-1)(b-1)x = (b-1)(a-1)x = 0$$

and

$$(a-1)^p x = (b-1)^p x = 0$$

for all $x \in M_{2n+1}$, which is immediate. This is sufficient to show that we have a representation of G since the equations $(a-1)(b-1) = (b-1)(a-1)$ and $(a-1)^p = 0 = (b-1)^p$ are defining relations for kG as a k-algebra with generators a, b and, in particular, they imply the relations, which define G as a group.

We now show that M_{2n+1} is indecomposable.

Proposition 11.4.2. *The quotient* $\operatorname{End}_{kG}(M_{2n+1})/\operatorname{Rad}\operatorname{End}_{kG}(M_{2n+1})$ *has dimension 1. Thus,* $\operatorname{End}_{kG}(M_{2n+1})$ *is a local ring and* M_{2n+1} *is indecomposable.*

Proof. We will show that $\operatorname{End}_{kG}(M_{2n+1})/I$ has dimension 1 for a certain nilpotent ideal I. Such an ideal I must be contained in the radical, being nilpotent, and the fact that it has codimension 1 will then force it to equal the radical. This will prove the result.

Observe that

$$\operatorname{Soc}(M_{2n+1}) = \operatorname{Rad}(M_{2n+1}) = kv_0 + \cdots kv_n.$$

The ideal I in question is $\operatorname{Hom}_{kG}(M_{2n+1}, \operatorname{Soc}(M_{2n+1}))$. This squares to zero since if $\phi : M_{2n+1} \to \operatorname{Soc}(M_{2n+1})$ then $\operatorname{Rad}(M_{2n+1}) \subseteq \operatorname{Ker}\phi$ and so $\phi \operatorname{Soc}(M_{2n+1}) = 0$.

We now show that I has codimension 1. It is easy to get an intuitive idea of why this is so, but not so easy to write it down in technically correct language. The intuitive idea is that the basis elements of M_{2n+1} lie in a string as shown diagrammatically, each element related to those on either side by the action of $a-1$ and $b-1$. Modulo elements of I, any endomorphism must send this string to a linear combination of strings of elements that are similarly related. If an endomorphism shifts the string either to the left or the right, then part of the string must fall off the end of the module, or in other words be sent to zero. The connection between adjacent basis elements then forces the whole shift to be zero, so that no shift to the left or right is possible. From this, we deduce that, modulo I, endomorphisms are scalar multiples of the identity.

We now attempt to write down this intuitive idea in formal terms. If ϕ is any endomorphism of M_{2n+1} then $\phi(\operatorname{Soc}(M_{2n+1})) \subseteq \operatorname{Soc} M_{2n+1}$ so ϕ induces an endomorphism $\bar{\phi}$ of $M_{2n+1}/\operatorname{Soc}(M_{2n+1})$. We show that $\bar{\phi}$ is necessarily a scalar multiple of the identity. To establish this, we will exploit the equations

$$(a-1)u_i = (b-1)u_{i-1} \quad \text{when } 2 \le i \le n$$

and also the fact that $(a-1)$ and $(b-1)$ both map $ku_1 + \cdots + ku_n$ injectively into $\operatorname{Soc}(M_{2n+1})$. Applying ϕ to the last equations we have, for $2 \le i \le n$,

$$\phi((a-1)u_i) = (a-1)\phi(u_i) = \phi((b-1)u_{i-1}) = (b-1)\phi(u_{i-1}).$$

It follows from this that $\bar{\phi}$ is completely determined once we know $\phi(u_1)$, since then $(a-1)\phi(u_2) = (b-1)\phi(u_1)$ determines $\bar{\phi}(u_2)$ by injectivity of $a-1$ on the span of the u_i, $(a-1)\phi(u_3) = (b-1)\phi(u_2)$ determines $\bar{\phi}(u_3)$ similarly, and so on.

Suppose that

$$\phi(u_1) \equiv \lambda_1 u_1 + \cdots + \lambda_r u_r \pmod{\operatorname{Soc}(M_{2n+1})}$$

where the λ_i are scalars and $\lambda_r \neq 0$. Multiplying both sides by $b-1$ and using the equations $(a-1)\phi(u_i) = (b-1)\phi(u_{i-1})$ as before, as well as injectivity of multiplication by $a-1$ on the span of the u_i, we see that

$$\phi(u_2) \equiv \lambda_1 u_2 + \cdots + \lambda_r u_{r+1} \pmod{\operatorname{Soc}(M_{2n+1})}$$

and inductively that

$$\phi(u_{n-r+1}) \equiv \lambda_1 u_{n-r+1} + \cdots + \lambda_r u_n \pmod{\operatorname{Soc}(M_{2n+1})}.$$

If it were the case that $r > 1$ then the equation

$$(b-1)\phi(u_{n-r+1}) = (a-1)\phi(u_{n-r+2}) = \lambda_1 v_{n-r+1} + \cdots + \lambda_r v_n$$

would have no solution, since no such vector where the coefficient of v_n is nonzero lies in the image of $a-1$.

We conclude that $r = 1$ and $\phi(u_i) \equiv \lambda_1 u_i \pmod{\operatorname{Soc}(M_{2n+1})}$ for some scalar λ_1, for all i with $1 \leq i \leq n$. Thus,

$$\phi - \lambda_1 1_{M_{2n+1}} : M_{2n+1} \to \operatorname{Soc}(M_{2n+1})$$

and so $\operatorname{End}_{kG}(M_{2n+1})/\operatorname{Hom}_{kG}(M_{2n+1}, \operatorname{Soc}(M_{2n+1}))$ has dimension 1. □

We can now establish the following.

Theorem 11.4.3 (D. G. Higman). *Let k be a field of characteristic p. Then kG has finite representation type if and only if Sylow p-subgroups of G are cyclic.*

Proof. By Proposition 11.4.1, it suffices to show that if P is a p-group then kP has finite representation type if and only if P is cyclic. We have seen in Theorem 6.1.2 that kP has finite representation type when P is cyclic. If P is not cyclic then P has the group $C_p \times C_p$ as a homomorphic image. (This may be proved using the fact that if $\Phi(P)$ is the Frattini subgroup of P then $P/\Phi(P) \cong (C_p)^d$ for some d and that P can be generated by d elements. Since P cannot be generated by a single element, $d \geq 2$ and so $(C_p)^2$ is an image of P.) The infinitely many nonisomorphic indecomposable $k[C_p \times C_p]$-modules become nonisomorphic indecomposable kP-modules via the quotient homomorphism, and this establishes the result. □

Even when the representation type is infinite, the arguments that we have been using still yield the following result.

Theorem 11.4.4. *Let k be a field of characteristic p. For any finite group G, the number of isomorphism classes of indecomposable kG-modules that are projective relative to a cyclic subgroup is finite.*

11.5 Infinite Representation Type and the Representations of $C_2 \times C_2$

We now turn to group rings of infinite representation type, namely, the group rings in characteristic p for which the Sylow p-subgroups of the group are not cyclic. We might expect that, even though the technical difficulties may be severe, a classification of indecomposable modules is about to be revealed. Perhaps the surprising thing about infinite representation type is that in some sense, most of the time, a classification of indecomposable modules is not merely something that is beyond our technical capabilities, it is rather something that may never be possible in any meaningful sense, because of inherent aspects of the problem.

Discussion of such matters raises the question of what we mean by a classification. There are many instances of classification in mathematics, but whichever one we are considering, we might reasonably understand that it is a description of some objects that is simpler than the objects themselves and that allows us to identify in a reasonable way some significant aspects of the situation. It would be inadequate to say that we had parametrized indecomposable modules by using the set of indecomposable modules itself to achieve the parametrization, because this would provide no simplification. It would also be inadequate to put the set of isomorphism classes of indecomposable modules in bijection with some abstract set of the same cardinality. Although we could certainly do this, and for some people it might be a classification, it would provide no new insight. Those who maintain that it is always possible to classify perhaps have such possibilities in mind. In reality, it might be very difficult, or impossible, to classify objects in any meaningful sense because there are simply too many of them to classify: if we were to put the objects in a list in a book, the list would simply be too long and lacking in structure to have any meaning. Of course, just because we cannot see how to make sense of a set of objects does not mean it cannot be done. However, it remains the case that for most group algebras in positive characteristic no one has been able to provide any reasonable classification of the indecomposable modules. Furthermore, we will present reasons why it would be hard to do so.

Infinite representation type divides up into two possibilities: tame and wild. For the tame group algebras, we can (in principle) classify the indecomposable modules. For the wild algebras, no one can see how to do it. Before we address these general questions, we will describe the indecomposable representations of $C_2 \times C_2$ over a field k of characteristic 2, since this exemplifies tame representation type. In constructing infinitely many indecomposable modules for $C_p \times C_p$, we already constructed some of the indecomposable $k[C_2 \times C_2]$-modules, but now we complete the picture. As before we let $G = C_2 \times C_2 = \langle a \rangle \times \langle b \rangle$ and we exhibit the modules diagrammatically by the action of $a - 1$ and $b - 1$ on a basis. Here are the indecomposable modules:

$$kG = \left(\begin{array}{ccc} & \bullet & \\ \swarrow & & \searrow \\ \bullet & & \bullet \\ \searrow & & \swarrow \\ & \bullet & \end{array}\right), \quad \text{and for } n \geq 1$$

$$W_{2n+1} = \left(\begin{array}{ccccccccc} \bullet & & & & \bullet & & & & \bullet \\ & \searrow & & \swarrow & & \searrow & \cdots & \swarrow & \\ & & \bullet & & & & \bullet & & \end{array}\right)$$

$$W_1 = M_1 = (\bullet)$$

$$M_{2n+1} = \left(\begin{array}{ccccccccc} & & \bullet & & & & \bullet & & \\ & \swarrow & & \searrow & & \swarrow & \cdots & \searrow & \\ \bullet & & & & \bullet & & & & \bullet \end{array}\right)$$

$$E_{f,n} = \left(\begin{array}{ccccccccc} & & \bullet & & & & \bullet & & \\ & \swarrow & & \searrow & & \swarrow & \cdots & \swarrow & \searrow \\ \bullet & & & & \bullet & & \bullet & & \end{array}\right)$$

$$E_{0,n} = \left(\begin{array}{ccccccccc} & & \bullet & & & & \bullet & & \\ & \swarrow & & \searrow & & \swarrow & \cdots & \swarrow & \\ \bullet & & & & \bullet & & \bullet & & \end{array}\right)$$

$$E_{\infty,n} = \left(\begin{array}{ccccccccc} \bullet & & & & \bullet & & \bullet & & \\ & \searrow & & \swarrow & & \searrow & \cdots & \searrow & \\ & & \bullet & & & & & & \bullet \end{array}\right)$$

In these diagrams each node represents a basis element of a vector space, a southwest arrow $\swarrow$ emanating from a node indicates that $a - 1$ sends that basis element to the basis element at its tip, and similarly a southeast arrow $\searrow$ indicates the action of $b - 1$ on a basis element. Where no arrow in some direction emanates from a node, the corresponding element $a - 1$ or $b - 1$ acts as zero.

The even-dimensional indecomposable representations $E_{f,n}$ require some further explanation. They are parametrized by pairs (f, n) where $f \in k[X]$ is an irreducible monic polynomial and $n \geq 1$ is an integer. Let the top row of nodes in the diagram correspond to basis elements $u_1, \ldots, u_n$, and the bottom row to basis elements $v_1, \ldots, v_n$. Let

$$(f(X))^n = X^{mn} + a_{mn-1}X^{mn-1} + \cdots + a_0.$$

The right-most arrow $\searrow$ starting at u_{mn} that has no terminal node is supposed to indicate that $(b - 1)u_{mn} = v_{mn+1}$ where

$$v_{mn+1} = -a_{mn-1}v_{mn-1} - \cdots - a_1 v_2 - a_0 v_1,$$

so that with respect to the given bases $b-1$ has matrix

$$\begin{bmatrix} 0 & 1 & \cdots & 0 \\ \vdots & \ddots & \ddots & \vdots \\ 0 & & 0 & 1 \\ -a_0 & & \cdots & -a_{mn-1} \end{bmatrix}.$$

The significance of this matrix is that it is an indecomposable matrix in rational canonical form with characteristic polynomial f^n.

Theorem 11.5.1. *Let k be a field of characteristic 2. The $k[C_2 \times C_2]$-modules shown are a complete list of indecomposable modules.*

Proof. We describe only the strategy of the proof, and refer to Exercises 3, 18, 19, and 20 at the end of this section and [3] for more details. The first step is to use the fact that the regular representation is injective, with simple socle spanned by $\sum_{g\in G} g$. If U is an indecomposable module for which $(\sum_{g\in G} g)U \neq 0$ then there is a vector $u \in U$ with $(\sum_{g\in G} g)u \neq 0$. The homomorphism $kG \to U$ specified by $x \mapsto xu$ is a monomorphism, since if its kernel were nonzero it would contain $\sum_{g\in G} g$, but this element does not lie in the kernel. Since kG is injective, the submodule kGu is a direct summand of U, and hence $U \cong kG$ since U is indecomposable. From this, we deduce that apart from the regular representation, every indecomposable module is annihilated by $\sum_{g\in G} g$, and hence is a module for the ring $kG/(\sum_{g\in G} g)$, which has dimension 3 and is isomorphic to $k[\alpha, \beta]/(\alpha^2, \alpha\beta, \beta^2)$, where α corresponds to $a-1 \in kG$ and β to $b-1 \in kG$.

Representations of this ring are the same thing as the specification of a vector space U with a pair of linear endomorphisms $\alpha, \beta : U \to U$ that annihilate each other and square to zero. The classification of such pairs of matrices up to simultaneous conjugacy of the matrices (which is the same as isomorphism of the module) was achieved by Kronecker in the nineteenth century, and he obtained the indecomposable forms that we have listed. □

The modules M_{2n+1} and W_{2n+1} have become known as *string modules* and the $E_{f,n}$ as *band modules*, in view of the form taken by the diagrams that describe them. More complicated classifications, but along similar lines, have been achieved for representations of the dihedral, semidihedral, and generalized quaternion 2-groups in characteristic 2. For dihedral 2-groups, all the modules apart from the regular representation are string modules or band modules (see [18]).

Provided the field k is infinite, $k[C_2 \times C_2]$ has infinitely many isomorphism types of indecomposable modules in each dimension larger than 1. They can nevertheless be grouped into finitely many families, as we have seen intuitively in their diagrammatic description. As a more precise version of this idea, consider the infinite-dimensional $k[C_2 \times C_2]$-module M with diagram

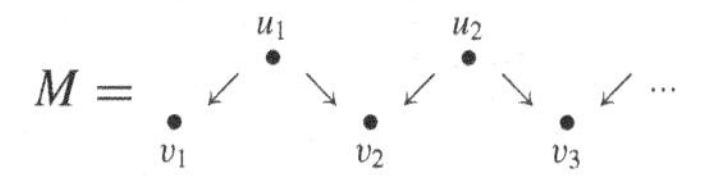

and basis $u_1, u_2, \ldots, v_1, v_2, \ldots$ This module has an endomorphism η that shifts each of the two rows one place to the right, specified by $\eta(u_i) = u_{i+1}$ and $\eta(v_i) = v_{i+1}$, so that M becomes a $(k[C_2 \times C_2], k[X])$-bimodule, where the indeterminate X acts via η. As $k[X]$-modules, we have $M \cong k[X] \oplus k[X]$. Given an irreducible polynomial $f \in k[X]$ and an integer $n \geq 1$, we may construct the $k[C_2 \times C_2]$-module $M \otimes_{k[X]} k[X]/(f^n)$, which is a module isomorphic to $(k[X]/(f^n))^2$ as a $k[X]$-module, and which is acted on by $k[C_2 \times C_2]$ as a module isomorphic to Ef, n. This construction accounts for all but finitely many of the indecomposable $k[C_2 \times C_2]$-modules in each dimension.

With our understanding improved by the last example, we now divide infinite representation type into two kinds: tame and wild. Let A be a finite-dimensional algebra over an infinite field k. We say A has *tame representation type* if it has infinite type and for each dimension d there are finitely many $(A, k[X])$-bimodules M_i that are free as $k[X]$-modules so that all but finitely many of the indecomposable A-modules of dimension d have the form $M_i \otimes_{k[X]} k[X]/(f^n)$ for some irreducible polynomial f and integer n. If the bimodules M_i can be chosen independently of d (as happens with representations of $C_2 \times C_2$), we say that A has *domestic representation type*, and otherwise, it is *nondomestic*.

We say that the finite-dimensional algebra A has *wild representation type* if there is a finitely generated $(A, k\langle X, Y\rangle)$-bimodule M that is free as a right $k\langle X, Y\rangle)$-module, such that the functor $M \otimes_{k\langle X,Y\rangle} -$ from finite-dimensional $k\langle X, Y\rangle$-modules to finite-dimensional A-modules preserves indecomposability and isomorphism classes. Here $k\langle X, Y\rangle$ is the free algebra on two noncommuting variables, having as basis the non-commutative monomials $X^{a_1} Y^{b_1} X^{a_2} Y^{b_2} \cdots$ where $a_i, b_i \geq 0$.

In view of the following theorem, it would have been possible, over an algebraically closed field, to define wild to be everything that is not finite or tame. We state the next three results without proof, since they take us outside the scope of this book.

Theorem 11.5.2 (Drozd [11]; Crawley–Boevey [9]). *Let A be a finite-dimensional algebra over an algebraically closed field. Then A has either finite, tame, or wild representation type.*

When A has wild representation type, the functor $M \otimes_{k\langle X,Y\rangle} -$ appearing in the definition of wild type has as its image a subcategory of the category of A-modules with indecomposable modules in bijection with those of $k\langle X, Y\rangle$, suggesting that a classification of indecomposable A-modules would imply one of indecomposable $k\langle X, Y\rangle$-modules. On the other hand, Brenner [8] has shown that, given any finite-dimensional algebra A, there is a full subcategory of the category of $k\langle X, Y\rangle$-modules naturally equivalent to the category of A-modules, and also a finite-dimensional $k\langle X, Y\rangle$-module B so that $\mathrm{End}_{k\langle X,Y\rangle}(B) = A$. This suggests that classifying indecomposable $k\langle X, Y\rangle$-modules is at least as difficult as classifying all finite-dimensional local algebras, as well as their indecomposable representations. In quantifying how difficult these problems are, Prest has shown in [15] that the theory of finite-dimensional $k\langle X, Y\rangle$-modules is undecidable, meaning that there exists a sentence in the language of finite-dimensional $k\langle X, Y\rangle$-modules that cannot be decided by any Turing machine.

In the context of group algebras, the division into finite, tame, and wild representation type is given by the following theorem, in which we include the result of D.G. Higman already proven.

Theorem 11.5.3 (Bondarenko, Brenner, Drozd, Higman, Ringel). *Let k be an infinite field of characteristic p and let G be a finite group with Sylow p-subgroup P. Then kG has finite representation type if and only if P is cyclic, and tame representation type if and only if $p = 2$ and P is dihedral, semidihedral, or generalized quaternion. In all other cases, kG has wild representation type.*

We include $C_2 \times C_2$ as a dihedral group in this theorem. The first step in the proof of this theorem we have already seen, and it is to identify the group algebras of finite representation type. Next, certain group algebras were established as being wild. This is implied by the following result; for this implication see [15]. There is another account in [17], using a differently worded definition of wild representation type.

Theorem 11.5.4 (Brenner [7]). *Let P be a finite p-group having either $C_p \times C_p$ (p odd), $C_2 \times C_4$ or $C_2 \times C_2 \times C_2$ as a homomorphic image, let k be a field of characteristic p, and let E be any finite-dimensional algebra over k. Then there exists a finite-dimensional kP-module M such that $\mathrm{End}_{kP}(M)$ has a nilpotent ideal J and a subalgebra E' isomorphic to E, with the property that the quotient map sends E' isomorphically to $\mathrm{End}_{kP}(M)/J$.*

A theorem of Blackburn [6, p. 74] implies that if P is a 2-group that is not cyclic and does not have $C_2 \times C_4$ or $C_2 \times C_2 \times C_2$ as a homomorphic image, then P is dihedral, semidihedral, or generalized quaternion. Groups with these as Sylow 2-subgroups were the only groups whose representation type was in question at this point. The representation type in these cases has been decided by classifying explicitly the indecomposable modules, showing that it is tame, and it was done by Bondarenko, Drozd, and Ringel. A later approach can be found in the work of Crawley–Boevey.

11.6 Vertices, Sources, and Green Correspondence

Having just given an impression of the difficulty of classifying indecomposable modules in general, we now explain some positive techniques that are available to understand them better.

Theorem 11.6.1. *Let R be a field or a complete discrete valuation ring, and let U be an indecomposable RG-module.*

(1) *There is a unique conjugacy class of subgroups Q of G that are minimal subject to the property that U is Q-projective.*
(2) *Let Q be a minimal subgroup of G such that U is Q-projective. There is an indecomposable RQ-module T that is unique up to conjugacy by elements of $N_G(Q)$ such that U is a summand of $T\uparrow_Q^G$. Such a T is necessarily a summand of $U\downarrow_Q^G$.*

Proof. (1) We offer two proofs of this result, one employing module theoretic techniques, and the other a ring-theoretic approach. Both proofs exploit similar ideas, in that the Mackey formula is a key ingredient.

First proof: we start by supposing that U is both H-projective and K-projective where H and K are subgroups of G. Then U is a summand of $U\downarrow_H^G\uparrow_H^G$ and also of $U\downarrow_K^G\uparrow_K^G$, so it is also a summand of

$$\begin{aligned} U\downarrow_H^G\uparrow_H^G\downarrow_K^G\uparrow_K^G &= \bigoplus_{g\in[K\backslash G/H]} ({}^g((U\downarrow_H^G)\downarrow_{K^g\cap H}^H))\uparrow_{K\cap {}^gH}^K)\uparrow_K^G \\ &= \bigoplus_{g\in[K\backslash G/H]} ({}^g(U\downarrow_{K^g\cap H}^G))\uparrow_{K\cap {}^gH}^G \end{aligned}$$

using transitivity of restriction and induction. Hence U must be a summand of some module induced from one of the groups $K\cap {}^gH$. If both H and K happen to be minimal subject to the condition that U is projective relative to these groups, we deduce that $K\cap {}^gH = K$, so $K\subseteq {}^gH$. Similarly, $H\subseteq {}^{g'}K$ for some g' and so H and K are conjugate.

Second proof: we start the same way and suppose that U is both H-projective and K-projective. We may write $1_U = \mathrm{tr}_H^G\alpha = \mathrm{tr}_K^G\beta$ for certain $\alpha\in\mathrm{End}_{RH}(U)$

and $\beta \in \mathrm{End}_{RK}(U)$. Now

$$\begin{aligned}1_U &= (\mathrm{tr}_H^G \alpha)(\mathrm{tr}_K^G \beta) = \mathrm{tr}_K^G((\mathrm{tr}_H^G \alpha)\beta)\\ &= \mathrm{tr}_K^G(\mathrm{tr}_H^G(\alpha\beta))\\ &= \sum_{g\in[K\backslash G/H]} \mathrm{tr}_{K\cap {}^gH}^G(c_g\alpha\beta).\end{aligned}$$

Since U is indecomposable its endomorphism ring is local and so some term $\mathrm{tr}_{K\cap {}^gH}^G(c_g\alpha\beta)$ must lie outside the unique maximal ideal of $\mathrm{End}_{RG}(U)$ and must be an automorphism. This implies that $\mathrm{tr}_{K\cap {}^gH}^G : \mathrm{End}_{R[K\cap {}^gH]}(U) \to \mathrm{End}_{RG}(U)$ is surjective, since the image of $\mathrm{tr}_{K\cap {}^gH}^G$ is an ideal, and so U is $K \cap {}^gH$-projective.

We now deduce as in the first proof that if K and H are minimal subgroups relative to which U is projective, then H and K are conjugate.

(2) Let Q be a minimal subgroup relative to which U is projective. We know that U is a summand of $U \downarrow_Q^G\uparrow_Q^G$ and hence it is a summand of $T \uparrow_Q^G$ for some indecomposable summand T of $U \downarrow_Q^G$. Suppose that T' is another indecomposable module for which U is a summand of $T' \uparrow_Q^G$. Now T is a summand of

$$T' \uparrow_Q^G\downarrow_Q^G = \bigoplus_{g\in[Q\backslash G/Q]} ({}^g(T' \downarrow_{Q^g\cap Q})) \uparrow_{Q\cap {}^gQ}^Q$$

and hence a summand of some $({}^g(T' \downarrow_{Q^g\cap Q})) \uparrow_{Q\cap {}^gQ}^Q$. For this element g, we deduce that U is $Q \cap {}^gQ$-projective and by minimality of Q, we have $Q = Q \cap {}^gQ$ and $g \in N_G(Q)$. Now T is a summand of ${}^gT'$, and since both modules are indecomposable, we have $T = {}^gT'$. We deduce from the fact that T is a summand of $U \downarrow_Q^G$ that $T' = {}^{g^{-1}}T$ must be a summand of $({}^{g^{-1}}U) \downarrow_Q^G$ and hence of $U \downarrow_Q^G$, since ${}^{g^{-1}}U \cong U$ as RG-modules. □

A minimal subgroup Q of G relative to which the indecomposable module U is projective is called a *vertex* of U, and it is defined up to conjugacy in G. We write $\mathrm{vtx}(U)$ to denote a subgroup Q that is a vertex of U. An RQ-module T for which U is a summand of $T \uparrow_Q^G$ is called a *source* of U and, given the vertex Q, it is defined up to conjugacy by elements of $N_G(Q)$.

We record some immediate properties of the vertex of a module.

Proposition 11.6.2. *Let R be a field of characteristic p or a complete discrete valuation ring with residue field of characteristic p.*

(1) The vertex of every indecomposable RG-module is a p-group.
(2) An indecomposable RG-module is projective if and only if it is free as an R-module, and its vertex is 1.
(3) A vertex of the trivial RG-module R is a Sylow p-subgroup of G.

Proof. (1) We know from Proposition 11.3.5 that every module is projective relative to a Sylow p-subgroup, and so vertices must be p-groups.

(2) If an indecomposable module is projective it is a summand of RG, which is induced from 1, so it must be free as an R-module and have vertex 1. Conversely, if U has vertex 1, it is a summand of $U \downarrow_1^G \uparrow_1^G$, so if U is free as an R-module, it is a summand of $R \uparrow_1^G = RG$, and hence is projective.

(3) Let Q be a vertex of R and P a Sylow p-subgroup of G containing Q. Then R is a summand of $R \uparrow_Q^G$, so $R \downarrow_P^G$ is a summand of $R \uparrow_Q^G \downarrow_P^G = \bigoplus_{g \in [P \backslash G/Q]} R \uparrow_{P \cap {}^g Q}^P$ and hence is a summand of $R \uparrow_{P \cap {}^g Q}^P$ for some $g \in G$. We claim that for every subgroup $H \leq P$, $R \uparrow_H^P$ is an indecomposable RP-module. From this, it will follow that $R = R \uparrow_{P \cap {}^g Q}^P$ and that $Q = P$. The only simple RP-module is the residue field k with the trivial action (in case R is a field already, $R = k$), and $\operatorname{Hom}_{RP}(R \uparrow_H^P, k) = \operatorname{Hom}_{RH}(R, k) \cong k$ is a space of dimension 1. This means that $R \uparrow_H^P$ has a unique simple quotient, and hence is indecomposable. □

We extract the final claim from the proof of Proposition 11.6.2(3).

Corollary 11.6.3. *Let R be a field of characteristic p or a complete discrete valuation ring with residue field of characteristic p. Let P be a p-group and let H be a subgroup of P. Then $R \uparrow_H^P$ is an indecomposable RP-module.*

The vertices of indecomposable modules partition the collection of all indecomposable RG-modules into subclasses, namely, for each p-subgroup H the modules with vertex H. The class of modules with vertex 1 that are free over R consists of the projective modules, and in this case, we have identified the isomorphism types of the individual modules. Identifying the isomorphism types of modules with cyclic vertex is also possible in many instances and by Theorem 11.4.4, there are only finitely many of them (Exercise 10 at the end of this chapter). Aside from this, every other class contains infinitely many isomorphism types, and they may be beyond our capabilities to classify.

We now bring in Green correspondence. This allows us to reduce many questions about indecomposable modules to a situation where the vertex of the module is a normal subgroup. The theory will be used when we come to consider blocks in Chapter 12, and it is very helpful in many other situations. The philosophy that many questions in modular representation theory of finite groups are determined by normalizers of p-subgroups is one of the most important themes in this area.

Green correspondence gives a bijection (denoted f in what follows, with inverse g) from isomorphism types of indecomposable RG-modules with vertex a given p-subgroup Q to isomorphism types of indecomposable $RN_G(Q)$-modules with vertex Q. It also gives an isomorphism between certain groups of homomorphisms, but we omit this part of the statement.

Theorem 11.6.4 (Green correspondence). *Let R be a field of characteristic p or a complete discrete valuation ring with residue field of characteristic p. Let Q be a p-subgroup of G and L a subgroup of G that contains the normalizer $N_G(Q)$.*

(1) Let U be an indecomposable RG-module with vertex Q. Then in any decomposition of $U \downarrow_L^G$ as a direct sum of indecomposable modules there is a unique indecomposable summand $f(U)$ with vertex Q. Writing $U \downarrow_L^G = f(U) \oplus X$, each summand of X is projective relative to a subgroup of the form $L \cap {}^xQ$ where $x \in G - L$.

(2) Let V be an indecomposable RL-module with vertex Q. Then in any decomposition of $V \uparrow_L^G$ as a direct sum of indecomposable modules there is a unique indecomposable summand $g(V)$ with vertex Q. Writing $V \uparrow_L^G = g(V) \oplus Y$, each summand of Y is projective relative to a subgroup of the form $Q \cap {}^xQ$ where $x \in G - L$.

(3) In the notation of parts (1) and (2), we have $gf(U) \cong U$ and $fg(V) \cong V$.

As a preliminary to the proof, notice that if $x \in G - L$ then $Q \cap {}^xQ$ is a strictly smaller group than Q, since $N_G(Q) \subseteq L$ so x does not normalize Q. On the other hand, $L \cap {}^xQ$ might be a group of the same size as Q, and in that case m it is conjugate to Q in G (by the element x). However, it will be important in step 1 of the proof to know that $L \cap {}^xQ$ cannot be conjugate to Q in L. The argument is that if $L \cap {}^xQ = {}^zQ$ for some $z \in L$ then $z^{-1}x \in N_G(Q) \subseteq L$ so $x \in zL = L$, which contradicts $x \in G - L$.

Proof. We will prove (2) before (1). Let V be an indecomposable RL-module with vertex Q.

Step 1. We show that in any decomposition as a direct sum of indecomposable RL-modules, $V \uparrow_L^G \downarrow_L^G$ has a unique summand with vertex Q, the other summands being projective relative to subgroups of the form $L \cap {}^xQ$ with $x \notin L$. To show this, let T be a source for V, so that $T \uparrow_Q^L \cong V \oplus Z$ for some RL-module Z. Put

$$V \uparrow_L^G \downarrow_L^G = V \oplus V'$$
$$Z \uparrow_L^G \downarrow_L^G = Z \oplus Z'$$

for certain RL-modules V' and Z'. Then

$$\begin{aligned} T \uparrow_Q^G \downarrow_L^G &= V \uparrow_L^G \downarrow_L^G \oplus Z \uparrow_L^G \downarrow_L^G \\ &= V \oplus V' \oplus Z \oplus Z' \\ &= \bigoplus_{x \in [L \backslash G/Q]} ({}^x(T \downarrow^Q_{L^x \cap Q})) \uparrow^L_{L \cap {}^x Q} . \end{aligned}$$

There is one summand in the last direct sum with $x \in L$, and it is isomorphic to $T \uparrow_Q^L = V \oplus Z$. The remaining summands are all induced from the subgroups $L \cap {}^x Q$ with $x \notin L$, and it follows that all indecomposable summands of V' and Z' are projective relative to these subgroups. This in particular implies the assertion we have to prove in this step, since such subgroups cannot be conjugate to Q by elements of L.

Step 2. We show that in any decomposition as a direct sum of indecomposable modules, $V \uparrow_L^G$ has a unique indecomposable summand with vertex Q and that the remaining summands are projective relative to subgroups of the form $Q \cap {}^x Q$ where $x \notin L$. To show this, write $V \uparrow_L^G$ as a direct sum of indecomposable modules and pick an indecomposable summand U for which $U \downarrow_L^G$ has V as a summand. This summand U must have vertex Q; for it is projective relative to Q, since V is, and if U were projective relative to a smaller group then V would be also, contradicting the fact that Q is a vertex of V. This shows that the direct sum decomposition of $V \uparrow_L^G$ has at least one summand with vertex Q.

Let U' be another summand of $V \uparrow_L^G$. Then $U' \downarrow_L^G$ must be a summand of V', in the notation of Step 1, and every indecomposable summand of $U' \downarrow_L^G$ is projective relative to a subgroup $L \cap {}^y Q$ with $y \notin L$. Since U' is a summand of $T \uparrow_Q^G$ it is projective relative to Q, and hence has a vertex Q' that is a subgroup of Q. Since $L \supseteq Q'$ it follows that $U' \downarrow_L^G$ has an indecomposable summand that on restriction to Q' has a source of U' as a summand, and so Q' is a vertex of this summand. It follows that some L-conjugate of Q' must be contained in one of the subgroups $L \cap {}^y Q$ with $y \notin L$. In other words, ${}^z Q' \subseteq L \cap {}^y Q$ for some $z \in L$. Thus, $Q' \subseteq {}^{z^{-1}y} Q$ where $x = z^{-1} y \notin L$. This shows that $Q' \subseteq Q \cap {}^x Q$ and completes the proof of assertion (2) of this theorem.

Step 3. We establish assertion (1). Suppose that U is an indecomposable RG-module with vertex Q. Letting T be a source of U, there is an indecomposable summand V of $T \uparrow_Q^L$ for which U is a summand of $V \uparrow_L^G$. This is because U is a summand of $T \uparrow_Q^G = (T \uparrow_Q^L) \uparrow_L^G$. This RG-module V must have vertex Q, since it is projective relative to Q, and if it were projective relative to a smaller subgroup then so would U be. Now $U \downarrow_L^G$ is a direct summand of $V \uparrow_L^G \downarrow_L^G$, and by Step 1 this has just one direct summand with vertex Q, namely, V. In fact, $U \downarrow_L^G$ must have an indecomposable summand that on further restriction to Q has T as a summand, and this summand has vertex Q. It follows that this summand must be isomorphic to V, and in any expression for $U \downarrow_L^G$ as a direct sum of indecomposable modules, one summand is isomorphic to V and the rest are projective relative to subgroups of the form $L \cap {}^xQ$ with $x \notin L$. This completes the proof of assertion (1) of the theorem.

Step 4. The final assertion of the theorem follows from the first two and the fact that U is isomorphic to a summand of $U \downarrow_L^G \uparrow_L^G$ and V is isomorphic to a summand of $V \uparrow_L^G \downarrow_L^G$. □

The use of Green correspondence is to allow us to understand the indecomposable modules for G in terms of the indecomposable modules for a subgroup of the form $N_G(Q)$ where Q is a p-group. The easiest situation where we may apply this result is when the characteristic of k is p and Q is a Sylow p-subgroup of G of order p. The detailed structure of the modules in this situation depends on the "Brauer tree" of the block to which they belong, which is outside the scope of this book. We can, however, use Green correspondence together with the information we already have about groups with a normal Sylow p-subgroup to say how many indecomposable modules there are.

We will write $l_k(G)$ for the number of isomorphism classes of simple kG-modules. By Theorem 7.3.9, this equals the number of isomorphism classes of indecomposable projective kG-modules and when k is a splitting field it equals the number of p-regular conjugacy classes, by Theorem 9.3.6. We will consider $l_k(N_G(Q)) = l_k(Q \rtimes K)$ where K is a group of order prime to p, and this equals $l_k(K)$, by Corollary 6.2.2.

Corollary 11.6.5. *Let k be a field of characteristic p and let G be a group with a Sylow p-subgroup Q of order p. Then the number of indecomposable kG-modules is $(p-1)l_k(N_G(Q)) + l_k(G)$.*

Proof. The Schur–Zassenhaus Theorem implies that $N_G(Q) = Q \rtimes K$ for some subgroup K of order prime to p, and $l_k(N_G(Q)) = l_kK$. By Corollary 11.2.2, there are $pl_k(N_G(Q))$ indecomposable $kN_G(Q)$-modules and $l_k(N_G(Q))$ of these are projective, so $(p-1)l_k(N_G(Q))$ indecomposable modules have vertex Q. By Green correspondence, this equals the number of indecomposable kG-modules with vertex Q. The remaining indecomposable kG-modules are projective, and there are $l_k(G)$ of them, giving $(p-1)l_k(N_G(Q)) + l_k(G)$ indecomposable modules in total. □

Example 11.6.6. When $G = S_4$ and $p = 3$, we determined the simple and projective $\mathbb{F}_3S_4$-modules in Example 10.1.5 and there are four of them. If Q is a Sylow 3-subgroup then $N_{S_4}(Q) \cong S_3$, which has two simple modules in characteristic 3. Therefore, the number of indecomposable $\mathbb{F}_3S_4$-modules is $(3-1)2+4=8$. However, examination of the Cartan matrix of $\mathbb{F}_3S_4$ shows that it is a direct sum of two blocks of defect zero and a Nakayama algebra isomorphic to $\mathbb{F}_3S_3$, so we did not need to invoke Green correspondence to obtain this result: we could have used Proposition 11.2.1 instead.

Example 11.6.7. When p is odd the group $SL(2,p)$ has order $p(p^2-1)(p-1)$ and has the subgroup Q of matrices $\begin{bmatrix} 1 & \alpha \\ 0 & 1 \end{bmatrix}$ with $\alpha \in \mathbb{F}_p$ as a Sylow p-subgroup with normalizer the matrices $\begin{bmatrix} \beta & \alpha \\ 0 & \beta^{-1} \end{bmatrix}$ with $\alpha, \beta \in \mathbb{F}_p$ and $\beta \neq 0$. This can be seen by direct calculation. Thus, $N(Q) \cong C_p \rtimes C_{p-1}$ and $l_k(N(Q)) = p-1$ for any field k of characteristic p. We have seen in Chapter 6, Exercise 25 that $SL(2,p)$ has at least p-simple modules in characteristic p, and in fact this is the complete list. It follows that $kSL(2,p)$ has $(p-1)^2+p = p^2-p+1$ indecomposable modules.

The case of $PSL(2,p) = SL(2,p)/\{\pm I\}$ when p is odd is similar. Now $N(Q) = C_p \rtimes C_{(p-1)/2}$ has $(p-1)/2$ simple modules, so the number of indecomposable modules is

$$\frac{(p-1)^2}{2} + \frac{p+1}{2} = \frac{p^2-p+2}{2}.$$

This applies, for instance, to the alternating group $A_5 \cong PSL(2,5)$ in characteristic 5, which has $\frac{5^2-5+2}{2} = 11$ indecomposable modules.

Further calculations with these modules will be found in the exercises.

We present in Theorem 11.6.9 a refinement of one part of Green correspondence, known as the Burry–Carlson–Puig Theorem, whose proof illustrates a number of techniques that are regularly used in this theory. The approach is ring-theoretic, and it is interesting to compare this with the module theoretic approach of our proof of the Green correspondence. Before that we single out a result that we will use a number of times.

Lemma 11.6.8. *(1) Suppose that B is a local ring and that $\{I_j \mid j \in J\}$ is a family of ideals of B. If $1 \in \sum_{j\in J} I_j$ then $1 \in I_j$ for some j, so that $I_j = B$.*

(2) (Rosenberg's Lemma) Suppose that B is an R-algebra that is finitely generated as an R-module, where R is a complete discrete valuation ring or a field. Let e be a primitive idempotent in B and suppose that $\{I_j \mid j \in J\}$ is a family of ideals of B. If $e \in \sum_{j\in J} I_j$ then $e \in I_j$ for some j.

Proof. (1) If $1 \notin I_j$ for all j then every I_j is contained in the unique maximal ideal of B and so $1 \notin \sum_{j\in J} I_j$ since 1 does not lie in the maximal ideal.

(2) The only idempotents in eBe are 0 and e since e is primitive. By Proposition 11.1.4 since eBe is finitely generated as an R-module, it is a local ring. It contains the family of ideals $\{eI_je \mid j \in J\}$ and e lies in their sum. Therefore by part (1), e lies in one of the ideals. □

Theorem 11.6.9 (Burry–Carlson–Puig). *Let R be a field of characteristic p or a complete discrete valuation ring with residue field of characteristic p. Let V be an indecomposable RG-module, Q a p-subgroup of G, and H a subgroup of G containing $N_G(Q)$. Suppose $V\downarrow_H^G = M \oplus N$ where M is indecomposable with vertex Q. Then V has vertex Q, from which it follows that M is the Green correspondent of V.*

Proof. We only need to show that V has vertex Q because the conclusion about Green correspondence then follows from Theorem 11.6.4. We will use Higman's criterion to verify this, as well as the Mackey formula, the fact that the image of the relative trace map between endomorphism rings is an ideal, and Rosenberg's Lemma.

For each subgroup K of G we will write $E_K(V)$ for the endomorphism ring $\mathrm{Hom}_{RK}(V, V)$, because this simplifies the notation. For each subgroup K of H the subset $\mathrm{tr}_K^H E_K(V)$ is an ideal of $E_H(V)$. We will put

$$J := \sum_{g \notin H} \mathrm{tr}^H_{H \cap {}^gQ} E_{{}^gQ}(V),$$

and this is an ideal of $E_H(V)$.

We show that for all $\alpha \in E_Q(V)$,

$$\mathrm{res}^G_H \mathrm{tr}^G_Q(\alpha) \equiv \mathrm{tr}^H_Q(\alpha) \pmod{J}.$$

This is a consequence of the Mackey formula, for

$$\mathrm{res}^G_H \mathrm{tr}^G_Q(\alpha) = \sum_{g \in [H\backslash G/Q]} \mathrm{tr}^H_{H \cap {}^gQ} \mathrm{res}^{{}^gQ}_{H \cap {}^gQ} c_g(\alpha)$$

and all of the terms in the sum lie in J except for the one represented by 1 which gives $\mathrm{tr}^H_Q(\alpha)$.

Let $e \in E_H(V)$ be the idempotent that is projection onto M. We claim that $e \in \mathrm{tr}^H_Q E_Q(V)$ and also that $e \notin J$. These come about because Q is a vertex of M and e is the identity in the ring $eE_H(V)e$, which may be identified with the local ring $E_H(M)$. Thus $e \in \mathrm{tr}^H_Q E_Q(M) \subseteq \mathrm{tr}^H_Q E_Q(V)$ since M is projective relative to Q. If e were to lie in J, which is a sum of ideals $\mathrm{tr}^H_{H \cap {}^gQ} E_{{}^gQ}(V)$ with $g \notin H$, it would lie in one of them by Rosenberg's Lemma, and hence M would be projective relative to $H \cap {}^gQ$ for some $g \notin H$. The vertex Q of M must be conjugate in H

to a subgroup of $H \cap {}^gQ$, so that ${}^gQ = {}^hQ$ for some $h \in H$. This implies that $g^{-1}h \in N_G(Q) \subseteq H$, so $g \in H$, a contradiction. Therefore $e \notin J$.

Let ϕ be the composite ring homomorphism $E_G(V) \xrightarrow{\text{res}^G_H} E_H(V) \to E_H(V)/J$ where the second morphism is the quotient map. Since it is the image of a local ring, $\phi(E_G(V))$ is also a local ring. It contains the ideal $\phi(\text{tr}^G_Q E_Q(V))$. We claim that this ideal contains the nonzero idempotent $e + J$. This is because we can write $e = \text{tr}^H_Q(\alpha)$ for some $\alpha \in E_Q(V)$ so that

$$\phi \circ \text{tr}^G_H(e) = \phi\text{tr}^G_H\text{tr}^H_Q(\alpha) = \phi\text{tr}^G_Q(\alpha) \equiv \text{tr}^H_Q(\alpha) = e \pmod J$$

by an earlier calculation.

Any ideal of a local ring containing a nonzero idempotent must be the whole ring, so that $\phi(\text{tr}^G_Q E_Q(V)) = \phi(E_G(V))$. Again since $E_G(V)$ is local, $\text{tr}^G_Q E_Q(V) = E_G(V)$.

We have just shown that V is projective relative to Q, so Q contains a vertex of V. The vertex cannot be smaller than Q because otherwise, by a calculation using the Mackey formula, M would have vertex smaller than Q. We conclude that Q is a vertex of V. □

11.7 The Heller Operator

The Heller or syzygy operator Ω provides a way to construct new indecomposable modules from old ones, obtaining modules $\Omega^i M$ that have importance in the context of homological algebra. Moving beyond the scope of this text, the stable module category of a group algebra has the structure of a triangulated category, and the shift operator is Ω^{-1}. Although we cannot describe that here, it is still important to know about the Heller operator simply as a useful tool in handling indecomposable modules. It will be used in Exercises 14 and 15 at the end of this chapter.

Suppose that R is either a field or a complete discrete valuation ring. Given an RG-module M we define ΩM to be the kernel of the projective cover $P_M \to M$, so that there is a short exact sequence $0 \to \Omega M \to P_M \to M \to 0$. Since projective covers are unique up to isomorphism of the diagram, ΩM is well defined up to isomorphism. We can immediately see that $\Omega M = 0$ if and only if M is projective, so that the operator Ω is not invertible on all modules. On the other hand if we exclude certain modules from consideration then we do find that Ω is invertible. In case R is a field, all we need to do is exclude the projective modules, which are also injective. We may define for any module N a module $\Omega^{-1}N$ to be the cokernel of the injective hull of N, so that there is a short exact sequence $0 \to N \to I_N \to \Omega^{-1}N \to 0$. In the other situation where R is a complete discrete valuation ring we restrict attention to RG-lattices. In this situation there is always a "relative injective hull" $N \to I_N$ for any RG-lattice N. It may be obtained as the dual of the projective cover $P_{N^*} \to N^*$, identifying N with its double dual and I_N with $(P_{N^*})^*$. Again we define $\Omega^{-1}N$ to be the cokernel, so that there is a short exact sequence $0 \to N \to I_N \to \Omega^{-1}N \to 0$. Note that since $(RG)^* \cong RG$ as RG-modules, the module I_N is in fact projective, as well as having an injective property relative to RG-lattices.

We state the next result for RG-lattices. If R happens to be a field, an RG-lattice is the same as a finitely generated RG module.

Proposition 11.7.1. *Let R be a field or a complete discrete valuation ring and G a finite group.*

(1) Let $0 \to U \to V \to W \to 0$ be a short exact sequence of RG-lattices. If W has no nonzero projective summand and $V \to W$ is a projective cover then $U \to V$ is a (relative) injective hull. Similarly, if U has no nonzero projective summand and $U \to V$ is a (relative) injective hull then $V \to W$ is a projective cover.

(2) For any RG-lattice M we have $(\Omega^{-1}M)^* \cong \Omega(M^*)$.

(3) If M is an RG-lattice with no nonzero projective summands then

$$\Omega^{-1}\Omega(M) \cong M \cong \Omega\Omega^{-1}(M).$$

(4) When M_1 and M_2 are RG-lattices we have

$$\Omega(M_1 \oplus M_2) \cong \Omega M_1 \oplus \Omega M_2$$

and

$$\Omega^{-1}(M_1 \oplus M_2) \cong \Omega^{-1}M_1 \oplus \Omega^{-1}M_2.$$

(5) Let M be an RG-lattice with no nonzero projective summands. Then M is indecomposable if and only if ΩM is indecomposable, if and only if $\Omega^{-1}M$ is indecomposable.

Proof. (1) The short exact sequence splits as a sequence of R modules, because W is projective as an R-module, and so the dual sequence $0 \to W^* \to V^* \to U^* \to 0$ is exact. Suppose that $V \to W$ is a projective cover and W has no nonzero projective summand. Then V^* is projective, and if $V^* \to U^*$ is not a projective cover then by Proposition 7.3.3 we may write $V^* = X \oplus Y$ where $X \to U^*$ is a projective cover and Y maps to zero. This would mean that $W^* \cong \Omega(U^*) \oplus Y$ where Y is projective, and on dualizing we deduce that W has a nonzero projective summand, which is not the case. Hence $V^* \to U^*$ is a projective cover, so that $U \to V$ is a (relative) injective hull. The second statement follows from the first on applying it to the dual sequence $0 \to W^* \to V^* \to U^* \to 0$.

(2) The dual of the sequence $0 \to \Omega(M^*) \to P_{M^*} \to M^* \to 0$ that computes $\Omega(M^*)$ is $0 \to M \to (P_{M^*})^* \to \Omega(M^*)^* \to 0$ and it computes $\Omega^{-1}(M)$. Thus $\Omega^{-1}(M) \cong \Omega(M^*)^*$ and dualizing again gives the result.

(3) These isomorphisms follow immediately from (1) since the same sequence that constructs ΩM also constructs $\Omega^{-1}\Omega(M)$ and the sequence that constructs $\Omega^{-1}M$ also constructs $\Omega\Omega^{-1}(M)$.

(4) This comes from the fact that the projective cover of $M_1 \oplus M_2$ is the direct sum of the projective covers of M_1 and M_2, and similarly with (relative) injective hulls.

(5) If $\Omega(M)$ were to decompose then so would $M \cong \Omega^{-1}\Omega(M)$ by (4), so the indecomposability of M implies the indecomposability of ΩM. The reverse implication and the equivalence with the indecomposability of $\Omega^{-1}M$ follow similarly. □

We see from this that Ω permutes the isomorphism types of indecomposable RG-lattices, with inverse permutation Ω^{-1}. Let us write Ω^i for the ith power of Ω when $i > 0$ and the $-i$th power of Ω^{-1} when $i < 0$. When $i = 0$ we put $\Omega^0 M = M$.

We sketch the role of Ω in homological algebra. A key notion in homological algebra is that of a *projective resolution* of a module M. This is a sequence of projective modules

$$\cdots \xrightarrow{d_3} P_2 \xrightarrow{d_2} P_1 \xrightarrow{d_1} P_0 \xrightarrow{d_0} 0 \xrightarrow{d_{-1}} 0 \cdots$$

that is exact everywhere except at P_0, where its homology (in this case the cokernel of the map d_1) is M. When R is a field or a complete discrete valuation ring we can construct always a *minimal* projective resolution, which has the further defining property that each map $P_i \to \operatorname{Ker} d_{i-1}$ is a projective cover. We see by induction that for a minimal projective resolution, $\Omega^i M \cong \operatorname{Ker} d_{i-1}$ when $i \geq 1$.

We will use properties of Ω in the exercises as part of a proof that the list of indecomposable modules for $k[C_2 \times C_2]$ is complete, where k is a field of characteristic 2. Before we leave Ω we mention its properties under the Kronecker product.

Proposition 11.7.2. *Let R be a field or a complete discrete valuation ring and G a finite group. For any RG-lattice M we have $M \otimes_R \Omega^i R \cong \Omega^i M \oplus Q_i$ for each i, where Q_i is a projective RG-module.*

Proof. Let $\cdots \to P_2 \to P_1 \to P_0 \to 0$ be a projective resolution of k. We claim that the sequence $\cdots \to M \otimes_R P_2 \to M \otimes_R P_1 \to M \otimes_R P_0 \to 0$ is exact except in position 0. This is because every map in the resolution must split as a map of R-modules since M is projective as an R-module. The homology in position 0 is $M \otimes_R k \cong M$ for the same reason. Furthermore, all of the modules $M \otimes_R P_i$ are projective RG-modules by Proposition 8.1.4. Thus we have a projective resolution of M, but there is no reason why it should be minimal. Assuming inductively that the kernel of the map in position $i-1$ has the form $\Omega^i M \oplus Q_i$ for some projective module Q_i (and this is true when $i = 0$), we deduce from Proposition 7.3.3(1) that the kernel in position i has the form $\Omega^{i+1} M \oplus Q_{i+1}$, which completes the proof by induction on i when $i \geq 0$. When $i < 0$ we can deduce the result by duality using part (2) of Proposition 11.7.1. □

We see from this result that up to isomorphism the modules $\Omega^i R$ where $i \in \mathbb{Z}$, together with the projective modules, are closed under the operations of taking indecomposable summands of tensor products. As a final comment we mention that Ω preserves vertices and Green correspondents of indecomposable modules. We leave this to Exercise 13 at the end of this chapter.

11.8 Some Further Techniques with Indecomposable Modules

There are many further techniques for handling indecomposable modules that have been developed, but at this point they start to go beyond the basic account we are attempting in this book and move into more specialist areas. Here is a pointer to some further topics. Where no other reference is given, and account of the topic can be found in the book by Benson [3].

- **Green's Indecomposability Theorem.** This states for a p-group, working with an algebraically closed field, that indecomposable modules remain indecomposable under induction. More generally the result is true, over an arbitrary field, for modules that are "absolutely indecomposable." This means for p-groups (over algebraically closed fields) that if a module has a certain subgroup as a vertex then the module is induced from that subgroup. It also has the consequence for arbitrary groups that if an indecomposable module has a certain subgroup as a vertex then its dimension is divisible by the index of the vertex in a Sylow p-subgroup.
- **Blocks with cyclic defect.** We have not yet defined the defect of a block (it is done in the next section) but this concept includes the case of all representations of groups with cyclic Sylow p-subgroups. Each block of cyclic defect is described by a tree, called the Brauer tree, and from this tree it is possible to give a complete listing of indecomposable modules together with their structure. An account of this theory can be found in the book by Alperin [2].

• **Auslander–Reiten theory.** The theory of almost split sequences (also known as Auslander–Reiten sequences) and the Auslander–Reiten quiver have been one of the fundamental tools in the abstract representation theory of algebras since they were introduced. They provide a way to describe some structural features of indecomposable modules even in situations of wild representation type, giving information that is more refined than the theory of vertices, for example. In the context of group representations they have been used in several ways, and most notable is the theory developed by Erdmann [12] in classifying the possible structures of blocks of tame representation type.

• **The Green ring.** This is a Grothendieck group of modules that has a basis in bijection with the isomorphism classes of indecomposable modules, and it is a useful home for constructions that extend classical results about the character ring. Notable are induction theorems of Conlon and Dress, generalizing the theorems for the character ring of Artin and Brauer. They have had various applications including a technique for computing group cohomology.

• **Diagrams for modules.** In this chapter and in some other places we have drawn diagrams with nodes and edges to indicate the structure of representations, but nowhere have these diagrams been defined in any general way. Evidently this is a useful approach to representations, and the reader may wonder what kind of theory these diagrams have. The trouble is that it is not easy to make a general definition that is sufficiently broad to apply to all modules that might arise. If we do give a general definition we find that the diagrams can be so complicated that they are no easier to understand than other ways of looking at representations. The reader can find more about this in the work of Alperin, Benson, Carlson, and Conway [1, 4, 5].

11.9 Summary of Chapter 11

Throughout, R is a complete discrete valuation ring with residue field of characteristic p (or a field of characteristic p) and k is a field of characteristic p.

- The indecomposable summands in any direct sum decomposition of a finitely generated RG-module have isomorphism type and multiplicity determined independently of the particular decomposition (the Krull–Schmidt Theorem).
- Every RG-module is projective relative to a Sylow p-subgroup. An RG-module is projective if and only if it is projective on restriction to a Sylow p-subgroup.
- The representation type of kG is characterized in terms of the Sylow p-subgroups of G. It is finite if and only if Sylow p-subgroups are cyclic. It is tame if and only if $p = 2$ and Sylow 2-subgroups are dihedral, semidihedral or generalized quaternion. Otherwise the representation type is wild.

- When G has a normal cyclic Sylow p-subgroup, kG is a Nakayama algebra and the indecomposable modules are completely described.
- When k is a field of characteristic 2 the indecomposable representations of $k[C_2 \times C_2]$ are classified as string or band modules. Modules for other tame 2-groups can be classified over k. It is not reasonable to expect to classify modules for an algebra of wild representation type.
- Every indecomposable RG-module has a vertex and a source. A vertex is always a p-subgroup. A module is projective if and only if it has vertex 1 and is R-free. There are only finitely many isomorphism types of indecomposable kG-modules with cyclic vertex. A vertex of the trivial module is a Sylow p-subgroup.
- Green correspondence gives a bijection between isomorphism types of indecomposable RG-modules with vertex D and indecomposable $R[N_G(D)]$-modules with vertex D. It allows us to give a complete listing of indecomposable kG-modules with cyclic vertex
- The Heller operator Ω provides a bijection from the set of isomorphism classes of nonprojective indecomposable modules to itself. It preserves vertices, sources and Green correspondents.

11.10 Exercises for Chapter 11

We will assume throughout these exercises that the ground ring R is either a complete discrete valuation ring with residue field of characteristic p, or a field k of characteristic p, so that the Krull–Schmidt Theorem holds.

1. Write out proofs of the following assertions. They refer to subgroups $H \leq K \leq G$ and $J \leq G$, an RG-module U and an RK-module V.

(a) If U is H-projective then U is K-projective.
(b) If U is H-projective and W is an indecomposable summand of $U \downarrow_J^G$ then W is $J \cap {}^gH$-projective for some element $g \in G$. Deduce that there is a vertex of W that is contained in a subgroup $J \cap {}^gH$.
(c) If U is a summand of $V \uparrow_K^G$ and V is H-projective then U is H-projective.
(d) For any $g \in G$, U is H-projective if and only if ${}^gU$ is ${}^gH$-projective.
(e) If U is H-projective and W is any RG-module then $U \otimes W$ is H-projective.

2. Let A be a ring with a 1.

(a) Suppose that U is an A-module with two decompositions $U = U_1 \oplus U_2 = V_1 \oplus V_2$ as left A-modules corresponding to idempotent decompositions $1_U = e_1 + e_2 = f_1 + f_2$ in $\text{End}_A(U)$. Show that $U_1 \cong V_1$ and $U_2 \cong V_2$ if and only if $f_1 = \alpha e_1 \alpha^{-1}$ for some $\alpha \in \text{Aut}_A(U)$.

(b) Suppose that $e, f \in A$ are idempotent elements. Show that $Ae \cong Af$ and $A(1-e) \cong A(1-f)$ as left A-modules if and only if $f = \alpha e \alpha^{-1}$ for some unit $\alpha \in A^\times$.

(c) Suppose that A is Noetherian as a left module over itself and suppose $I \lhd A$ is a 2-sided ideal of A with $I \subseteq \text{Rad}\, A$. Let $e, f \in A$ be idempotents for which $e + I = f + I$. Show that $f = \alpha e \alpha^{-1}$ for some unit $\alpha \in A^\times$. [Use part (b) and the uniqueness of projective covers.]

3. Let U be an indecomposable module for a finite-dimensional algebra A over a field. Assuming that U is not simple, show that $\text{Soc}\, U \subseteq \text{Rad}\, U$. Deduce that if U has Loewy length 2 then $\text{Soc}\, U = \text{Rad}\, U$.

4. Let Q be a subgroup of G. Suppose that U is an indecomposable RG-module that is Q-projective and that $U \downarrow_Q^G$ has an indecomposable summand that is not projective relative to any proper subgroup of Q. Show that Q is a vertex of U.

5. Suppose that Q is a vertex of an indecomposable RG-module U and that H is a subgroup of G that contains Q.

(a) For each subgroup $Q' \subseteq H$ that is conjugate in G to Q, show that $U \downarrow_H^G$ has an indecomposable summand with vertex Q'. Deduce that if $U \downarrow_H^G$ is indecomposable then subgroups of H that are conjugate in G to Q are all conjugate in H.

(b) Show that there is an indecomposable RH module V with vertex Q so that U is a direct summand of $V \uparrow_H^G$.

6. Suppose that H is a subgroup of G and that V is an indecomposable RH-module with vertex Q, where $Q \leq H$. Show that $V \uparrow_H^G$ has an indecomposable direct summand with vertex Q. Show that for every p-subgroup Q of G there is an indecomposable RG-module with vertex Q.

7. Let H be a subgroup of G.

(a) Show that $R \uparrow_H^G$ has an indecomposable summand for which a Sylow p-subgroup of H is a vertex. In particular, R has vertex a Sylow p-subgroup of G.

(b) Let U be an indecomposable summand of $R \uparrow_H^G$. Show that the source of U is the trivial module (for the subgroup that is the vertex of U).

[Because of this, the indecomposable summands of permutation modules (over a field) are sometimes called *trivial source modules*, and also p-permutation modules when the field k has characteristic p.]

8. Let U be an indecomposable trivial source RG-module with vertex Q (see Exercise 7). Show that the Green correspondent $f(U)$ with respect to $N_G(Q)$ is a projective module for $R[N_G(Q)/Q]$, made into a $k[N_G(Q)]$-module via inflation along the quotient map $N_G(Q) \to N_G(Q)/Q$. Show, conversely, that these inflated projectives are a complete list of the Green correspondents of trivial source modules. Deduce that the number of trival source RG-modules equals the sum, over conjugacy classes of p-subgroups Q of G, of $l_k(N_G(Q))$, where $l_k(H)$ denotes the number of isomorphism classes of simple kH-modules.

9. Given an indecomposable RG-module U, let X be the set of pairs (Q, T) such that Q is a vertex of U and T is a source of U with respect to Q. For each $g \in G$ define ${}^g(Q, T) := ({}^gQ, {}^gT)$. Show that this defines a permutation action of G on X, and that it is transitive.

10. Let H be a noncyclic p-subgroup of G and k a field of characteristic p. Show that there are infinitely many isomorphism types of indecomposable kG-modules with vertex H.

11. Let U be an indecomposable kG-module where k is a field. Assuming that U is not projective, show that $U/\operatorname{Rad} U \cong \operatorname{Soc} \Omega U$.

12. Let k be a field of characteristic p and suppose that U is a kG-module with the property that for every proper subgroup $H < G$, $U \downarrow_H^G$ is a projective kH-module.

(a) Show that if G is a cyclic p-group of order $> p$ then U must be a projective kG-module.
(b) Show by example that if $G = C_2 \times C_2$ is the Klein four-group and $p = 2$ then U need not be a projective kG-module.

13. Let U be an indecomposable RG-module with vertex Q, let $H \supseteq N_G(Q)$ and let $V = f(U)$ be the RH-module that is the Green correspondent of U.

(a) Show that U and ΩU have the same vertex.
(b) Show that $f(\Omega U) \cong \Omega(f(U))$ and $g(\Omega V) \cong \Omega(g(V))$.
(c) Show that $f(U^*) \cong f(U)^*$ and $g(V^*) \cong g(V)^*$.

14. The group $G := PSL(2, 5)$ (which is isomorphic to A_5) has three simple modules over $k = \mathbb{F}_5$, of dimensions 1, 3 and 5, and has Cartan matrix (with the simples taken in the order just given)

$$\begin{bmatrix} 2 & 1 & 0 \\ 1 & 3 & 0 \\ 0 & 0 & 1 \end{bmatrix}.$$

You may assume that a Sylow 5-subgroup is cyclic of order 5 and that its normalizer H is dihedral of order 10.

(a) Show that Ω has two orbits, each of length 4, on the nonprojective kD_{10}-modules. Deduce that for kG there are 8 indecomposable nonprojective modules in two Ω orbits of length 4.
(b) Show that apart from a block of defect zero, each indecomposable projective module for kG has Loewy length 3, and that its radical series equals its socle series.
(c) Show that each nonprojective indecomposable module for kG has Loewy length at most 2.
(d) Identify the socle and the radical quotient of each of the 8 indecomposable nonprojective kG-modules.

15. The group $G := PSL(2,7)$ (which is isomorphic to $GL(3,2)$) has four simple modules over $k = \mathbb{F}_7$, of dimensions 1, 3, 5, and 7, and has Cartan matrix (with the simples taken in the order just given)

$$\begin{bmatrix} 2 & 0 & 1 & 0 \\ 0 & 3 & 1 & 0 \\ 1 & 1 & 2 & 0 \\ 0 & 0 & 0 & 1 \end{bmatrix}.$$

You may assume that a Sylow 7-subgroup is cyclic of order 7 and that its normalizer H is a non-abelian group $\langle x, y \mid x^7 = y^3 = 1,\ yxy^{-1} = x^2 \rangle$ of order 21.

(a) Show that Ω has three orbits, each of length 6, on the nonprojective kH-modules. Deduce that for kG there are 18 indecomposable nonprojective modules in three Ω orbits of length 6.
(b) Show that apart from a block of defect zero, each indecomposable projective module for kG has Loewy length 3, and that its radical series equals its socle series.
(c) Show that each nonprojective indecomposable module for kG has Loewy length at most 2.
(d) Identify the socle and the radical quotient of each of the 18 indecomposable nonprojective kG-modules.

16. Let k be a field of characteristic p and suppose that $G = H \rtimes K$ where $H = \langle x \rangle \cong C_{p^n}$ is cyclic of order p^n and K is a group of order relatively prime to p (as in Corollary 11.2.2).

(a) Let U_r be the indecomposable kH-module of dimension r, $1 \le r \le p^n$. If $J \le H$, show that J is a vertex of U_r if and only if $r = |H : J|q$ where q is prime to p. [Use the fact, shown in Exercise 24 of Chapter 6, that indecomposable kJ-modules induce to indecomposable kH-modules.]

(b) Using the description of projective kG-modules in Proposition 8.3.3 and the description of indecomposables in Proposition 11.2.1, show that if V is an indecomposable kG-module then $V \downarrow_H^G$ is a direct sum of copies of a single indecomposable kH-module U_r, for some r, the number of copies being $\dim S$ for some simple kK-module S. Let $J \le H$ be a vertex of V. Show that $r = |H : J|q$ for some number q prime to p. Assuming that k is a splitting field for G, show that $\dim V = |H : J|q'$ for some number q' prime to p.

17. Let H be a subgroup of G and let U, V be RG-modules. We say that an RG-module homomorphism $f : U \to V$ *factors through* an H-projective module if there is an H-projective RG-module M and RG-module homomorphisms $U \xrightarrow{\theta} M \xrightarrow{\phi} V$ so that $f = \phi\theta$.

(a) Similarly to Proposition 11.3.4, show that the following conditions are equivalent:
   (1) f factors through an H-projective module,
   (2) f factors through an induced module $N \uparrow_H^G$ for some RH-module N,
   (3) f factors as $U \xrightarrow{\eta} U \downarrow_H^G \uparrow_H^G \xrightarrow{\phi} V$ for some RG-module homomorphism ϕ, where η is the map constructed after Corollary 11.3.3,
   (4) f factors as $U \xrightarrow{\theta} V \downarrow_H^G \uparrow_H^G \xrightarrow{\epsilon} V$ for some RG-module homomorphism θ, where ϵ is the map constructed after Corollary 11.3.3,
   (5) f lies in the image of $\mathrm{tr}_H^G : \mathrm{Hom}_{RH}(U, V) \to \mathrm{Hom}_{RG}(U, V)$.

(b) When R is a field or a complete discrete valuation ring, use Exercise 7 from Chapter 8 to show that the space of homomorphisms $U \to V$ that factor through a projective has dimension equal to the multiplicity of P_R as an RG-module summand of $\mathrm{Hom}_R(U, V)$.

18. Let k be a field of characteristic 2, and let U be an indecomposable $k[C_2 \times C_2]$-module that is not simple or projective. By considering the Loewy length of U, $\dim \mathrm{Soc}\, U$ and $\dim U/\mathrm{Soc}\, U$ show that one of the following three possibilities must occur:

(a) $\dim \Omega U < \dim U < \dim \Omega^{-1} U$, or
(b) $\dim \Omega U > \dim U > \dim \Omega^{-1} U$, or
(c) $\dim \Omega U = \dim U = \dim \Omega^{-1} U$.

In cases (a) and (b) show that $U = \Omega^i(k)$ for some i. Deduce that $\dim U$ is odd and $|\dim \operatorname{Soc} U - \dim U/\operatorname{Soc} U| = 1$.

In case (c) show that $\dim \operatorname{Soc} U = \dim U/\operatorname{Soc} U$ so that $\dim U$ is even.

19. Consider one of the even-dimensional $k[C_2 \times C_2]$-modules $E_{f,n}$ (in the notation used before Theorem 11.5.1) where k is a field of characteristic 2, and suppose that $f \neq 0, \infty$. The goal of this question is to show that $E_{f,n}$ is indecomposable. Define $u_{mn+1} = -a_{mn-1}u_{mn-1} - \cdots - a_1u_2 - a_0u_1$ and let $\eta : E_{f,n} \to E_{f,n}$ be the linear map specified by $\eta(u_i) = u_{i+1}$, $\eta(v_i) = v_{i+1}$ where $1 \leq i \leq mn$.

(a) Show that $\eta \in \operatorname{End}_{k[C_2\times C_2]}(E_{f,n})$.
(b) Show that the subalgebra $k[\eta]$ of $\operatorname{End}_{k[C_2\times C_2]}(E_{f,n})$ generated by η is isomorphic to $k[X]/(f^n)$.
(c) Show that

$$k[\eta] + \operatorname{Hom}_{k[C_2\times C_2]}(E_{f,n}, \operatorname{Soc}(E_{f,n})) = \operatorname{End}_{k[C_2\times C_2]}(E_{f,n}).$$

(d) Deduce that $E_{f,n}$ is an indecomposable $k[C_2 \times C_2]$-module.

20. Let $G = \langle a\rangle \times \langle b\rangle = C_2 \times C_2$ and let U be an even-dimensional indecomposable kG-module where k is a field of characteristic 2. Suppose that multiplication by $a - 1$ induces an isomorphism $U/\operatorname{Rad}(U) \to \operatorname{Soc}(U)$.

(a) Show that there is an action of the polynomial ring $k[X]$ on U so that $X(a-1)u = (b-1)u = (a-1)Xu$ for all $u \in U$. Show that $U \cong k\langle a\rangle \otimes_k (U/\operatorname{Rad}(U))$ as $k\langle a\rangle \otimes_k K[X]$-modules.
(b) Show that invariant subspaces of U as a $k\langle a\rangle \otimes_k K[X]$-module are also invariant subspaces of U as a kG-module. Show that $U/\operatorname{Rad}(U)$ is an indecomposable $k[X]$-module and deduce that $U/\operatorname{Rad}(U) \cong k[X]/(f^n)$ as $k[X]$-modules for some irreducible polynomial f and integer n.
(c) Prove that $U \cong E_{f,n}$ as kG-modules.

21. Let (F, R, k) be a p-modular system in which R is complete. Let L be an RG-lattice.

(a) Show that L is an indecomposable RG-module if and only if $L/\pi L$ is an indecomposable kG-module. [Use Proposition 9.5.2 and Corollary 11.1.5.]

(b) Assuming that L is an indecomposable RG-lattice, show that the vertices of L and $L/\pi L$ are the same, and that if T is a source of L then $T/\pi T$ is a source of $L/\pi L$.

(c) Again assuming that L is indecomposable, let Q denote a vertex of L and let H be a subgroup of G containing $N_G(Q)$. Show that $f(L)/\pi f(L) \cong f(L/\pi L)$ where f denotes the Green correspondence with respect to H. Prove a similar formula for the reverse Green correspondence map g.

12
Blocks

Block theory is one of the deepest parts of the representation theory of finite groups. In this chapter we can only scratch the surface of the sophisticated constructions and techniques that have been developed. Our goal is to serve the needs of the nonspecialist who might attend a talk on this topic and wish to understand something. For such purposes it is useful to know the basic definitions, to have an idea of the defect group of a block and its relation to vertices of indecomposable modules in the block, and to have seen the different methods used in block theory—sometimes ring theoretic, sometimes module theoretic. We will describe these things, and at that stage we will be able to see that the blocks of defect zero introduced in Chapter 9 do indeed deserve the name. We will demonstrate the equivalence of several different definitions of the defect group of a block, given in terms of the theory of vertices, the relative trace map, and the Brauer morphism. We will conclude with a discussion of the Brauer correspondent and Brauer's First Main Theorem, which establishes a correspondence between blocks with a certain defect group D and blocks of $N_G(D)$ with defect group D.

One of the confusing things about blocks of finite groups is that there often seems to be more than one definition of the same concept. Different authors use different approaches, and then may have more than one definition in mind at the same time. The common ground is a finite group G and a splitting p-modular system (F, R, k), where R is a discrete valuation ring with field of fractions F of characteristic zero, and with residue field k of characteristic p. Whatever a p-block of G is, it determines and is determined by any of the following data:

- an equivalence class of kG-modules, sometimes simple, sometimes indecomposable, sometimes arbitrary,
- an equivalence class of RG-modules,

- a primitive central idempotent in kG, or in RG,
- an equivalence class of primitive idempotents in kG, or in RG,
- an equivalence class of FG-modules,
- an indecomposable 2-sided direct summand of kG,
- an indecomposable 2-sided direct summand of RG,
- a division of the Cartan matrix of kG into block diagonal form.

We will explain the relationship between these different aspects of a p-block of a finite group.

12.1 Blocks of Rings in General

We make the definition that a *block* of a ring A with identity is a primitive idempotent in the center $Z(A)$. A block is thus a primitive central idempotent (rather than a central primitive idempotent, which is a more restrictive condition). We will show in this section that blocks determine, and are determined by, the modules that "belong" to them. We eventually characterize blocks of finite-dimensional algebras in terms of block-diagonal decompositions of the Cartan matrix. First we recall, without proof, Proposition 3.6.1.

Proposition 12.1.1. *Let A be a ring with identity. Decompositions*

$$A = A_1 \oplus \cdots \oplus A_r$$

as direct sums of 2-sided ideals A_i biject with expressions

$$1 = e_1 + \cdots + e_r$$

as a sum of orthogonal central idempotent elements, where e_i is the identity element of A_i and $A_i = Ae_i$. The A_i are indecomposable as rings if and only if the e_i are primitive central idempotent elements. If every A_i is indecomposable as a ring then the A_i, and also the primitive central idempotents e_i, are uniquely determined as subsets of A, and every central idempotent can be written as a sum of certain of the e_i.

This decomposition of an algebra as a direct sum of ideals also gives rise to a decomposition of its modules.

Proposition 12.1.2. *Let A be a ring with identity, let $1 = e_1 + \cdots + e_n$ be a sum of orthogonal idempotents of $Z(A)$ and let U be an A-module. Then $U = e_1U \oplus \cdots \oplus e_nU$ as A-modules. Thus if U is indecomposable we have $e_iU = U$ for precisely one i, and $e_jU = 0$ for $j \neq i$. These summands of U satisfy* $\operatorname{Hom}_A(e_iU, e_jU) = 0$ *if $i \neq j$.*

Proof. Every element u of U can be written $u = e_1u + \cdots + e_nu$, so that $U = e_1U + \cdots + e_nU$. The sum is direct because if $u \in e_iU \cap \sum_{j \neq i} e_jU$ then $e_iu = u = 0$, since e_i acts as the identity on summand i and as zero on the other summands. If $f : e_iU \to e_jU$ is a homomorphism then $f(u) = e_jf(e_iu) = e_je_if(u) = 0$ for all u in U, if $i \neq j$. □

We will say that an A-module U *belongs to* (or *lies in*) the block e if $eU = U$. As an immediate consequence of Proposition 12.1.2 we have the following.

Corollary 12.1.3. *Let A be a ring with identity. An A-module belongs to a block e if and only if each of its direct summands belongs to e. An indecomposable A-module U belongs to a unique block of A, and it is characterized as the block e for which $eU \neq 0$. The modules that belong to a block determine that block.*

Proof. This is immediate from Proposition 12.1.2. For the last sentence we observe that each block e does have some modules that belong to it, namely, the Ae-modules regarded as A-modules via the projection to Ae. □

Example 12.1.4. When A is a finite-dimensional semisimple algebra over a field, the blocks correspond to the matrix summands of A, each block being the idempotent that is the identity element of a matrix summand. Each simple module lies in its own block, and so there is (up to isomorphism) only one indecomposable module in each block. When $A = \mathbb{C}G$ is a group algebra over $\mathbb{C}$ we have seen in Theorem 3.6.2 a formula in terms of characters for the primitive central idempotent in $\mathbb{C}G$ corresponding to each simple complex representation.

We now characterize the blocks of an algebra in module theoretic terms by means of an equivalence relation on the simple modules. The next lemmas prepare for this.

Lemma 12.1.5. *Let e be a block of a ring A with identity. If $0 \to U \to V \to W \to 0$ is a short exact sequence of A-modules then V belongs to e if and only if U and W belong to e.*

In other words, every submodule and factor module of a module that belongs to e also belong to e, and an extension of two modules that belong to e also belongs to e. Thus, by Proposition 12.1.2, all factors of submodules of an indecomposable module lie in the same block.

Proof. A module belongs to e if and only if multiplication by e is an isomorphism of that module. This property holds for V if and only if it holds for U and W. □

Lemma 12.1.6. *Let A be a ring with identity. Let C_1, C_2 be two sets of simple A-modules with the property that $C_1 \cup C_2$ contains all isomorphism types of simple modules, $C_1 \cap C_2 = \emptyset$, and for all $S \in C_1$, $T \in C_2$ there is no nonsplit extension $0 \to S \to V \to T \to 0$ or $0 \to T \to V \to S \to 0$. Then every finite length module M can be written $M = U_1 \oplus U_2$ where U_1 has all its composition factors in C_1 and U_2 has all its composition factors in C_2. The submodules U_1 and U_2 are the unique largest submodules of M with all composition factors in C_1 and C_2, respectively.*

Proof. It is possible to find a composition series $0 = M_0 \subset M_1 \subset \cdots \subset M_n = M$ so that for some i, the composition factors of M_i all lie in C_1 and the composition factors of M/M_i all lie in C_2. This is so because, starting with any composition series, if we ever find a composition factor M_j/M_{j-1} in C_2 with M_{j+1}/M_j in C_1 then the short exact sequence

$$0 \to M_j/M_{j-1} \to M_{j+1}/M_{j-1} \to M_{j+1}/M_j \to 0$$

splits, and so we can find another composition series with the order of the factors M_j/M_{j-1} and M_{j+1}/M_j interchanged. Repeating this, we can move M_j/M_{j-1} and the other composition factors from C_2 above all composition factors in C_1. We take $U_1 = M_i$ once this has been done. Reversing the roles of C_1 and C_2, we find U_2 similarly. Now $M = U_1 \oplus U_2$ because $U_1 \cap U_2 = 0$, having all its composition factors in both C_1 and C_2, and $U_1 + U_2$ has composition length equal to that of M, so $U_1 + U_2 = M$.

If U_1' is a submodule of M with all its composition factors in C_1 then $U_1 + U_1'$ has all its composition factors in C_1, so has composition length at most the length of U_1 by the Jordan–Hölder theorem. From this, it follows that $U_1' \subseteq U_1$. The argument showing maximality of U_2 is similar. □

We put an equivalence relation on the set of simple modules of an algebra A: define $S \sim T$ if either $S \cong T$ or there is a list of simple A-modules $S = S_1, S_2, \ldots, S_n = T$ so that for each $i = 1, \ldots, n-1$, the modules S_i and S_{i+1} appear in a nonsplit short exact sequence of A-modules $0 \to U \to V \to W \to 0$ with $\{U, W\} = \{S_i, S_{i+1}\}$. It is immediate that if S and T belong to different equivalence classes then every sequence $0 \to S \to V \to T \to 0$ must split, and the equivalence classes are the smallest sets of modules with this property.

Proposition 12.1.7. *Let A be a finite-dimensional algebra over a field k. The following are equivalent for simple A-modules S and T:*

(1) S and T lie in the same block.
(2) There is a list of simple A-modules $S = S_1, S_2, \ldots, S_n = T$ so that S_i and S_{i+1} are both composition factors of the same indecomposable projective module, for each $i = 1, \ldots, n-1$.
(3) $S \sim T$.

Proof. The implication (2) $\Rightarrow$ (1) follows because all composition factors of an indecomposable (projective) module all belong to the same block, by Lemma 12.1.5.

The implication (3) $\Rightarrow$ (2) follows because if $0 \to U \to V \to W \to 0$ is a non-split short exact sequence with U and W simple then V is a homomorphic image of the projective cover P_W of W, so that U and V are composition factors of the same indecomposable projective module.

To prove that (1) imples (3), suppose that S and T lie in the same block. Partition the simple A-modules as $\mathcal{C}_1 \cup \mathcal{C}_2$ where $\mathcal{C}_1$ consists of the simple modules equivalent to S and $\mathcal{C}_2$ consists of the remaining simple modules. As in Lemma 12.1.6 write $A = A_1 \oplus A_2$ where all composition factors of A_1 are in $\mathcal{C}_1$ and all composition factors of A_2 are in $\mathcal{C}_2$. These submodules of A are left ideals, and they are also right ideals because if $x \in A$ then $A_i x$ is a homomorphic image of A_i under right multiplication by x, so has composition factors in $\mathcal{C}_i$, and is therefore contained in A_i, by Lemma 12.1.6. Thus, A_1 is a direct sum of blocks of A, and since T lies in the same block as S, T is a composition factor of A_1. Therefore, $S \sim T$. □

The effect of this result is that the division of the simple A-modules into blocks can be achieved in a purely combinatorial fashion, knowing the Cartan matrix of A. This is the content of the next corollary, in which the term "block" is used in two different ways. The connection with the block matrix decomposition of the Cartan matrix is probably the origin of the use of the term in representation theory.

Corollary 12.1.8. *Let A be a finite-dimensional algebra over a field k. On listing the simple A-modules so that modules in each block occur together, the Cartan matrix of A has a block diagonal form, with one block matrix for each block of the group. Up to permutation of simple modules within blocks and permutation of the blocks, this is the unique decomposition of the Cartan matrix into block diagonal form with the maximum number of block matrices.*

Proof. Given any matrix, we may define an equivalence relation on the set of rows and columns of the matrix by requiring that a row be equivalent to a column if and only if the entry in that row and column is nonzero, and extending this by transitivity to an equivalence relation. In the case of the Cartan matrix, the row indexed by a simple module S is in the same equivalence class as the column indexed by P_S, because S is a composition factor of P_S. If we order the rows and columns of the Cartan matrix so that the rows and columns in each equivalence class come together, the matrix is in block diagonal form, with square blocks, and this is the unique such expression with the maximal number of blocks (up to permutation of the blocks and permutation of rows and columns within a block). By Proposition 12.1.7(2), the matrix blocks biject with the blocks of A. □

Example 12.1.9. We have seen from Theorem 9.6.1 and the remarks afterward that a block of defect zero my be regarded as a simple projective kG-module, where k is a field of characteristic p. Since projective modules are also injective, such a simple projective module lies in its own equivalence class under $\sim$, so there is only one simple module in this block, and in fact only one indecomposable module in this block.

12.2 p-Blocks of Groups

When considering blocks of group algebras, we are really only interested in the blocks of RG and kG, where (F, R, k) is a splitting p-modular system. This is because FG is semisimple and the theory of blocks of FG is nothing more than what has already been described in previous chapters. We show that the blocks of RG and of kG biject under reduction modulo the maximal ideal (π) of R.

Proposition 12.2.1. *Let G be a finite group and (F, R, k) a p-modular system in which R is complete.*

(1) Reduction modulo (π) gives a surjective ring homomorphism $Z(RG) \to Z(kG)$. It induces a bijection

$$\{\text{idempotents of } Z(RG)\} \leftrightarrow \{\text{idempotents of } Z(kG)\}.$$

Each idempotent of $Z(kG)$ lifts uniquely to an idempotent of $Z(RG)$, and each idempotent of $Z(RG)$ reduces uniquely to an idempotent of $Z(kG)$. Under this process, primitive idempotents correspond to primitive idempotents.

(2) An RG-module U belongs to a block $e \in Z(RG)$ if and only if $U/\pi U$ belongs to the image $\overline{e}$ of e in $Z(kG)$.

Proof. (1) The conjugacy class sums $\sum_{g\sim x} g$ (i.e., the sum of all elements g conjugate to x) form a basis for $Z(RG)$ over R, and over k they form a basis for $Z(kG)$, by Lemma 3.4.2. Reduction modulo (π) sends one basis to the other, which proves surjectivity.

The correspondence of idempotents comes partly from a lifting argument that we have seen before in Proposition 9.4.3. Since $\pi^{n-1}Z(RG)/\pi^n Z(RG)$ is a nilpotent ideal in $Z(RG)/\pi^n Z(RG)$, it follows that we may lift any idempotent

$e_{n-1} + \pi^{n-1}Z(RG)$ to an idempotent $e_n + \pi^n Z(RG)$, thereby obtaining from any idempotent $e_1 + \pi Z(RG) \in Z(kG)$ a Cauchy sequence $e_1, e_2, \ldots$ of elements of $Z(RG)$ whose limit is an idempotent e which lifts e_1, that is, which reduces modulo π to e_1. This shows that reduction modulo π gives a surjective map between the sets of idempotents.

To see that this map on idempotents is injective, suppose that idempotents e and f both reduce to the same idempotent $\overline{e} = \overline{f}$. The idempotent ef also reduces to $\overline{ef} = \overline{e}$ and hence $e(1-f)$ is an idempotent that reduces to 0. But $RGe = RGef \oplus RGe(1-f)$ is a summand of RG, so that $RGe(1-f)$ is a free R-module which, under reduction modulo π of RG, reduces to zero. It follows that $e(1-f) = 0$ so that $e = ef$. Similarly, $f = ef$ and $f = e$.

We have also seen before an argument that primitive idempotents correspond to primitive idempotents under lifting and reduction modulo π. In the present situation, lifting and reduction of idempotents preserve sums and orthogonality: if e and f are idempotents of $Z(RG)$ with $ef = 0$ then $\overline{ef} = 0$. Equally, if e and f are idempotents of $Z(RG)$ with $\overline{ef} = 0$ then $ef = 0$ because it is the unique idempotent that reduces to $\overline{ef}$, and by injectivity of reduction modulo π on idempotents. Thus, if $e = e_1 + e_2$ is a sum of orthogonal idempotents of $Z(kG)$ with lifts $\hat{e}, \hat{e}_1, \hat{e}_2$ in $Z(RG)$ then $\hat{e}_1 + \hat{e}_2$ lifts e, so equals $\hat{e}$ by uniqueness, and $\hat{e}_1\hat{e}_2$ lifts $0 = e_1 e_2$ so is zero. Thus, $\hat{e} = \hat{e}_1 + \hat{e}_2$ is a sum of orthogonal idempotents, so that $\hat{e}$ is not primitive. It is immediate that reduction modulo π also preserves sums of orthogonal idempotents. Hence both lifting and reduction modulo π preserve primitivity. □

Because of this, it is the same thing to study the blocks of RG and of kG since they correspond to each other under reduction modulo π. The modules belonging to blocks of RG and of kG also correspond under reduction modulo π and lifting (when that is possible). If U is a kG-module, we may regard it also as an RG-module via the surjection $RG \to kG$ and if e is a block of RG with image the block $\overline{e} \in Z(kG)$, there is no difference between saying that U belongs to e or that U belongs $\overline{e}$.

We may also partition the simple FG-modules into blocks in a way consistent with the blocks for RG and kG, as we now describe. This is not the same as the partition by blocks of FG, which is trivial. Regarding RG as a subset of FG, a primitive central idempotent e of RG is also a central idempotent of FG. We say that an FG module U *belongs* to e if $eU = U$. Evidently, each simple FG-module belongs to a unique block. We see that if U is an RG-module and $U_0 \subset U$ is any R-form of U (i.e., a full RG-lattice in U) then U_0 belongs to e if and only if U belongs to e.

We define a *p-block* of G to be the specification of a block e of RG, understanding also the corresponding block of kG, the modules which belong to these blocks, the FG modules which belong to these blocks and also the ring direct summands eRG of RG and $\bar{e}kG$ of kG. The block to which the trivial module R belongs is called the *principal block*.

Example 12.2.2. When G is a p-group and k is a field of characteristic p the regular representation kG is indecomposable as a module (Corollary 6.3.7) and hence as a ring, so that the identity element of kG is a primitive central idempotent (Proposition 12.1.1). There is only one p-block in this situation.

Example 12.2.3. We have seen in Theorem 9.6.1 that a block of defect zero for G over a splitting p-modular system (F, R, k) corresponds to having 2-sided direct summand of RG that is a matrix algebra $M_n(R)$, and also a matrix summand $M_n(k)$ of kG that is the reduction modulo π of the summand of RG. Thus, a block of defect zero is a p-block in the sense just defined. There is a unique simple kG-module in this block, and it is projective. Since extensions between copies of this module must split, we see that it satisfies the criterion of Proposition 12.1.7 to be the only simple module in its block. It lifts to a unique RG-lattice, and all RG-sublattices of it are isomorphic to it. We showed that a single simple $\mathbb{C}G$-module lies in such a block and that the corresponding primitive central idempotent of $\mathbb{C}G$ in fact lies in RG. We see now why the word "block" appears in the term "block of defect zero," but we do not yet know what "defect zero" means.

Example 12.2.4. When $G = S_3$ in characteristic 2, there are two blocks, since the simple module of degree 2 is a block of defect zero, and the only other simple module is the trivial module, which lies in the principal block. This has been seen in Example 7.2.2, Example 9.5.7, and Theorem 9.6.1. The projective cover of the trivial module as an RG-module has character equal to the sum of the characters of the trivial representation and the sign representation, and so the principal block of RG acts as the identity on these two ordinary representations, meaning that these ordinary characters lie in the principal block. The other ordinary character, of degree 2, lies in the block of defect zero, and so there are precisely two ordinary characters in the principal 2-block.

In characteristic 3, S_3 has only one block since there are two simple modules over a field of characteristic 3 (trivial and sign) and the sign representation appears as a composition factor of the projective cover of the trivial module (Example 9.5.7). This means that the sign representation belongs to the principal 3-block.

Example 12.2.5. Suppose that $G = K \rtimes H$ where K has order prime to p and H is a p-group, so that G has a normal p-complement and is *p-nilpotent*. We saw in Theorem 8.4.1 that this is precisely the situation in which each indecomposable projective module has only one isomorphism type of composition factor. Another way of expressing this is to say that the Cartan matrix is diagonal. Thus, the groups for which each p-block contains just one simple module in characteristic p are characterized as the p-nilpotent groups.

We conclude this section by describing explicitly the primitive central idempotents for a p-nilpotent group.

Proposition 12.2.6. *Let $G = K \rtimes H$ be a p-nilpotent group, where K has order prime to p and H is a p-group. Let k be a field of characteristic p. Each block of kG lies in kK, and is the sum of a G-conjugacy class of blocks of kK.*

Proof. Observe that if $1 = e_1 + \cdots + e_n$ is the sum of blocks of kK then for each i and $g \in G$ the conjugate ge_ig^{-1} is also a block of kK. For this, we verify that this element is idempotent, and also that it is central in kK, which is so since if $x \in K$ then $xge_ig^{-1} = g(g^{-1}xg)e_ig^{-1} = ge_i(g^{-1}xg)g^{-1} = ge_ig^{-1}x$. Furthermore, ge_ig^{-1} is primitive in $Z(kK)$ since if it were the sum of two orthogonal central idempotents, on conjugating back by g^{-1} we would be able to deduce that e_i is not primitive either.

The blocks of kK are uniquely determined, and it follows that $ge_ig^{-1} = e_j$ for some j. Thus, G permutes the blocks of kK. Notice also that, since kK is semisimple, each e_i corresponds to a unique simple kK-module on which it acts as the identity, and that e_i acts as zero on the other simple kK-modules.

For each fixed i, the element $f = \sum_{e=ge_ig^{-1}} e \in Z(kK)$ is idempotent. It acts as the identity precisely on the simple kK-modules that are G-conjugate to the simple module determined by e_i. It is central in kG since if $x \in G$ then $xfx^{-1} = \sum_{e=ge_ig^{-1}} xex^{-1} = f$, the sum again being over the elements in the G-orbit of e_i. We now show that f is primitive in $Z(kG)$. Suppose instead that $f = f_1 + f_2$ is a sum of orthogonal idempotents in $Z(kG)$. Then there are nonisomorphic simple kG-modules U_1 and U_2 with $f_1U_1 = U_1$ and $f_2U_2 = U_2$. Since $ff_1 = f_1$ and $ff_2 = f_2$, we have $fU_1 = U_1$ and $fU_2 = U_2$.

By Clifford's Theorem, $U_1 \downarrow_K^G \cong S_1^a \oplus \cdots \oplus S_t^a$ for some integer a, where $S_1, \ldots, S_t$ are G-conjugate simple kK-modules. Since $f \in kK$, we have $fS_i = S_i$ for all i, and this identifies these modules exactly as the G-orbit of simple kK modules on which f acts as the identity. By a similar argument, $U_2 \downarrow_K^G \cong S_1^b \oplus \cdots \oplus S_t^b$ for some integer b, where the S_i are the same modules. Since K consists of the p-regular elements of G it follows that the Brauer characters of

U_1 and U_2 are scalar multiples of one another. Since Brauer characters of non-isomorphic simple modules are linearly independent we deduce that $U_1 \cong U_2$, a contradiction. This shows that f is primitive in $Z(kG)$. □

Example 12.2.7. We may see the phenomenon described in the last result in many examples, of which the smallest nontrivial one is $G = S_3 = K \rtimes H$ where $K = \langle(1, 2, 3)\rangle$ and $H = \langle(1, 2)\rangle$, taking p to be 2 and $k = \mathbb{F}_4$. By Theorem 3.6.2 (modified for the case of positive characteristic by Exercise 8 of Chapter 10) the blocks of kK are

$$\begin{aligned} e_1 &= () + (1, 2, 3) + (1, 3, 2), \\ e_2 &= () + \omega(1, 2, 3) + \omega^2(1, 3, 2), \text{ and} \\ e_3 &= () + \omega^2(1, 2, 3) + \omega(1, 3, 2), \end{aligned}$$

where ω is a primitive cube root of 1 in $\mathbb{F}_4$. (In case the reader expects some factors $\frac{1}{3}$, note that $3 = 1$ in characteristic 2. In the next expression -1 has been written as 1 and 2 as 0.) In the action of G on these idempotents there are two orbits, namely, $\{e_1\}$ and $\{e_2, e_3\}$. The blocks of kG are thus e_1 and $e_2 + e_3 = (1, 2, 3) + (1, 3, 2)$. These idempotents have already appeared in Example 7.2.2, where it was calculated that $kGe_1 = \begin{smallmatrix}1\\1\end{smallmatrix} = P_1$ and $kG(e_2 + e_3) = 2 \oplus 2$ is the direct sum of two simple projective kG-modules. This structure has also been considered in Exercise 5 of Chapter 8. From Theorem 9.6.1, we know the second summand is a block of defect zero, and $kG(e_2 + e_3) \cong M_2(k)$ as rings. We also see that $kGe_1 \cong kC_2 \cong k[X]/(X^2)$ as rings. An approach to this can be found in Exercise 12 of Chapter 8.

12.3 The Defect of a Block: Module Theoretic Methods

The defect group of a p-block of G is a p subgroup of G that measures how complicated the block is. If the defect group has order p^d, we say that the block has defect d. In very rough terms, the larger d is, the more complicated the block.

In view of the multiple meanings of the term "block," it will be no surprise that there is more than one approach to the definition of a defect group of a block. We will start with a module theoretic approach pioneered by J. A. Green, and after that relate it to a ring theoretic approach that goes back to Brauer.

There are two ways to obtain the regular representation from the group ring RG. One is to regard RG as a left RG-module via multiplication from the left. The second way is to use the left action of G on itself in which an element $g \in G$ acts by multiplication from the right by g^{-1}. The module we obtain in this way is isomorphic to the regular representation obtained via left multiplication.

Because these two actions commute with each other we may combine them, and regard kG as a representation of $G \times G$ with an action given by

$$(g_1, g_2)x = g_1 x g_2^{-1} \quad \text{where } g_1, g_2 \in G, x \in kG.$$

It will be important to consider the diagonal embedding of $\delta : G \to G \times G$ specified by $\delta(g) = (g, g)$. Via this embedding we obtain yet another action of G on RG, which is the action given by conjugation: $g \cdot x = gxg^{-1}$.

Proposition 12.3.1. *Let R be a commutative ring with a 1.*

(1) *The submodules of RG, regarded as a module for $R[G \times G]$, are precisely the 2-sided ideals of RG.*
(2) *Any decomposition of RG as a direct sum of indecomposable $R[G \times G]$-modules is a decomposition as a direct sum of blocks.*
(3) *Regarded as a representation of $G \times G$, RG is a transitive permutation module in which the stabilizer of 1 is $\delta(G)$. Thus $RG \cong R \uparrow_{\delta(G)}^{G \times G}$ as $R[G \times G]$-modules and every summand of RG as an $R[G \times G]$-module is $\delta(G)$-projective.*

Proof. (1) The 2-sided ideals of RG are precisely the R-submodules of RG that are closed under multiplication from the left and from the right by G. These is equivalent to being an $R[G \times G]$-submodule.

(2) This follows from (1) and Proposition 12.1.1.

(3) Evidently $G \times G$ permutes the basis of RG consisting of the group elements. We only need check that $\mathrm{Stab}_{G \times G}(1) = \{(g_1, g_2) \mid g_1 1 g_2^{-1} = 1\} = \delta(G)$. □

Corollary 12.3.2. *Let R be a discrete valuation ring with residue field of characteristic p (or a field of characteristic p). Let G be a finite group and let e be a block of RG. Then, regarded as a $R[G \times G]$-module, the summand eRG has a vertex of the form $\delta(D)$ where D is a p-subgroup of G. Such a subgroup D is uniquely defined to within conjugacy in G.*

Proof. By Proposition 12.3.1, eRG is indecomposable as an $R[G \times G]$-module, and since it is $\delta(G)$-projective it has a vertex contained in $\delta(G)$. Such a subgroup is necessarily a p-group, so has the form $\delta(D)$ for some p-subgroup D of G where $\delta(D)$ is determined up to conjugacy in $G \times G$. If D_1 is another subgroup of G for which $\delta(D_1)$ is a vertex of eRG then $\delta(D_1) = {}^{(g_1, g_2)}\delta(D)$ for some $g_1, g_2 \in G$, and so for all $x \in D$, $({}^{g_1}x, {}^{g_2}x) \in \delta(D_1)$. Thus, ${}^{g_1}x \in D_1$ for all $x \in D$ and since D and D_1 have the same order it follows that $D_1 = {}^{g_1}D$. □

Let R be a discrete valuation ring with residue field of characteristic p and let $e \in Z(RG)$ be a p-block of G. We define a subgroup D of G to be a *defect group* of the block e if $\delta(D)$ is a vertex of eRG as an $R[G \times G]$-module. According to Corollary 12.3.2, D is defined up to conjugacy in G and is a p-group. If $|D| = p^d$, we say that d is the *defect* of e, but often we abuse this terminology and say simply that the defect of the block is D. According to these definitions, a block of defect 0 is one whose defect group is the identity subgroup. It is not immediately apparent that this definition of a block of defect zero coincides with the previous one, and this will have to be proved. There are many characterizations of the defect group of a block, and we will see some of them in Theorem 12.4.5, Theorem 12.5.2, and Corollary 12.5.5.

We could also have defined a defect group of a p-block of G using the field k instead of the discrete valuation ring R: it is a subgroup D of G so that $\delta(D)$ is a vertex of $\overline{e}kG$ as a $k[G \times G]$-module. This gives the same conjugacy class of subgroups D, since reduction modulo π preserves vertices of indecomposable modules. This was shown in Exercise 21 of Chapter 11.

We now start to investigate the kinds of p-subgroups of G that can be defect groups and we will see that the possibilities are restricted.

Theorem 12.3.3 (Green). *Let e be a p-block of G with defect group D, and let P be a Sylow p-subgroup of G that contains D. Then $D = P \cap {}^gP$ for some element $g \in C_G(D)$.*

Proof. Let P be a Sylow p-subgroup of G containing D. We consider

$$\begin{aligned} RG \downarrow^{G\times G}_{P\times P} &= R \uparrow^{G\times G}_{\delta(G)} \downarrow^{G\times G}_{P\times P} \\ &= \bigoplus_{y\in[P\times P\backslash G\times G/\delta(G)]} ({}^y(R \downarrow^{\delta(G)}_{(P\times P)^y\cap\delta(G)})) \uparrow^{P\times P}_{(P\times P)\cap{}^y\delta(G)} \\ &= \bigoplus_{y\in[P\times P\backslash G\times G/\delta(G)]} R \uparrow^{P\times P}_{(P\times P)\cap{}^y\delta(G)} . \end{aligned}$$

Now each $R \uparrow^{P\times P}_{(P\times P)\cap{}^y\delta(G)}$ is indecomposable since $P \times P$ is a p-group, as shown in Corollary 11.6.3. The summand eRG of RG, on restriction to $P \times P$, has a summand that on further restriction to $\delta(D)$ has a source of eRG as a summand. This summand of $eRG \downarrow^{G\times G}_{P\times P}$ also has vertex $\delta(D)$ (an argument that is familiar from the proof of the Green correspondence 11.6.4), and must have the form $R \uparrow^{P\times P}_{(P\times P)\cap{}^y\delta(G)}$ for some $y \in G \times G$. Thus $\delta(D)$ is conjugate in $P \times P$ to the subgroup $(P \times P) \cap {}^y\delta(G)$, so $\delta(D) = {}^z((P \times P) \cap {}^y\delta(G))$ for some $z \in P \times P$.

The elements $1 \times G = \{(1,t) \mid t \in G\}$ form a set of coset representatives for $\delta(G)$ in $G \times G$, and so we may assume $y = (1,t)$ for some $t \in G$. Write $z = (r,s)$, where $r, s \in P$. Now

$$\begin{aligned}(P \times P) \cap {}^{(1,t)}\delta(G) &= \{(x, {}^t x) \mid x \in P \text{ and } {}^t x \in P\}\\ &= \{(x, {}^t x) \mid x \in P \cap P^t\}\\ &= {}^{(1,t)}\delta(P \cap P^t)\end{aligned}$$

and $\delta(D) = {}^{(r,s)(1,t)}\delta(P \cap P^t)$. The projection onto the first coordinate here equals $D = {}^r(P \cap P^t) = {}^rP \cap {}^{rt^{-1}}P = P \cap {}^{rt^{-1}}P$ since $r \in P$. At this point, the proof is complete, apart from the fact that rt^{-1} might not centralize D. Now ${}^{(1,t^{-1})(r^{-1},s^{-1})}\delta(D) \subseteq \delta(G)$, so that ${}^{r^{-1}}x = {}^{t^{-1}s^{-1}}x$ for all $x \in D$, and $rt^{-1}s^{-1} \in C_G(D)$. Since $s \in P$ we have $D = P \cap {}^{rt^{-1}s^{-1}}P$ and this completes the proof. □

Corollary 12.3.4. *If e is a p-block of G with defect group D then*

$$D = O_p(N_G(D)) \supseteq O_p(G),$$

where $O_p(G)$ denotes the largest normal p-subgroup of G.

Proof. We start by observing that $O_p(G)$ is the intersection of the Sylow p-subgroups of G; for the intersection of the Sylow p-subgroups is a normal p-subgroup since the Sylow p-subgroups are closed under conjugation, and on the other hand, $O_p(G) \subseteq P$ for some Sylow p-subgroup P, and hence ${}^g(O_p(G)) = O_p(G) \subseteq {}^gP$ for every element $g \in G$. Since ${}^gP$ accounts for all Sylow p-subgroups, by Sylow's Theorem, it follows that $O_p(G)$ is contained in their intersection.

We see immediately that $D \supseteq O_p(G)$ since D is the intersection of two Sylow p-subgroups, and hence contains the intersection of all Sylow p-subgroups.

To prove that $D = O_p(N_G(D))$, let P be a Sylow p-subgroup of G that contains a Sylow p-subgroup of $N_G(D)$. Such a P necessarily has the property that $P \cap N_G(D)$ is a Sylow p-subgroup of $N_G(D)$. Now $D = P \cap {}^gP$ for some $g \in C_G(D)$ and so in particular $g \in N_G(D)$. Thus, ${}^gP \cap N_G(D) = {}^g(P \cap N_G(D))$ is also a Sylow p-subgroup of $N_G(D)$, and $D = (P \cap N_G(D)) \cap {}^g(P \cap N_G(D))$ is the intersection of two Sylow p-subgroups of $N_G(D)$. Thus, $D \supseteq O_p(N_G(D))$. But on the other hand, D is a normal p-subgroup of $N_G(D)$, and so is contained in $O_p(N_G(D))$. Thus, we have equality. □

The condition on a subgroup D of G that appeared in Corollary 12.3.4, namely, that $D = O_pN_G(D)$, is quite restrictive. Such subgroups are called *p-radical subgroups*. They play an important role in the study of conjugation within G, a topic which goes by the name of *fusion*. It has applications in many directions aside from block theory, including group theoretic classification questions (such as the classification of finite simple groups) and topological questions (such as the study of group cohomology and classifying spaces). The terminology p-radical comes from the fact that when G is a finite group of Lie type in characteristic p (such as $SL(n, p^r)$ and other such groups) the p-radical subgroups are precisely the unipotent radicals of parabolic subgroups. Thus, the definition of p-radical subgroup extends the notion from groups of Lie type in defining characteristic to all finite groups. The reader should be warned that the term "p-radical" has also been used in different senses, one of which is described in the book by Feit [13]. The subgroups we are calling p-radical have also been called p-stubborn subgroups by some authors.

It is immediate that Sylow p-subgroups of a group are p-radical, as is $O_p(G)$. An exercise in group theory (presented as Exercise 2) shows that other p-radical subgroups must lie between these two extremes. It is always the case that G has a p-block with defect group a Sylow p-subgroup (the principal block has this property, as will be seen in Corollary 12.4.7) but it need not happen that $O_p(G)$ is the defect group of any block. An example of this is S_4 in characteristic 2: from the Cartan matrix computed in Chapter 10 and from Corollary 12.1.8, we see that the only block is the principal block; this also follows from Corollary 12.5.8.

12.4 The Defect of a Block: Ring Theoretic Methods

The advantage of the module theoretic approach to the defect group of a block is that it allows us to exploit the theory of vertices and sources already developed. In view of the fact that blocks are rings, it is not surprising that there is also a ring theoretic approach. We use it in the next results, showing in Corollary 12.4.6 that modules for a block with defect group D are projective relative to D. As a consequence we will see that defect groups of the principal block are Sylow p-subgroups (Corollary 12.4.7) and also that blocks of defect zero are correctly named (Corollary 12.4.8). Before that we obtain further characterizations of the defect group.

Our main tool will be the relative trace map. We have already been using the properties of this map in the context of endomorphism rings of modules and we will now do something very similar in the context of blocks. It is helpful to introduce a more general axiomatic setting that includes these examples.

Let R be a commutative ring with a 1 and G a group. We define a *G-algebra* over R to be an R-algebra A together with an action of G on A by R-algebra automorphisms. Thus, for each $g \in G$ and $a \in A$, there is defined an element

${}^g a \in A$ so that the mapping $a \mapsto {}^g a$ is R-linear, ${}^g(ab) = {}^g a {}^g b$ always and ${}^g 1 = 1$. A homomorphism of G-algebras $\phi : A \to B$ is defined to be an algebra homomorphism so that $\phi({}^g a) = {}^g \phi(a)$ always holds.

Given an RG-module U, we have already been using the G-algebra structure on $\mathrm{End}_R(U)$ given by $({}^g f)(u) = gf(g^{-1}u)$ whenever $f \in \mathrm{End}_R(U)$, $g \in G$, $u \in U$. Another example of a G-algebra is the group ring RG, where for $x \in RG$ and $g \in G$ we define ${}^g x = gxg^{-1}$. In fact, the RG-module structure on U is given by an algebra homomorphism $RG \to \mathrm{End}_R(U)$, and it is a homomorphism of G-algebras. The same holds for any block B, and for B-modules: since B is a summand of RG as a representation of $G \times G$, B is preserved under the G-algebra action on RG and so becomes a G-algebra in its own right. If U is a B-module, we obtain a G-algebra homomorphism $B \to \mathrm{End}_R(U)$, since this factors as $B \to RG \to \mathrm{End}_R(U)$ where the first map is inclusion (not an algebra homomorphism, but the action of G is preserved).

Whenever A is a G-algebra we have algebras of fixed points A^H for each subgroup $H \leq G$, and as before the relative trace map $\mathrm{tr}_H^G : A^H \to A^G$, the inclusion $\mathrm{res}_H^G : A^G \to A^H$ and conjugation $c_g : A^H \to A^{{}^g H}$ given by $a \mapsto {}^g a$. They satisfy the properties already established in Lemma 11.3.1 as well as the analogues of Lemma 11.3.2 and Corollary 11.3.3 which we present now.

Proposition 12.4.1. *Let A be a G-algebra and $H \leq G$.*

(1) *If $a \in A^G$, $b \in A^H$, we have $a(\mathrm{tr}_H^G b) = \mathrm{tr}_H^G(ab)$ and $(\mathrm{tr}_H^G b)a = \mathrm{tr}_H^G(ba)$.*
(2) *The image of $\mathrm{tr}_H^G : A^H \to A^G$ is an ideal of A^G. The inclusion $\mathrm{res}_H^G : A^G \to A^H$ is a ring homomorphism.*
(3) *If $\phi : A \to B$ is a homomorphism of G-algebras then $\phi(\mathrm{tr}_H^G(b)) = \mathrm{tr}_H^G(\phi(b))$.*

Proof. The calculations are the same as for Lemma 11.3.2 and Corollary 11.3.3, and part (3) is immediate. □

For most of the arguments, we will present it will be sufficient to consider G-algebras, but sometimes a stronger property is needed which block algebras possess. We define an *interior G-algebra* over R (a notion due to L. Puig) to be an R-algebra A together with a group homomorphism $u : G \to A^\times$ where $A^\times$ denotes the group of units of A. As an example, the group algebra RG is itself an interior G-algebra where u is the inclusion of G as a subset of RG. Whenever A is an interior G-algebra and $\phi : A \to B$ is an algebra homomorphism (sending 1_A to 1_B) we find that B becomes an interior G-algebra via the homomorphism $\phi u : G \to B^\times$. Thus, if B is a block of G the algebra homomorphism $RG \to B$ makes B into an interior G-algebra. As a further example, if U is any RG-module its structure is determined by an algebra homomorphism

$RG \to \mathrm{End}_R(U)$ that expresses the action of RG. In fact, to specify a module action of G on U is the same as specifying the structure of an interior G-algebra on $\mathrm{End}_R(U)$.

Given an interior G-algebra A, we obtain the structure of a G-algebra on A by letting each $g \in G$ act as $a \mapsto {}^g a := u(g)au(g^{-1})$. We see immediately that our three examples of the G-algebra structure on RG, on a block, and on $\mathrm{End}_R(U)$ are obtained in this way.

Why should we consider interior G-algebras? The reason here is that if A is an interior G-algebra and U is an A-module then we can recover the structure of U as an RG-module from the composite homomorphism $G \xrightarrow{u} A \to \mathrm{End}_R(U)$, and without the extra property of an interior G-algebra we cannot do this. This property is used in the next lemma, which will be used in Corollary 12.4.6.

Proposition 12.4.2. *Let A be an interior G-algebra over R, let U be an A-module and let H be a subgroup of G. Suppose that $1_A = \mathrm{tr}_H^G a$ for some element $a \in A^H$. Then, regarded as an RG-module, U is H-projective.*

Proof. The representation of A on U is given by a G-algebra homomorphism $\phi : A \to \mathrm{End}_R(U)$. This homomorphism is a homomorphism of G-algebras and by Proposition 12.4.1, we have $1_U = \phi(1_A) = \phi(\mathrm{tr}_H^G a) = \mathrm{tr}_H^G \phi(a)$. Thus, by Higman's criterion (Proposition 11.3.4(6)), U is H-projective. □

Our goal now is Theorem 12.4.5, which makes a connection between the module theoretic approach to the defect group as we have defined it, and the ring theoretic approach. The proof will follow from the next two results. As before, $\delta(G) = \{(g, g) \mid g \in G\}$.

Lemma 12.4.3. *Let U be an $R[G \times G]$-module that is $\delta(H)$-projective. Then $U \downarrow_{\delta(G)}^{G\times G}$ is $\delta(H)$-projective as a representation of $\delta(G)$.*

Proof. The hypothesis says that U is a summand of some module $V \uparrow_{\delta(H)}^{G\times G}$, so $U \downarrow_{\delta(G)}^{G\times G}$ is a summand of

$$V \uparrow_{\delta(H)}^{G\times G} \downarrow_{\delta(G)}^{G\times G} = \bigoplus_{x \in [\delta(G)\backslash G\times G/\delta(H)]} ({}^x(V \downarrow_{\delta(G)^x \cap \delta(H)}^{\delta(H)})) \uparrow_{\delta(G)\cap {}^x\delta(H)}^{\delta(G)} .$$

In this formula, we may write each $x \in G \times G$ as $x = (a, b) = (a, a)(1, a^{-1}b)$ and now

$$\begin{aligned}\delta(G) \cap {}^x\delta(H) &= {}^{(a,a)}(\delta(G) \cap {}^{(1,a^{-1}b)}\delta(H)) \\ &= {}^{(a,a)}\{(h, {}^{a^{-1}b}h) \mid h = {}^{a^{-1}b}h\} \\ &= {}^{(a,a)}\delta(C_H(a^{-1}b)) \le {}^{(a,a)}\delta(H).\end{aligned}$$

It follows that every summand of the decomposition of $V \uparrow_{\delta(H)}^{G\times G}\downarrow_{\delta(G)}^{G\times G}$ is projective relative to a $\delta(G)$-conjugate of $\delta(H)$, which is the same as being $\delta(H)$-projective. Thus, $U \downarrow_{\delta(G)}^{G\times G}$ is $\delta(H)$-projective. □

The point about these lemmas is that we are relating the structure of a block algebra eRG as a representation of $G \times G$ to its structure as a representation of G via the isomorphism $G \to \delta(G)$. Note in this context that if U is an RG-module and $\alpha \in \mathrm{End}_{RH}(U)$ then $\mathrm{tr}_H^G\alpha$ and $\mathrm{tr}_{\delta(H)}^{\delta(G)}\alpha$ are exactly the same thing, from the definitions. This is because G is taken to act on $\mathrm{End}_R(G)$ via δ.

Lemma 12.4.4. *Let $e \in Z(RG)$ be a central idempotent of RG and $H \leq G$. Then eRG is projective relative to $\delta(H)$ as an $R[G \times G]$-module if and only if $e = \mathrm{tr}_H^G a$ for some element $a \in (eRG)^H$.*

Proof. Suppose first that eRG is $\delta(H)$-projective as an $R[G \times G]$-module. Then by Lemma 12.4.3 $(eRG) \downarrow_{\delta(G)}^{G\times G}$ is $\delta(H)$-projective. By Higman's criterion (Proposition 11.3.4(6)), there is an endomorphism α of eRG as a $R[\delta(H)]$-module so that the identity morphism can be written $1_{eRG} = \mathrm{tr}_{\delta(H)}^{\delta(G)}\alpha = \mathrm{tr}_H^G\alpha$. Now

$$\begin{aligned} e &= 1_{eRG}(e) \\ &= (\mathrm{tr}_{\delta(H)}^{\delta(G)}\alpha)(e) \\ &= \sum_{g\in[G/H]} {}^g\alpha(e) \\ &= \sum_{g\in[G/H]} {}^g(\alpha({}^{g^{-1}}e)) \\ &= \sum_{g\in[G/H]} g(\alpha(g^{-1}eg))g^{-1} \\ &= \sum_{g\in[G/H]} g\alpha(e)g^{-1} \\ &= \mathrm{tr}_H^G(\alpha(e)), \end{aligned}$$

using the fact in the middle that e is central. We take $a = \alpha(e)$ and the proof in this direction is complete.

Conversely, suppose $e = \mathrm{tr}_H^G a$ for some $a \in (eRG)^H$. Thus, $hah^{-1} = a$ for all $h \in H$ and $e = \sum_{g\in[G/H]} gag^{-1}$. Now $\alpha : eRG \to eRG$ specified by $\alpha(x) = ax$ is an endomorphism of $R[\delta(H)]$-modules (which is the same thing as an endomorphism of RH-modules), since

$$\alpha((h,h)x) = ahxh^{-1} = h(h^{-1}ah)xh^{-1} = haxh^{-1} = (h,h)\alpha(x).$$

We claim that $\mathrm{tr}_{\delta(H)}^{\delta(G)}\alpha = 1_{eRG}$. For any $x \in eRG$, we have

$$\begin{aligned}
(\mathrm{tr}_{\delta(H)}^{\delta(G)}\alpha)(x) &= \sum_{g\in[G/H]} (g,g),\alpha((g^{-1},g^{-1})x) \\
&= \sum_{g\in[G/H]} (g,g)\alpha(g^{-1}xg) \\
&= \sum_{g\in[G/H]} (g,g)ag^{-1}xg \\
&= \sum_{g\in[G/H]} gag^{-1}xgg^{-1} \\
&= \sum_{g\in[G/H]} gag^{-1}x \\
&= ex \\
&= x.
\end{aligned}$$

This shows that $eRG \downarrow_{\delta(G)}^{G\times G}$ is $\delta(H)$-projective, and hence, since eRG is $\delta(G)$-projective by Proposition 12.3.1, it implies that eRG is $\delta(H)$-projective. □

Putting the last results together, we obtain a proof of the following characterization of the defect group in terms of the effect of the relative trace map on the interior G algebra eRG. This characterization could have been used as the definition of the defect group.

Theorem 12.4.5. *Let R be a discrete valuation ring with residue field of characteristic p, let e be a block of RG and D a p-subgroup of G. Then D is a defect group of e if and only if D is a minimal subgroup with the property that $\mathrm{tr}_D^G : (eRG)^D \to (eRG)^G$ is surjective. Equivalently, D is a minimal subgroup with the property that $e = \mathrm{tr}_D^G a$ for some $a \in (eRG)^D$.*

Proof. From the definition, D is a defect group of e if and only if D is a minimal subgroup of G such that eRG is $\delta(D)$-projective. By Lemma 12.4.4, it is equivalent to say that D is minimal such that e can be written $e = \mathrm{tr}_{\delta(D)}^{\delta(G)} a$ for some $a \in (eRG)^D$. With the understanding that G acts on eRG via $\delta(G)$, we may write this as $e = \mathrm{tr}_D^G a$. Since $\mathrm{tr}(eRG)^D$ is an ideal of $(eRG)^G$, it is equivalent to say that $\mathrm{tr}_D^G : (eRG)^D \to (eRG)^G$ is surjective. □

Corollary 12.4.6. *Let e be a block of RG with defect group D. Then every eRG-module is projective relative to D and hence has a vertex that is a subgroup of D.*

Proof. We see from Theorem 12.4.5 that it is possible to write $e = \mathrm{tr}_D^G a$ for some $a \in (eRG)^D$. Thus, by Proposition 12.4.2, every eRG-module is projective relative to D. □

Note that it follows by the remarks about Green's indecomposability theorem at the end of Chapter 11 that if $|G| = |D|p^a q$ where q is relatively prime to p, then p^a divides the dimension of every FG-module and kG-module in the block.

It is a fact (which we will not prove in general) that every block has an indecomposable module with vertex exactly the defect group D, so that the defect group may be characterized as the unique maximal vertex of modules in the block. In the case of the principal block, the trivial module provides an example of such a module and it allows us to deduce the next corollary.

Corollary 12.4.7. *The defect groups of the principal p-block are the Sylow p-subgroups of G.*

Proof. A defect group is a p-subgroup of G, and it must contain a vertex of the trivial module, which is a Sylow p-subgroup by Proposition 11.6.2 part (3). □

We are now in a position to show that our previous use of the term "block of defect zero" in Chapter 9 is consistent with the definitions of this chapter. A block of defect zero was taken to be a block with a representation satisfying any of the equivalent conditions of Theorem 9.6.1. These were seen to be equivalent to the condition that over the field k the block has a simple projective representation, and also equivalent to the condition that over k the block is a matrix algebra.

Corollary 12.4.8. *Let k be a field of characteristic p that is a splitting field for G. A block of kG has defect zero in the sense of this chapter if and only if it has defect zero in the sense of Chapter 9.*

Proof. If the block has defect zero in the sense of this chapter, its defect group is 1 and by Corollary 12.4.6, every module in the block is 1-projective, or in other words projective. Thus, the block has a simple projective module and by Theorem 9.6.1 and the comment immediately after its proof, the block has defect zero in the sense considered there.

Conversely, suppose the block has defect zero in the sense of Chapter 9, so that ekG is a matrix algebra over k. We will show that ekG is projective as a $k[G \times G]$-module, and will make use of the isomorphism $k[G \times G] \cong kG \otimes_k kG$. Let $* : kG \to kG$ be the algebra antiisomorphism that sends each group element g to its inverse, so that $(\sum \lambda_g g)^* = \sum \lambda_g g^{-1}$. An element x in the second kG factor in the tensor product acts on ekG as right multiplication by x^*, and it follows that $e \otimes e^*$ acts as the identity on ekG (Since $e^{**} = e$). Now $e \otimes e^*$ is a central idempotent in $kG \otimes_k kG$ that generates the 2-sided ideal $ekG \otimes_k e^*kG = ekG \otimes_k (ekG)^*$. This is the tensor product of two matrix algebras, since the image of a matrix algebra under an antiisomorphism is a matrix algebra. Such a tensor product is again a matrix algebra, for if E_{ij} and F_{kl} are two matrix algebra bases (consisting of the matrices that are nonzero in only one place, where the entry is 1), then the tensors $E_{ij} \otimes_k F_{kl}$ are a basis for the tensor product that multiply together in the manner of a matrix algebra basis. We see that the block $ekG \otimes_k e^*kG$ of $kG \otimes_k kG$ is semisimple, and so ekG is a projective $k[G \times G]$-module. This shows that the defect group of ekG is 1, and so ekG has defect zero as defined in this chapter. □

12.5 The Brauer Morphism

We will use the Brauer morphism in the next section in the proof of Brauer's First Main Theorem. Before that, we will use it in this section to give a characterization of the defect group of a block (Theorem 12.5.2) and in proving Corollary 12.5.8, which says that if G has a normal p-subgroup whose centralizer is a p-group then kG only has one block.

The Brauer morphism was originally defined by Brauer in the specific context of idempotents in group rings, but it has subsequently been realized that the same construction is important more widely. It applies whenever we have a structure with mappings like restriction, conjugation, and the relative trace map satisfying the usual identities, including the Mackey formula. We first introduce the Brauer morphism in (not quite such) a general context and then make the connection with the map as Brauer conceived it.

Throughout this section we will work over a field k of characteristic p. When U is a kG-module and $K \le H$ are subgroups of G, we have already made use of the inclusion of fixed points $\mathrm{res}^H_K : U^H \to U^K$, the relative trace map $\mathrm{tr}^H_K : U^K \to U^H$ and the conjugation map $c_g : U^H \to U^{gH}$ for each $g \in G$ specified by $c_g(x) = gx$. We define for each subgroup $H \le G$ the *Brauer quotient*

$$\overline{U}(H) = U^H / \sum_{K<H} \mathrm{tr}^H_K(U^K).$$

We write $K < H$ to mean that K is a proper subgroup of H, excluding the possibility that $K = H$. The Brauer quotient may also be defined over a discrete valuation ring, but in that case the definition we have given needs to be modified by factoring out $(\pi)U$ as well as the other terms. We will not consider that generality here.

We define the *Brauer morphism* $\mathrm{Br}_H^G : U^G \to \overline{U}(H)$ to be the composite

$$U^G \xrightarrow{\mathrm{res}_H^G} U^H \to \overline{U}(H)$$

where the second map is the quotient homomorphism. Here U^G is simply a vector space but $\overline{U}(H)$ has further structure: it is a $k[N_G(H)]$-module with the action determined by the maps $c_g : U^H \to U^{{}^gH}$. If $g \in N_G(H)$ then ${}^gH = H$ so c_g preserves H-fixed points, and since $c_g \mathrm{tr}_K^G U^K = \mathrm{tr}_{{}^gK}^H c_g U^K = \mathrm{tr}_{{}^gK}^H U^{{}^gK}$ we see that c_g permutes the terms in the sum being factored from U^H to produce $\overline{U}(H)$, so c_g has a well-defined action on $\overline{U}(H)$. We see also that the image of Br_H^G lies in the fixed points under the action of $N_G(H)$. If U happens furthermore to be a G-algebra then $\sum_{K<H} \mathrm{tr}_K^H(U^K)$ is an ideal of U^H by Proposition 12.4.1, the Brauer quotient $\overline{U}(H)$ is an $N_G(H)$-algebra, and the Brauer morphism is a ring homomorphism.

The Brauer quotient provides a means to express properties of the relative trace map. We already have a characterization in Theorem 12.4.5 of the defect group of a block in terms of this map. We now provide a corresponding characterization in terms of the Brauer quotient. This is preceded by a more technical lemma.

Lemma 12.5.1. *Let U be an kG-module where k is a field of characteristic p and let H and J be subgroups of G.*

(1) If $\overline{U}(H) \neq 0$ then H is a p-group.
(2) If $\mathrm{Br}_H^G(\mathrm{tr}_J^G(a)) \neq 0$ for some element $a \in U^J$ then H is conjugate to a subgroup of J.

Proof. (1) For any group H, if P is a Sylow p-subgroup of H then $\mathrm{tr}_P^H : U^P \to U^H$ is surjective, because any $u \in U^H$ can be written $u = \frac{1}{|H:P|}\mathrm{tr}_P^H u$. Thus if H is not a p-group then P is a proper subgroup of H and so $\sum_{K<H} \mathrm{tr}_K^H(U^K) = U^H$ and $\overline{U}(H) = 0$.

(2) $\mathrm{Br}_H^G(\mathrm{tr}_J^G(a))$ is the image in $\overline{U}(H)$ of $\mathrm{res}_H^G \mathrm{tr}_J^G(a) = \sum_{g \in [H\backslash G/J]} \mathrm{tr}_{H \cap {}^gJ}^H(ga)$. If this is not zero in $\overline{U}(H)$ then, for some term in the sum, $H \cap {}^gJ$ must not be a proper subgroup of H, or in other words $H \subseteq {}^gJ$. □

The next result provides two more characterizations of the defect group of a block. These are obtained by applying what we have just done to kG, regarded as a representation of G via the conjugation action, so that $(kG)^G$ is the center of kG. Compare parts (2) and (3) of this result with the characterization from Theorem 12.4.5 that a defect group D is a minimal subgroup such that $e = \mathrm{tr}_D^G a$ for some $a \in (ekG)^H$.

Theorem 12.5.2. *Let k be a field of characteristic p, e a block of kG, and D a subgroup of G. The following are equivalent.*

(1) e has defect group D.
(2) $e = \mathrm{tr}_D^G(a)$ for some element $a \in (kG)^D$ and $\mathrm{Br}_D^G(e) \neq 0$.
(3) D is a maximal subgroup of G such that $\mathrm{Br}_D^G(e) \neq 0$.

We already know that D is uniquely defined up to conjugacy by condition (1), so it is also uniquely defined up to conjugacy by conditions (2) and (3).

Proof. (1) $\Rightarrow$ (2) If e has defect group D then D is minimal among groups for which $e = \mathrm{tr}_D^G(a)$ for some $a \in (kG)^D$ by Theorem 12.4.5. Thus certainly $e = \mathrm{tr}_D^G(a)$. Suppose that $\mathrm{Br}_D^G(e) = 0$. Then $e \in \sum_{K<D} \mathrm{tr}_K^D(kG)^K$ from the definition of Br, and we may write $e = \sum_{K<D} \mathrm{tr}_K^D(u_K)$ where $u_K \in (kG)^K$. Note that although this expression only suggests that $e \in (kG)^D$, in fact $e \in (kG)^G$. Now

$$\begin{aligned} e &= ee \\ &= (\mathrm{tr}_D^G(a))(\sum_{K<D} \mathrm{tr}_K^D(u_K)) \\ &= \mathrm{tr}_D^G(a \sum_{K<D} \mathrm{tr}_K^D(u_K)) \\ &= \mathrm{tr}_D^G \sum_{K<D} \mathrm{tr}_K^D(au_K) \\ &= \sum_{K<D} \mathrm{tr}_K^G(au_K). \end{aligned}$$

Thus $e \in \sum_{K<D} \mathrm{tr}_K^G(kG)^K$ and since e is a primitive idempotent, $e \in \mathrm{tr}_K^G(kG)^K$ for some $K < D$, by Rosenberg's Lemma 11.6.8 and Corollary 11.3.3. This contradicts the minimal property of D and so the supposition $\mathrm{Br}_D^G(e) = 0$ was false.

(2) $\Rightarrow$ (3) Suppose that $e = \mathrm{tr}_D^G(a)$ for some $a \in (kG)^D$ and $\mathrm{Br}_D^G(e) \neq 0$. By Lemma 12.5.1, if K is any subgroup for which $\mathrm{Br}_K^G(e) \neq 0$ then K is a subgroup of a conjugate of D, and this shows that D is a maximal subgroup of G such that $\mathrm{Br}_D^G(e) \neq 0$ (and also that it is unique up to conjugacy).

(3) $\Rightarrow$ (1) Suppose that condition (3) holds and let D_1 be a defect group of e. By the implication (1) $\Rightarrow$ (3) we know that D_1 is a maximal subgroup for which $\mathrm{Br}_{D_1}(e) \neq 0$ and, by the comment at the end of the proof of (2) $\Rightarrow$ (3) it is unique up to conjugacy among such subgroups. Since D also has this property, D and D_1 are conjugate, and D is a defect group. □

The relative trace map and the Brauer morphism have the convenient theoretical properties we have just described, but so far it does not appear to be easy to calculate with them in specific cases. We remedy this situation by showing that the Brauer morphism for the module kG acted upon by conjugation has an interpretation in terms of subgroups of G and group elements. We start with a general lemma about permutation modules. Since we will apply this in the situation of a subgroup H of G we denote our group by H. If Ω is an H-set we write $\tilde{\Omega} := \sum_{\omega\in\Omega} w$ as an element of the permutation module $k\Omega$. (Elsewhere we have written this element as $\overline{\Omega}$, but the bar notation is already in use here.)

Lemma 12.5.3. *Let Ω be an H-set and $k\Omega$ the corresponding permutation module.*

(1) The fixed point space $(k\Omega)^H$ has as a basis the H-orbit sums $\tilde{\Omega}_1, \dots, \tilde{\Omega}_n$ where $\Omega = \Omega_1 \sqcup \cdots \sqcup \Omega_n$ is a disjoint union of H-orbits.

(2) If H is a p-group and k is a field of characteristic p then $\sum_{K<H} \mathrm{tr}_K^H((k\Omega)^K)$ is the subspace of $(k\Omega)^H$ with basis the orbit sums $\tilde{\Omega}_i$ where $|\Omega_i| > 1$. Thus

$$(k\Omega)^H = k[\Omega^H] \oplus \sum_{K<H} \mathrm{tr}_K^H((k\Omega)^K).$$

The Brauer quotient $\overline{k\Omega}(H)$ may be identified with $k[\Omega^H]$, the span of the fixed points of H on Ω.

Proof. (1) For each transitive H-set Ω_i we know that $(k\Omega_i)^H$ is the 1-dimensional space spanned by $\tilde{\Omega}_i$ (Chapter 6, Exercise 8). From this it follows that if $\Omega = \Omega_1 \cup \cdots \cup \Omega_n$ where the Ω_i are the orbits of H on Ω then $(k\Omega)^H = (k\Omega_1)^H \oplus \cdots \oplus (k\Omega_n)^H$ has as a basis the orbit sums.

(2) Suppose H is a p-group and k has characteristic p. Since the relative trace map preserves direct sums it suffices to assume that H acts transitively on Ω, so $\Omega \cong H/J$ for some subgroup J. We have $\sum_{\omega\in\Omega} \omega = \mathrm{tr}_J^H(\omega_0)$ for any chosen $\omega_0 \in \Omega^J$. Thus $\sum_{K<H} \mathrm{tr}_K^H((k\Omega)^K)$ contains the span of the orbit sums for orbits of size larger than 1. On the other hand if $\Omega = \{\omega\}$ has size 1 and $K < H$ then $\mathrm{tr}_K^H \omega = |H : K|\omega = 0$ since H is a p-group so that $|H : K| = 0$ in k. This means that $\sum_{K<H} \mathrm{tr}_K^H((k\Omega)^K)$ equals the span of the orbit sums for orbits of size larger than 1, giving the claimed decomposition of $(k\Omega)^H$. □

Corollary 12.5.4. *Let G act on kG via conjugation and let H be a p-subgroup of G. Then $\overline{kG}(H) \cong k[C_G(H)]$. With this identification the Brauer morphism is a ring homomorphism $\mathrm{Br}_H^G : Z(kG) \to k[C_G(H)]^{N_G(H)}$ that truncates a group ring element $\sum_{g\in G} \lambda_g g$ to $\sum_{g\in C_G(H)} \lambda_g g$.*

Proof. In the conjugation action kG is a permutation module and the set of fixed points of H on G is $C_G(H)$. The Brauer morphism may be identified as inclusion of fixed points followed by projection onto the first factor in the decomposition $(kG)^H = k[C_G(H)] \oplus \sum_{K<H} \mathrm{tr}_K^H((kG)^K)$ given in Lemma 12.5.3. Under this identification the Brauer map is truncation of support to $C_G(H)$. □

In older treatments the Brauer map may be defined as the map $Z(kG) \to k[C_G(H)]$ that truncates a group ring element to have support on $C_G(H)$. A disadvantage of this direct approach is that it is less obvious that the Brauer morphism is a ring homomorphism and has the other properties we have described. It also ignores the larger context in which the Brauer quotient and morphism may be defined for any kG-module, and this has significance beyond the scope of this text. When the module is kG with the conjugation action, the interpretation of Br_H^G as truncation of the support of a group ring element to $C_G(H)$ does provide a concrete understanding of this homomorphism in the context of blocks. We immediately see the following, for example.

Corollary 12.5.5. *Let e be a block of kG. A defect group of e is a maximal p-subgroup D of G such that e has some part of its support in $C_G(D)$.*

Proof. This follows from part (3) of Theorem 12.5.2 and Corollary 12.5.4 since the condition given is exactly the requirement that $\mathrm{Br}_D^G(e) \neq 0$. □

Example 12.5.6. We illustrate with $G = S_3$ and $k = \mathbb{F}_2$ where we have blocks $e_1 = () + (1,2,3) + (1,3,2)$ and $e_2 = (1,2,3) + (1,3,2)$ (see Example 12.2.7). We already know from that example that e_1 is the principal block and e_2 is a block of defect zero. This means that e_1 has defect group a Sylow 2-subgroup $H = \langle(1,2)\rangle$, by Corollary 12.4.7, and e_1 has defect group 1. We can see this information in several different ways using the characterizations of the defect group. Since $C_G(H) = H$ we see that e_1 has part of its support in $C_G(H)$, but e_2 does not. According to Corollary 12.5.5 e_1 has defect group a 2-subgroup containing H, which must be H, and e_2 has defect group not containing H, and hence 1. These same calculations are encoded by the Brauer morphism, which takes values $\mathrm{Br}_H^G(e_1) = ()$, $\mathrm{Br}_H^G(e_2) = 0$, $\mathrm{Br}_1^G(e_1) = e_1$ and $\mathrm{Br}_1^G(e_2) = e_2$ showing again that H and 1 are (respectively) the largest subgroups Q (up to conjugacy) for which $\mathrm{Br}_Q^G(e_1) \neq 0$ and $\mathrm{Br}_Q^G(e_2) \neq 0$. We may also use the relative trace map to compute the defect groups of these blocks. We have by direct calculation $e_1 = \mathrm{tr}_H^G(e_1)$ and $e_2 = \mathrm{tr}_1^G((1,2,3))$. The latter shows that e_2 has defect group 1, by Theorem 12.4.5, or by Theorem 12.5.2 in combination with the information about the Brauer morphism. The expression for e_1 does not help us determine its defect group because every block can be expressed as a trace from a Sylow subgroup.

Notice that, as with Galois correspondence, for each containment of subgroups $K \leq H$ we have the reverse containment of fixed points $(kG)^K \supseteq (kG)^H$ and also a containment of centralizers $C_G(K) \supseteq C_G(H)$. This means that we obtain a commutative diagram

$$\begin{array}{ccccccc} \mathrm{Br}_H^G: & (kG)^G & \xrightarrow{\mathrm{res}_H^G} & (kG)^H & \longrightarrow & \overline{kG}(H) & = & kC_G(H) \\ & \| & & \downarrow \mathrm{res}_K^H & & & & \downarrow \iota \\ \mathrm{Br}_K^G: & (kG)^G & \xrightarrow{\mathrm{res}_K^G} & (kG)^K & \longrightarrow & \overline{kG}(K) & = & kC_G(K) \end{array}$$

where ι is the inclusion of the centralizer group rings, so that Br_K^G is the composite $\iota \circ \mathrm{Br}_H^G$. We see from this that if $\mathrm{Br}_H^G(x) \neq 0$ for some $x \in (kG)^G$ then $\mathrm{Br}_K^G(x) \neq 0$ for every $K \leq H$, since $\mathrm{Br}_H^G(x)$ equals $\mathrm{Br}_K^G(x)$ with its support truncated to $C_G(H)$. This observation provides a strengthening of part (3) of Theorem 12.5.2: if K is any subgroup of a defect group D of a block e of kG, then $\mathrm{Br}_K^G(e) \neq 0$.

The next result is important and interesting in its own right. It will be used when we come to describe the Brauer correspondence of blocks. The statement of this result is given in terms of $O_p(G)$, the largest normal p-subgroup of G. Although $O_p(G)$ could, in principle, be 1, the result gives no information in that case, so it is really a result about groups with a nontrivial normal p-subgroup.

Proposition 12.5.7. *Let $Q = O_p(G)$ be the largest normal p-subgroup of G. Then every block of G lies in $k[C_G(Q)]$ and is the sum of a G-orbit of blocks of $C_G(Q)$.*

Proof. We show first that $\sum_{K<Q} \mathrm{tr}_K^Q (kG)^K \subseteq \mathrm{Rad}((kG)^Q)$. Observe that if S is a simple kG-module then Q acts trivially on S by Corollary 6.2.2. Thus if $K < Q$, $a \in (kG)^K$, and $u \in S$ we have

$$\begin{aligned} \mathrm{tr}_K^Q(a) \cdot u &= \sum_{g \in [Q/K]} gag^{-1}u \\ &= \sum_{g \in [Q/K]} au \\ &= |Q : K| au = 0. \end{aligned}$$

Thus $\sum_{K<Q} \mathrm{tr}_K^Q (kG)^K$ annihilates every simple kG-module and so is contained in the radical of kG. It follows that $\sum_{K<Q} \mathrm{tr}_K^Q (kG)^K$ is a nilpotent ideal, and so is contained in $\mathrm{Rad}((kG)^Q)$.

If e is a central idempotent of kG then both e and $\mathrm{Br}_Q(e)$ may be regarded as idempotent elements of $(kG)^Q$ that map under the quotient homomorphism

$(kG)^Q \to \overline{kG}(Q) = k[C_G(Q)]$ to $\mathrm{Br}_Q(e)$. Since the kernel of this homomorphism is nilpotent it follows that e and $\mathrm{Br}_Q(e)$ are conjugate in $(kG)^Q$, by Exercise 2 of Chapter 11. (The argument is that the kernel I is contained in $\mathrm{Rad}((kG)^Q)$ so $(kG)^Q e$ and $(kG)^Q \mathrm{Br}_Q^G(e)$ are both projective covers of $\overline{kG}(Q) \cdot \mathrm{Br}_Q^G(e)$, and hence are isomorphic. For a similar reason $(kG)^Q(1-e)$ and $(kG)^Q(1 - \mathrm{Br}_Q^G(e))$ are isomorphic so there is an automorphism θ of $(kG)^Q$ sending $(kG)^Q e$ to $(kG)^Q \mathrm{Br}_Q^G(e)$ and $(kG)^Q(1-e)$ to $(kG)^Q(1 - \mathrm{Br}_Q^G(e))$. Now $\theta(x) = x\alpha$ for some unit $\alpha \in (kG)^Q$ and it has the property that $ue\alpha = u\alpha \mathrm{Br}_Q^G(e)$ for all $u \in (kG)^Q$. Hence $e = \alpha \mathrm{Br}_Q^G(e)\alpha^{-1}$.) Since e is central its only conjugate is e. Thus $e = \mathrm{Br}_Q(e)$, and this lies in $k[C_G(Q)]$.

We can now write $e = f_1 + \cdots + f_n$ as a sum of primitive central idempotents f_i of $k[C_G(Q)]$, and since e is stable under conjugation by G the $f_1, \ldots, f_n$ must be a union of G orbits in the conjugation action on the blocks of $k[C_G(Q)]$. However the sum of a single G orbit of the f_i is already a central idempotent of kG and e is the sum of such sums, so if e is a block it must be the sum of a single G-orbit of the f_i. □

Corollary 12.5.8. *If there is a normal p-subgroup Q of G for which $C_G(Q)$ is a p-group (for example, if $C_G(Q) \subseteq Q$) then G has only one p-block.*

Proof. Any block of kG must lie in $k[C_G(Q)]$ by Proposition 12.5.7; but this is the group ring of a p-group that has only one block, so kG also has only one block. □

Example 12.5.9. The preceding corollary implies, for example, that if K is a subgroup of $\mathrm{Aut}(Q)$ where Q is a p-group then the semidirect product $Q \rtimes K$ has only one p-block. Such is the case with the symmetric group S_4 at $p = 2$ on taking Q to be the normal Klein four-group, and there are many other examples of this phenomenon, including dihedral groups $C_p \rtimes C_2$ in characteristic p, the non-abelian group $C_7 \rtimes C_3$ of order 21 in characteristic 7, and so on.

12.6 Brauer Correspondence

We now define the Brauer correspondence of blocks. Our goal is Brauer's First Main Theorem, Theorem 12.6.4, which shows that blocks are parametrized by blocks of normalizers of p-subgroups of G. We will define the Brauer correspondent whenever b is a block of kJ, where J is a subgroup of G satisfying $HC_G(H) \subseteq J \subseteq N_G(H)$ for some p-subgroup H of G. In this situation the Brauer morphism is a ring homomorphism $\mathrm{Br}_H^G : Z(kG) \to (k[C_G(H)])^{N_G(H)}$, and the latter ring is contained in $Z(kJ)$ since $C_G(H) \subseteq J$, and elements that are fixed under $N_G(H)$ are also fixed under J. If $1 = e_1 + \cdots + e_n$ is the sum of blocks of kG then $1 = \mathrm{Br}_H^G(e_1) + \cdots + \mathrm{Br}_H^G(e_n)$ is a sum of orthogonal central idempotents of kJ. Thus if b is a block of kJ then

$$b = b\mathrm{Br}_H^G(e_1) + \cdots + b\mathrm{Br}_H^G(e_n)$$

is a decomposition of b as a sum of orthogonal idempotents in $Z(kJ)$, and since b is primitive in $Z(kJ)$ there is a unique block e_i of G so that $b\mathrm{Br}_H^G(e_i) = b$. We write b^G for this block e_i, and call it the *Brauer correspondent* of b. There may be several blocks b' of kJ with the same Brauer corresponding block of kG: $b'^G = b^G$. We see that the blocks of kG partition the blocks of kJ by this means.

There was a choice of p-subgroup H in the definition of b^G, since it would be possible to have another p-subgroup H_1 with $H_1 C_G(H_1) \subseteq J \subseteq N_G(H_1)$, and perhaps the definition of b^G would be different using H_1. In fact the choice of H does not matter, as we now show.

Proposition 12.6.1. *If b is a block of kJ, the definition of b^G is independent of the choice of the p-group H satisfying $HC_G(H) \subseteq J \subseteq N_G(H)$. The subgroups J that satisfy this condition for some p-subgroup H of G are characterized by the requirement that $C_G(O_p(J)) \subseteq J$. Furthermore, we may take $H = O_p(J)$ in the definition of b^G.*

Proof. Suppose that $HC_G(H) \subseteq J \subseteq N_G(H)$ for some p-subgroup H. Then H is a normal p-subgroup of J so we have $H \subseteq O_p(J)$ and $C_G(H) \supseteq C_J(O_p(J))$. Thus $C_G(O_p(J)) \subseteq J$. Conversely, if this is satisfied then taking $H_2 = O_p(J)$ we have $H_2 C_G(H_2) \subseteq J \subseteq N_G(H_2)$ and so the final statements are proved.

By Proposition 12.5.7 every central idempotent of kJ lies in $k[C_J(O_p(J))]$. For each block e_i of kG the idempotent $\mathrm{Br}_H^G(e_i)$ thus lies in $k[C_J(O_p(J))]$ and could have been computed by truncating the support of e_i to lie in $C_J(O_p(J))$ instead of $C_G(H)$. This identification of $\mathrm{Br}_H^G(e_i)$ is independent of the choice of H and hence so is b^G. □

Proposition 12.6.2. *Let H be a p-subgroup of G, J a subgroup of G with $HC_G(H) \subseteq J \subseteq N_G(H)$ and b a block of kJ. Then b^G has a defect group that contains a defect group of b.*

Proof. If $D \subseteq J$ is a defect group of b then $D \supseteq H$ by Corollary 12.3.4 since H is a normal p-subgroup of J, and so $C_G(H) \supseteq C_G(D) = C_J(D)$ since $J \supseteq C_G(H)$. Let us write $\mathrm{Br}_H^G(b^G) = b + b_1$ where b and b_1 are orthogonal central idempotents of kJ. We can now apply Br_D^J to this, and since this map truncates to $C_J(D)$ we get the same thing as if we had originally applied Br_D^G. Thus $\mathrm{Br}_D^G(b^G) = \mathrm{Br}_D^J(\mathrm{Br}_H^G(b^G)) = \mathrm{Br}_D^J(b) + \mathrm{Br}_D^J(b_1)$ is a sum of orthogonal idempotents in $k[C_J(D)]$. Since $\mathrm{Br}_D^J(b) \neq 0$ it follows that $\mathrm{Br}_D^G(b^G) \neq 0$, and hence D is contained in a defect group of b^G by Theorem 12.5.2. □

The following lemma is entirely technical and is used in the proof of Theorem 12.6.4.

Lemma 12.6.3. *Let H be a subgroup of G and U a kG-module. Then*

$$\mathrm{Br}_H^G \mathrm{tr}_H^G = \mathrm{tr}_H^{N_G(H)} \mathrm{Br}_H^H : U^H \to \overline{U}(H)^{N_G(H)}.$$

Proof. $\mathrm{Br}_H^G \mathrm{tr}_H^G(x)$ is the image in $\overline{U}(H)$ of

$$\mathrm{res}_H^G \mathrm{tr}_H^G(x) = \sum_{g \in [H \backslash G/H]} \mathrm{tr}_{H \cap {}^gH}^H({}^gx).$$

The only terms that contribute have $H \cap {}^gH = H$, which happens if and only if $g \in N_G(H)$, so the image equals the image of $\sum_{g \in [N_G(H)/H]} {}^gx = \mathrm{tr}_H^{N_G(H)} x$. □

The following is a basic version of Brauer's First Main Theorem.

Theorem 12.6.4 (Brauer's First Main Theorem). *Let k be field of characteristic p that is a splitting field for G and all of its subgroups and let D be a p-subgroup of G. The Brauer morphism induces a bijection between blocks of kG with defect group D and blocks of $kN_G(D)$ with defect group D with inverse given by the Brauer correspondent.*

Proof. Let us write N for $N_G(D)$. We prove that if e is a block of kG with defect group D then $\mathrm{Br}_D^G e$ is a block of kN with defect group D; and if b is a block of kN with defect group D then b^G is a block of kG with defect group D. Furthermore $(\mathrm{Br}_D^G e)^G = e$ and $\mathrm{Br}_D^G(b^G) = b$.

Let $b \in Z(kN)$ have defect D. Then by Theorem 12.4.5 $b = \mathrm{tr}_D^N(a)$ for some

$$a \in (kN)^D = k[C_G(D)] \oplus \sum_{K<D} \mathrm{tr}_K^D(kN)^K.$$

Since tr_D^N preserves both of the two summands on the right and $b \in k[C_G(D)]$ by Proposition 12.5.7, we may assume $a \in k[C_G(D)]$ and so $\mathrm{Br}_D^D(a) = a$. Thus $b = \mathrm{tr}_D^N \mathrm{Br}_D^D(a) \in \mathrm{tr}_D^N(k[C_G(D)])$.

Let $e = b^G \in Z(kG)$ be the Brauer correspondent of b, so that $b = b\mathrm{Br}_D^G(e)$ is a summand of $\mathrm{Br}_D^G(e)$ and e has a defect group $D_1 \supseteq D$ by Proposition 12.6.2. We will show that $D_1 = D$. Now, using Lemma 12.6.3,

$$\mathrm{Br}_D^G(\mathrm{tr}_D^G((kG)^D)) = \mathrm{tr}_D^N \mathrm{Br}_D^D((kG)^D) = \mathrm{tr}_D^N(k[C_G(D)])$$

and so, since Br_D^G is a ring homomorphism,

$$\mathrm{Br}_D^G(e\mathrm{tr}_D^G((kG)^D)) = \mathrm{Br}_D^G(e) \cdot \mathrm{Br}_D^G(\mathrm{tr}_D^G((kG)^D)) = \mathrm{Br}_D^G(e) \cdot \mathrm{tr}_D^N(k[C_G(D)]),$$

which contains b. Thus the ideal $e\mathrm{tr}_D^G((kG)^D)$ of $eZ(kG)$ is not nilpotent, since it has an image under a ring homomorphism that contains a nonzero idempotent. It follows that $e\mathrm{tr}_D^G((kG)^D) = eZ(kG)$, since $eZ(kG)$ is local by Proposition 11.1.4. This implies, since $\mathrm{tr}_D^G(e(kG)^D) = e\mathrm{tr}_D^G((kG)^D)$, that e lies in the image of tr_D^G and so has defect group contained in D. This completes the argument that the defect group of e equals D.

We have also just seen that $\mathrm{Br}_D^G(eZ(kG))$ contains b, and it also contains $\mathrm{Br}_D^G(e)$. It is an image of a local ring and hence is local and contains only one nonzero idempotent. It follows that $b = \mathrm{Br}_D^G(e)$. This shows us that $b \mapsto b^G$ is a one-to-one mapping from blocks of kN with defect group D to blocks of kG with defect group D, and that its inverse on one side is Br_D^G. We conclude by observing that this mapping is surjective. For, if $e \in Z(kG)$ is a block with defect group D then $\mathrm{Br}_D^G(e)$ is a sum of blocks b of kN for which $e = b^G$, and the blocks b have defect groups that are subgroups of D by Proposition 12.6.2. On the other hand every block of kN has defect group containing D, since D is a normal p-subgroup of N. It follows that $e = b^G$ for some block of kN with defect group D, showing that $b \mapsto b^G$ is surjective with the domain and codomain as specified earlier. This completes the proof. □

Example 12.6.5. When $G = S_3$ and $k = \mathbb{F}_2$ we have seen (in Example 12.2.7 and elsewhere) that $e_1 = () + (1,2,3) + (1,3,2)$ has defect group $\langle(1,2)\rangle$ and $e_2 = (1,2,3) + (1,3,2)$ has defect group 1. Here $\mathrm{Br}_{\langle(1,2)\rangle}^G(e_1) = () \in k\langle(1,2)\rangle$, so that the Brauer correspondent $()^G$ of the only idempotent in $k\langle(1,2)\rangle$ is e_1, this giving a bijection between the idempotents of S_3 and $k\langle(1,2)\rangle$ with defect group $\langle(1,2)\rangle$.

Example 12.6.6. Let $G = A_5$ and let k be a splitting field of characteristic 2. A Sylow 2-subgroup $P \cong C_2 \times C_2$ has $N_G(P) \cong A_4$, and we have seen in Exercise 2 from Chapter 8 that kA_4 has only one block. Thus there is only one block of kG with defect group P: it is the principal block. The subgroups C_2 have Sylow p-subgroups P as their normalizers and there are no blocks of kP with C_2 as defect group, since P is a 2-group. Hence there are no blocks of kG with this defect group. This confirms information we know from a different source, to the effect that C_2 cannot be a defect group by Corollary 12.3.4 because it is not O_2 of its normalizer. Finally there remain the blocks of defect zero of kA_5, which have defect group 1. There is in fact just one of these, as may be seen by inspecting the character table of A_5 for characters of degree divisible by 4.

12.7 Further Reading

Having stated a "basic" version of Brauer's First Main Theorem, the reader will naturally wonder what a less basic version looks like, and what other main theorems may be attributed to Brauer. An extended version of the first main theorem (with the notation in force in Theorem 12.6.4) says that blocks of $kN_G(D)$ with defect group D biject with $N_G(D)$-conjugacy classes of blocks b of $k[DC_G(D)/D]$ of defect zero, such that $|\,\mathrm{Stab}(b) : DC_G(D)|$ is not divisible by p, where $\mathrm{Stab}(b)$ is the stabilizer of b under conjugation by $N_G(D)$.

Brauer's Second Main Theorem can be stated in various ways, but in one version has the implication that Green correspondence is compatible with Brauer correspondence.

Brauer's third main theorem says that the Brauer correspondent of a block is the principal block if and only if that block is the principal block.

To read further about these, see the books by Benson [3], Alperin [2], and Thévenaz [21]. The account by Thévenaz goes into considerable detail and describes the theory of blocks that has been developed by Puig.

It is also important to know about the theory of blocks with a cyclic defect group. This theory describes completely the structure of the indecomposable projective modules, as well as their indecomposable representations, in terms of the combinatorial properties of a tree called the Brauer tree, computed from the decomposition matrix. A good place to start reading is the book by Alperin [2].

12.8 Summary of Chapter 12

- Blocks correspond to blocks of the Cartan matrix, indecomposable ring summands of kG or RG, primitive central idempotents in kG or RG, and certain equivalence classes of representations in characteristic 0 or in characteristic p.
- The defect group of a block may also be characterized in several ways, using the bimodule structure of the block as a ring summand, the relative trace map, the Brauer morphism, and the vertices of modules in the block.
- A defect group is always the intersection of two Sylow p-subgroups and is the largest normal p-subgroup of its normalizer.
- The principal block has Sylow p-subgroups as defect groups.
- The Brauer correspondent b^G provides a bijection between blocks of G and of $N_G(D)$ with defect group D.

12.9 Exercises for Chapter 12

We assume throughout that (F, R, k) is a complete splitting p-modular system for G.

1. (a) Show that if there exists an RG-module U for which both U and U^* belong to the same block of RG then, for every module V in that block, V^* also belongs to the same block.

(b) Find an example of a group algebra kG in characteristic p with a module U so that U and U^* lie in different blocks.

2. A subgroup Q of G is defined to be p-radical if and only if $Q = O_p(N_G(Q))$.

(a) Show that Q is p-radical if and only if $O_p(N_G(Q)/Q) = 1$.

(b) Suppose that Q_1 and Q_2 are p-radical subgroups of G. Show that if $N_G(Q_1) \supseteq N_G(Q_2)$ then $Q_1 \subseteq Q_2$.

(c) Show that if Q is a p-radical subgroup of G then $Q \supseteq O_p(G)$.

3. (a) Compute the 2-radical and 3-radical subgroups of S_4. For each 2-radical and 3-radical subgroup, determine whether there is a block of S_4 (in characteristic 2 or 3) whose defect group is that subgroup. [The Cartan matrices for S_4 were computed in Example 10.1.5.]

(b) Compute the 2-radical subgroups of A_5. For each 2-radical subgroup determine whether there is a 2-block of A_5 whose defect group is that subgroup. [The character table of A_5 was computed in Chapter 4, Exercise 5 and is repeated in Appendix B.]

(c) Compute the 2-radical and 7-radical subgroups of $GL(3, 2)$. For each 2-radical and 7-radical subgroup, determine whether there is a block of $GL(3, 2)$ (in characteristic 2 or 7) whose defect group is that subgroup. [Either use the results of Chapter 10 Exercises 1 and 2 or the tables in Appendix B. Use also the group theoretic information about $GL(3, 2)$ in Appendix B.]

4. Let A be an algebra and let U and V be A-modules that lie in different blocks of A. Show that

(a) $\operatorname{Hom}_A(U, V) = 0$ and

(b) every short exact sequence of A-modules $0 \to U \to W \to V \to 0$ is split.

5. Let (F, R, k) be a complete p-modular system and G a finite group. The decomposition number d_{TS} is the multiplicity of the simple kG-module S as a composition factor of the reduction modulo π of the simple FG-module T. Fix a p-block e of G.

(a) Show that the simple FG-modules belonging to e are precisely the simple FG-modules T for which there exists a simple kG-module S belonging to e with $d_{TS} \neq 0$.
(b) Show that the simple kG-modules belonging to e are precisely the simple kG-modules S for which there exists a simple FG-module T belonging to e with $d_{TS} \neq 0$.

6. Let G_1 and G_2 be finite groups.

(a) Show that the block idempotents of $R[G_1 \times G_2]$ are precisely the e_1e_2 where e_i is a block idempotent of RG_i, $i = 1, 2$. [The difficulty is to show that the central idempotent e_1e_2 is primitive.]
(b) Suppose that $D_i \leq G_i$ is a defect group of e_i, where $i = 1, 2$. Show that $D_1 \times D_2$ is a defect group of e_1e_2.

7. Let $G = A_4 = K \rtimes H$ where $K \cong C_2 \times C_2$ and $H \cong C_3$.

(a) Write down a complete list of the primitive central idempotents in $\mathbb{F}_3K$.
(b) Compute the orbits of G on the set of idempotents found in (a) and hence find the block idempotents of $\mathbb{F}_3G$.
(c) For each of the blocks of $\mathbb{F}_3G$ compute the effect of Br_1^G and Br_H^G. Hence compute the defect groups of each block by this means.

8. Let $G = K \rtimes H$ be a p-nilpotent group, so that H is a p-group and K has order prime to p. We saw in Proposition 12.2.6 that each block f of kG is the sum of a G-conjugacy class of blocks of kK, namely, $f = e_1 + \cdots + e_t$ where the e_i are blocks of kK forming a single orbit under the conjugation action of G. Since kK is semisimple each e_i corresponds to a unique simple kK-module T_i and by Theorem 8.4.1 there is a unique simple kG-module S belonging to f.

(a) Show that t is a power of p.
(b) Show that $S \downarrow_K^G \cong (T_1 \oplus \cdots \oplus T_t)^n$ as kK-modules, for some n.
(c) Show that a defect group of f is contained in $\mathrm{Stab}_H(e_1)$. [Use a characterization of the defect group in terms of the relative trace map.]

9. Let $G = Q_8 \rtimes C_3 \cong SL(2, 3)$. Write down the block idempotents for the three 3-blocks of this group. Verify, using the methods of this chapter, that two of them have defect group a Sylow 3-subgroup, and one has defect 0. [Use the character tables of G given in Appendix B.]

10. Let $G = (C_2 \times C_2 \times C_2) \rtimes C_7$ where the cycle of order 7 acts nontrivially on $C_2 \times C_2 \times C_2$. To construct this action, observe that $GL(3, 2)$ has order 168, so has a Sylow 7-subgroup of order 7, which necessarily must permute the seven nonidentity elements of $C_2 \times C_2 \times C_2$ transitively. Write down the block idempotents of G in characteristic 7, and determine their defect groups.

11. A group $G = K \rtimes H$ is said to be a *Frobenius group* if every nonidentity element of H acts (by conjugation) without fixed points on the nonidentity elements of K: for all nonidentity $h \in H$, $C_K(h) = \{1\}$. Let G be such a group.

(a) Show that if k is any nonidentity element of K then ${}^kH \cap H = 1$.
(b) Suppose that H is a Sylow p-subgroup of G. Show that the only possible defect groups of blocks are H and $\{1\}$, and that a block idempotent f has defect group H if and only if the coefficient of 1 in f is nonzero.

12. Let $e \in Z(kG)$ be a block idempotent and suppose the coefficient of 1 in e is nonzero. Show that the defect groups of e are the Sylow p-subgroups of G.

13. Let (F, R, k) be a splitting p-modular system for G. Let e be a block idempotent of kG with defect group D, and suppose $g \in G$ is an element that centralizes a p-power element of G that does not lie in any conjugate of D.

(a) Use the characterization of D in terms of the Brauer morphism to show that g does not lie in the support of e.
(b) Suppose furthermore that e is a block of defect 0 corresponding to an ordinary character χ. Show that $\chi(g) \in (\pi)$, the maximal ideal of R.

[Use the formula in Theorem 3.6.2 that gives the block idempotent in RG in terms of the characters of G, and that reduces to e.]

14. Let (F, R, k) be a splitting p-modular system for G. Show that a p-block of G has defect zero if and only if every RG lattice lying in the block is a projective RG-module.

10. [illegible] when the [illegible] is of order [illegible] on $C_2 \times C_2$. [illegible] (1.2.2) has order [illegible] [illegible] [illegible] seven [illegible]. Write down the block idempotents of G and [illegible] defect groups.

11. A [illegible] elements [illegible] fixed points [illegible] $(G) = [2]$ [illegible]

(a) [illegible]

(b) Suppose that [illegible] is a Sylow p-subgroup of G. Show that [illegible] defect group of [illegible] and [1] [illegible] block idempotent [illegible] if and only if the coefficient of 1 in e is nonzero.

12. [illegible] block idempotent and [illegible]

13. Let [illegible] be the [illegible] of G with defect group D and suppose x is an element [illegible] p-power [illegible]

(a) Use the [illegible] to show that [illegible] does not [illegible]

(b) Suppose further that [illegible] is a block of [illegible] corresponding to an [illegible] Show that [illegible] the [illegible] of G.

[illegible]

14. [illegible] Show that a p-block of [illegible]

Appendix A
Discrete Valuation Rings

Let F be a field. A (multiplicative) *valuation* on F is a mapping $\phi : F \to \mathbb{R}_{\geq 0}$ such that

- $\phi(a) = 0$ if and only if $a = 0$,
- $\phi(ab) = \phi(a)\phi(b)$ for all $a, b \in F$, and
- $\phi(a + b) \leq \phi(a) + \phi(b)$ for all $a, b \in F$.

In many texts, the theory is developed in terms of additive valuations. A suitable reference that uses the multiplicative language is [14].

Example A.1. No matter what the field F is, we always have the valuation

$$\phi(a) = \begin{cases} 0 & \text{if } a = 0 \\ 1 & \text{otherwise.} \end{cases}$$

This valuation is the *trivial valuation*, and we generally exclude it.

Example A.2. If F is any subfield of the field of complex numbers, we may take $\phi(a) = |a|$, the absolute value of a.

Example A.3. Let $F = \mathbb{Q}$ and pick a prime p. Every rational number $a \in \mathbb{Q}$ may be written $a = \frac{r}{s}$ where $(r, s) = 1$. We set

$$\nu_p(a) = \begin{cases} \infty & \text{if } a = 0, \\ \text{power to which } p \text{ divides } r & \text{if } p \mid r, \\ -\text{power to which } p \text{ divides } s & \text{otherwise,} \end{cases}$$

so that if $a \neq 0$ then

$$a = p^{\nu_p(a)} \frac{r'}{s'}$$

where $(r', p) = 1 = (s', p)$. Now let λ be any real number with $0 < \lambda < 1$ and put

$$\phi(a) = \begin{cases} \lambda^{\nu_p(a)} & \text{if } a \neq 0 \\ 0 & \text{if } a = 0. \end{cases}$$

Often λ is taken to be $\frac{1}{p}$, but the precise choice of λ does not affect the properties of the valuation. This valuation is called the *p-adic valuation* on $\mathbb{Q}$.

This last ϕ is an example of a valuation that satisfies the so-called *ultrametric inequality*

$$\phi(a+b) \leq \max\{\phi(a), \phi(b)\},$$

which, in the case of this example, comes down to the fact that if $p^n \mid a$ and $p^n \mid b$ where $a, b \in \mathbb{Z}$, then $p^n \mid (a+b)$. We say that ϕ is *non-Archimedean* if it satisfies the ultrametric inequality. The valuations in the third example are also *discrete*, meaning that $\{\phi(a) \mid a \in K,\ a \neq 0\}$ is an infinite cyclic group under multiplication. It is the case that discrete valuations are necessarily non-Archimedean.

We deduce from the axioms for a valuation that $\phi(1) = \phi(-1) = 1$. Using this, we see that every valuation ϕ gives rise to a metric $d(a, b) = \phi(a-b)$ on the field F. We say that two valuations are *equivalent* if and only if the metric spaces they determine are equivalent, that is, they give rise to the same topologies. In Example A.3, changing the value of λ between 0 and 1 gives an equivalent valuation.

Theorem A.4 (Ostrowski). *Up to equivalence, the nontrivial valuations on $\mathbb{Q}$ are the ones just described, namely, the usual absolute value and for each prime number a non-Archimedean valuation.*

It is a fact that if R is a ring of algebraic integers (or more generally a Dedekind domain) with quotient field F, the non-Archimedean valuations on R, up to equivalence, biject with the maximal ideals (which are the same as the nonzero prime ideals) of R.

Let ϕ be a non-Archimedean valuation on a field F. The set $R_\phi = \{a \in F \mid \phi(a) \leq 1\}$ is a ring, called the *valuation ring* of ϕ. Any ring arising in this way for some ϕ is called a *valuation ring*. If the valuation is discrete, the ring is called a *discrete valuation ring*. We set $P_\phi = \{a \in F \mid \phi(a) < 1\}$, and this is evidently an ideal of R_ϕ. For example, if $F = \mathbb{Q}$ and ϕ is the non-Archimedean valuation corresponding to the prime p then R_ϕ is the localization of $\mathbb{Z}$ at p, and P_ϕ is its unique maximal ideal.

Proposition A.5. *Let ϕ be a discrete valuation on a field F with valuation ring R_ϕ and valuation ideal P_ϕ.*

(1) *An element $a \in R_\phi$ is invertible if and only if $\phi(a) = 1$.*
(2) *P_ϕ is the unique maximal ideal of R_ϕ, consisting of the noninvertible elements.*
(3) *$P_\phi = (\pi)$ where π is any element such that $\phi(\pi)$ generates the value group of ϕ.*
(4) *Every element of R_ϕ is uniquely expressible $a = \pi^\nu a'$ where a' is a unit in R_ϕ. In this situation, $\phi(a) = \phi(\pi)^\nu$.*
(5) *The ideals of R_ϕ are precisely the powers $P_\phi^n = (\pi^n)$. Thus R_ϕ is a principal ideal domain.*
(6) *F is the field of fractions of R_ϕ.*

Proof. The proofs are all rather straightforward. If $ab = 1$ where $a, b \in R_\phi$ then $\phi(ab) = \phi(a)\phi(b) = \phi(1) = 1$ and since $\phi(a)$ and $\phi(b)$ are real numbers between 0 and 1 it follows that they equal 1. Conversely, if $\phi(a) = 1$ let b be the inverse of a in F. Since $\phi(b) = 1$ also it follows that $b \in R_\phi$ and a is invertible in R_ϕ. This proves part (1).

Now P_ϕ is seen to consist of the noninvertible elements of R_ϕ. It is an ideal, and it follows that it is the unique maximal ideal. Defining π to be an element for which $\phi(\pi)$ generates the value group of ϕ we see that $\pi \in P_\phi$. If $a \in P_\phi$ is any element then $\phi(a) = \phi(\pi)^\nu$ for some ν and so $\phi(\pi^{-\nu}a) = 1$. Thus, $\pi^{-\nu}a = a'$ for some unit $a' \in R_\phi$, and so $a = \pi^\nu a'$. This proves that $P_\phi = (\pi)$, and also the first statement of (4). The uniqueness of the expression in (4) comes from the fact that in any expression $a = \pi^\nu a'$ with a' a unit, necessarily ν is defined by $\phi(a) = \phi(\pi)^\nu$, and then a' is forced to be $\pi^{-\nu}a$.

To prove (5), if I is any ideal of R_ϕ we let n be minimal so that I contains a nonzero element a with $\phi(a) = \phi(\pi)^n$. Then by (4), we have $(a) = (\pi^n) \supseteq I$, and it follows that $I = (\pi^n)$.

As for (6), given any element $a \in F$, either $a \in R_\phi$ or $\phi(a) = \phi(\pi)^{-n}$ for some $n > 0$. In the second case, the element $a' = \pi^n a$ has $\phi(a') = 1$ so $a' \in R_\phi$ and now $a = \frac{a'}{\pi^n}$, showing that a lies in the field of fractions of R_ϕ in both cases. □

When we reduce representations from characteristic zero to positive characteristic we need to work with algebraic number fields, that is, field extensions of $\mathbb{Q}$ of finite degree. Let F be an algebraic number field, and R its ring of integers. We quote without proof some facts about this situation. A full account may be found in [14] and other standard texts on number theory. A *fractional ideal* in F is a finitely generated R-submodule I of F. For any such I, we put

$I^{-1} = \{x \in F \mid xI \subseteq F\}$. With this definition of inverse and with a multiplication defined the same way as the multiplication of ideals, the fractional ideals form a group, whose identity is R. Every fractional ideal may be written uniquely as a product $I = \mathfrak{p}_1^{a_1} \cdots \mathfrak{p}_t^{a_t}$ where $\mathfrak{p}_1, \ldots, \mathfrak{p}_t$ are maximal ideals of R and the a_i are nonzero integers (which may be positive or negative). Let us write $\nu_{\mathfrak{p}_i}(I) = a_i$, and let $0 < \lambda < 1$. Then for each maximal ideal $\mathfrak{p}$ of R, we obtain a discrete valuation on F by putting $\phi(a) = \lambda^{\nu_{\mathfrak{p}}(Ra)}$, which is called the $\mathfrak{p}$-*adic valuation* on F.

Proposition A.6. *Let F be an algebraic number field with ring of algebraic integers R, and let ϕ be the discrete valuation on F associated to a maximal ideal $\mathfrak{p}$ of R. Let $R_{\mathfrak{p}}$ be the valuation ring of ϕ with maximal ideal $P_{\mathfrak{p}}$. Then $R_{\mathfrak{p}}$ is the localization of R at $\mathfrak{p}$, and the inclusion $R \to R_{\mathfrak{p}}$ induces an isomorphism $R/\mathfrak{p} \cong R_{\mathfrak{p}}/P_{\mathfrak{p}}$.*

Proof. We assume the group structure of the set of fractional ideals. The localization of R at $\mathfrak{p}$ is

$$\{\frac{a}{b} \mid a, b \in R,\ b \notin \mathfrak{p}\}$$

and this is clearly contained in $R_{\mathfrak{p}}$. Conversely, if $\frac{a}{b} \in R_{\mathfrak{p}}$ then $\nu_{\mathfrak{p}}(a) \geq \nu_{\mathfrak{p}}(b)$ and we choose a field element $x \in \mathfrak{p}^{-\nu_{\mathfrak{p}}(b)} - \mathfrak{p}^{-\nu_{\mathfrak{p}}(b)+1}$. Writing $\frac{a}{b} = \frac{ax}{bx}$ expresses $\frac{a}{b}$ as a quotient of elements of R with $bx \notin \mathfrak{p}$, showing that $\frac{a}{b}$ lies in the localization.

The kernel of the composite $R \to R_{\mathfrak{p}} \to R_{\mathfrak{p}}/P_{\mathfrak{p}}$ is $R \cap P_{\mathfrak{p}} = \mathfrak{p}$. We show that this composite is surjective. We can write any element of $R_{\mathfrak{p}}/P_{\mathfrak{p}}$ as $\frac{a}{b} + P_{\mathfrak{p}}$ where $a, b \in R$ and $b \notin \mathfrak{p}$. Since $\mathfrak{p}$ is maximal in R, $R/\mathfrak{p}$ is a field and so there exists $c \in R$ with $bc - 1 \in \mathfrak{p}$. Now $\frac{a}{b} - ac = \frac{a}{b}(1 - bc) \in P_{\mathfrak{p}}$ and $\frac{a}{b} + P_{\mathfrak{p}} = ac + P_{\mathfrak{p}}$ is the image of $ac \in R$. These observations show that we have an isomorphism $R/\mathfrak{p} \cong R_{\mathfrak{p}}/P_{\mathfrak{p}}$. □

Given a valuation ϕ on F we may form the completion $\hat{F}$ as a metric space, that contains F in a canonical way. We state without proof that the completion $\hat{F}$ acquires a ring structure extending that of F, and that $\hat{F}$ is a field. The valuation ϕ extends uniquely to a valuation $\hat{\phi}$ on $\hat{F}$, and $\hat{F}$ is complete in the metric given by $\hat{\phi}$. If ϕ is non-Archimedean then so is $\hat{\phi}$, and if ϕ is discrete then so is $\hat{\phi}$, with the same value group. Thus in the case of a discrete valuation we have a valuation ring $\hat{R}_{\phi} = \{a \in \hat{F} \mid \hat{\phi}(a) \leq 1\}$ with unique maximal ideal $\hat{P}_{\phi} = \{a \in \hat{F} \mid \hat{\phi}(a) < 1\}$, and $\hat{P}_{\phi} = \hat{R}_{\phi}(\pi)$ since $\hat{\phi}(\pi) = \phi(\pi)$ generates the value group. (We should properly write $\hat{R}_{\hat{\phi}}$, etc., but this seems excessive.) The ideals of $\hat{R}_{\phi}$ are exactly the powers $\hat{P}_{\phi}^n$.

When ϕ is the p-adic valuation on $\mathbb{Q}$, the completion $\hat{\mathbb{Q}}$ is denoted $\mathbb{Q}_p$ and is called the *field of p-adic rationals*. The valuation ring of $\mathbb{Q}_p$ with respect to $\hat{\phi}$ is denoted $\mathbb{Z}_p$ and is called the *ring of p-adic integers*.

Lemma A.7. *Let ϕ be a discrete valuation on a field F with valuation ring R_ϕ. The inclusion $R_\phi \hookrightarrow \hat{R}_\phi$ induces an isomorphism $R_\phi/P_\phi^n \cong \hat{R}_\phi/\hat{P}_\phi^n$ for all n.*

Proof. Consider the composite homomorphism $R_\phi \to \hat{R}_\phi \to \hat{R}_\phi/\hat{P}_\phi^n$. Its kernel is P_ϕ^n and the desired isomorphism will follow if we can show that this homomorphism is surjective. To show this, given $a \in \hat{R}_\phi$ we know from the construction of the completion that there exists $b \in R_\phi$ with $\phi(b-a) < \phi(\pi)^n$, that is, $b - a \in \hat{P}_\phi^n$. Now b maps to $a + \hat{P}_\phi^n$. □

The completion $\hat{R}_\phi$ is, by definition, the set of equivalence classes of Cauchy sequences in R_ϕ. We comment that a sequence (a_i) of elements of R_ϕ is a Cauchy sequence if and only if for every n there exists a number N so that whenever $i, j > N$ we have $a_i - a_j \in P_\phi^n$, that is, $a_i \equiv a_j \pmod{P_\phi^n}$.

Lemma A.8. *Let ϕ be a discrete valuation on a field F with valuation ring R_ϕ, maximal ideal P_ϕ and completion $\hat{R}_\phi$. Any element of $\hat{R}_\phi$ is uniquely expressible as a series*

$$a = a_0 + a_1\pi + a_2\pi^2 + \cdots$$

where the a_i lie in a set of representatives $\mathcal{S}$ for R_ϕ/P_ϕ.

Proof. Let $a \in \hat{R}_\phi$. Since $\hat{R}_\phi/\hat{P}_\phi \cong R_\phi/P_\phi$, we have $a + \hat{P}_\phi = a_0 + \hat{P}_\phi$ for some uniquely determined $a_0 \in \mathcal{S}$. Now $a - a_0 \in \hat{P}_\phi$ so $a = a_0 + \pi b_1$ for some $b_1 \in \hat{R}_\phi$. Repeating this construction we write $b_1 = a_1 + \pi b_2$ with $a_1 \in \mathcal{S}$ uniquely determined, and in general $b_n = a_n + \pi b_{n+1}$ with $a_n \in \mathcal{S}$ uniquely determined. Now $a_0, a_0 + a_1\pi, a_0 + a_1\pi + a_2\pi^2, \ldots$ is a Cauchy sequence in R_ϕ whose limit is a, and we write this limit as the infinite series. □

The last result, combined with Proposition A.6, provides a very good way to realize the completion $\hat{R}_\phi$. For example, in the case of the p-adic valuation on $\mathbb{Q}$ we may take $\mathcal{S} = \{0, 1, \ldots, p-1\}$. The completion $\hat{\mathbb{Z}} = \mathbb{Z}_p$ may be realized as the set of infinite sequences $\cdots a_3a_2a_1a_0.$ of elements from $\mathcal{S}$ presented in positions to the left of a "point," analogous to the decimal point (which we write on the line, rather than raised above the line). Thus, a_0 is in the 1s position, a_1 is in the ps position, a_2 is in the p^2s position, and so on. Unlike decimal numbers these strings are potentially infinite to the left of the point, whereas decimal numbers are potentially infinite to the right of the point. Addition and multiplication of these strings are performed by means of the same algorithms (carrying values from one position to the next when p is exceeded, etc.) that are used with infinite decimals. Note that p-adic integers have the advantage over decimals that, whereas certain real numbers have more than one decimal representation, distinct p-adic expansions always represent distinct elements of $\mathbb{Z}_p$.

Exercises for Appendix A

1. With the description of the p-adic integers as the set of infinite sequences

$$\cdots a_3 a_2 a_1 a_0 \bullet$$

in positions to the left of a "point," where $a_i \in \{0, \ldots, p-1\}$, show that when $p = 2$, we have

$$-1 = \cdots \overline{1}111\bullet \quad \text{and}$$

$$\frac{1}{3} = \cdots \overline{10}101011\bullet$$

Find the representation of the fraction $1/5$ in the 2-adic integers. What fraction does $\cdots \overline{1100}110011\bullet$ represent?

2. Show that the field of p-adic rationals $\mathbb{Q}_p$ may be constructed as the set of sequences $\cdots a_3 a_2 a_1 a_0 \bullet a_{-1} a_{-2} \cdots a_{-n}$ that may be infinite to the left of the point, but must be finite to the right of the point, where $a_i \in \{0, \ldots, p-1\}$ for all i. Show that the field of rational numbers $\mathbb{Q}$ is the subset of these sequences that eventually recur.

Appendix B
Character Tables

We collect here the character tables that have been studied in this book. We use the notation $\zeta_n = e^{2\pi i/n}$ for a primitive nth root of unity. Where a group is isomorphic to one of the groups $SL(2, p)$ or $PSL(2, p)$ we have emphasized this, because there is a special construction of simple modules over $\mathbb{F}_p$, described in Chapter 6, Exercise 25.

Cyclic and Abelian Groups

We let $C_n = \langle x | x^n = 1 \rangle$.

Characteristic 0

C_n
Ordinary characters

g $\lvert C_G(g)\rvert$	1 n	x n	$\cdots$ $\cdots$	x^{n-1} n
$\chi_{\zeta_n^s}$ $(0 \le s \le n-1)$	1	ζ_n^s	$\cdots$	$\zeta_n^{s(n-1)}$

Notes for Cyclic and Abelian Groups

This table was described in Proposition 4.1.1. Ordinary character tables of abelian groups are obtained as tensor products of the character tables of the cyclic direct factor groups, according to Corollary 4.1.4. The representations in positive characteristic were described for abelian groups in Example 8.2.1.

The Symmetric Group S_3

This group is isomorphic to the dihedral group D_6 and also to $GL(2, 2) = SL(2, 2)$

Characteristic 0

S_3
Ordinary characters

g $\|C_G(g)\|$	() 6	(12) 2	(123) 3
χ_1	1	1	1
χ_{sign}	1	-1	1
χ_2	2	0	-1

Characteristic 2

$S_3 \cong SL(2, 2)$
Brauer simple $p = 2$

g $\|C_G(g)\|$	() 6	(123) 3
ϕ_1	1	1
ϕ_2	2	-1

$S_3 \cong SL(2, 2)$
Brauer projective $p = 2$

g $\|C_G(g)\|$	() 6	(123) 3
η_1	2	2
η_2	2	-1

$S_3 \cong SL(2, 2)$
Decomposition matrix $p = 2$

	ϕ_1	ϕ_2
χ_1	1	0
χ_{sign}	1	0
χ_2	0	1

$S_3 \cong SL(2, 2)$
Cartan matrix $p = 2$

	η_1	η_2
ϕ_1	2	0
ϕ_2	0	1

Characteristic 3

S_3
Brauer simple $p = 3$

g $\|C_G(g)\|$	() 6	(12) 2
ϕ_1	1	1
ϕ_{sign}	1	-1

S_3
Brauer projective $p = 3$

g $\|C_G(g)\|$	() 6	(12) 2
η_1	3	1
η_{sign}	3	-1

S_3
Decomposition matrix $p = 3$

	ϕ_1	ϕ_{sign}
χ_1	1	0
χ_{sign}	0	1
χ_2	1	1

S_3
Cartan matrix $p = 3$

	η_1	η_{sign}
ϕ_1	2	1
ϕ_{sign}	1	2

Notes for S_3

The ordinary character table of S_3 was one of the first constructed, in Example 3.1.2. The Brauer tables were constructed in Example 10.1.4.

The Dihedral and Quaternion Groups of Order 8

We put

$$D_8 = \langle x, y | x^4 = y^2 = 1,\ yxy^{-1} = x^{-1} \rangle$$
$$Q_8 = \langle x, y | x^4 = 1,\ x^2 = y^2,\ yxy^{-1} = x^{-1} \rangle.$$

Characteristic 0

In both cases, the character table is

D_8 and Q_8
Ordinary characters

g $\lvert C_G(g)\rvert$	1 8	x^2 8	x 4	y 4	xy 4
χ_1	1	1	1	1	1
χ_{1a}	1	1	1	−1	−1
χ_{1b}	1	1	−1	1	−1
χ_{1c}	1	1	−1	−1	1
χ_2	2	−2	0	0	0

Notes for D_8 and Q_8

To determine the conjugacy classes it is convenient simply to consider a list of the elements in the group $\{1, x, x^2, x^3, y, xy, x^2y, x^3y\}$ and explicitly calculate the effect of conjugacy. In both cases, the derived subgroup is $\langle x^2 \rangle$ with a $C_2 \times C_2$ quotient, so that the four 1-dimensional characters look the same in both cases. There remains a fifth character which is determined by orthogonality relations, so the character tables must be the same. In the case of D_8, the 2-dimensional character is also obtained from the natural construction of D_8 as a the group of symmetries of a square. In the case of Q_8, it is obtained from the action on the quaternion algebra, as in Chapter 2, Exercise 12. In both cases, the 2-dimensional character is induced from any linear character of a subgroup of index 2 that does not have x^2 in its kernel.

The Alternating Group A_4

Characteristic 0

A_4
Ordinary characters

g / $\|C_G(g)\|$	() / 12	(12)(34) / 4	(123) / 3	(132) / 3
χ_1	1	1	1	1
χ_{1a}	1	1	ζ_3	ζ_3^2
χ_{1b}	1	1	ζ_3^2	ζ_3
χ_3	3	-1	0	0

Characteristic 2

A_4
Brauer simple $p = 2$

g / $\|C_G(g)\|$	() / 12	(123) / 3	(132) / 3
ϕ_1	1	1	1
ϕ_{1a}	1	ζ_3	ζ_3^2
ϕ_{1b}	1	ζ_3^2	ζ_3

A_4
Brauer projective $p = 2$

g / $\|C_G(g)\|$	() / 12	(123) / 3	(132) / 3
η_1	4	1	1
η_{1a}	4	ζ_3	ζ_3^2
η_{1b}	4	ζ_3^2	ζ_3

A_4
Decomposition matrix $p = 2$

	ϕ_1	ϕ_{1a}	ϕ_{1b}
χ_1	1	0	0
χ_{1a}	0	1	0
χ_{1b}	0	0	1
χ_3	1	1	1

A_4
Cartan matrix $p = 2$

	η_1	η_{1a}	η_{1b}
ϕ_1	2	1	1
ϕ_{1a}	1	2	1
ϕ_{1b}	1	1	2

Characteristic 3

A_4
Brauer simple $p = 3$

g / $\|C_G(g)\|$	() / 12	(12)(34) / 4
ϕ_1	1	1
ϕ_3	3	-1

A_4
Brauer projective $p = 3$

g / $\|C_G(g)\|$	() / 12	(12)(34) / 4
η_1	3	3
η_3	3	-1

A_4
Decomposition matrix $p = 3$

	ϕ_1	ϕ_3
χ_1	1	0
χ_{1a}	1	0
χ_{1b}	1	0
χ_3	0	1

A_4
Cartan matrix $p = 3$

	η_1	η_3
ϕ_1	3	0
ϕ_3	0	1

Notes for A_4

The element (123) conjugates transitively the three nonidentity elements of order two in the unique Sylow 2-subgroup $\langle(12)(34), (13)(24)\rangle$. The remaining group elements have order three, conjugated in two orbits by the Sylow 2-subgroup, which equals the derived subgroup. There are three degree 1 characters. The remaining character can be found from the orthogonality relations; it can be constructed by inducing a nontrivial character from $\langle(12)(34), (13)(24)\rangle$; it is also the character of the realization of A_4 as the group of rotations of a regular tetrahedron.

The Dihedral and Quaternion Groups of Order 16

Put

$$D_{16} = \langle x, y | x^8 = y^2 = 1,\ yxy^{-1} = x^{-1} \rangle$$
$$Q_{16} = \langle x, y | x^8 = 1,\ x^4 = y^2,\ yxy^{-1} = x^{-1} \rangle.$$

Characteristic 0

In both cases, the character table is

D_{16} and Q_{16}
Ordinary characters

g $\lvert C_G(g)\rvert$	1 16	x^4 16	x^2 8	x 8	x^5 8	y 4	xy 4
χ_1	1	1	1	1	1	1	1
χ_{1a}	1	1	1	1	1	-1	-1
χ_{1b}	1	1	1	-1	-1	1	-1
χ_{1c}	1	1	1	-1	-1	-1	1
χ_{2a}	2	2	-2	0	0	0	0
χ_{2b}	2	-2	0	$\sqrt{2}$	$-\sqrt{2}$	0	0
χ_{2c}	2	-2	0	$-\sqrt{2}$	$\sqrt{2}$	0	0

Notes for the Dihedral and Quaternion Groups of Order 16

As with D_8 and Q_8, find the conjugacy classes by listing the elements. The quotient by $\langle x^4 \rangle$ is a copy of either D_8 or Q_8 (depending on the case), and we obtain the top 5 rows of the character table by lifting (or inflating) the characters from the quotient group. The final two characters are obtained by inducing the characters χ_{ζ_8} and $\chi_{\zeta_8^3}$ from the cyclic subgroup $\langle x \rangle$.

The Semidihedral Group of Order 16

The semidihedral group of order 16 has a presentation

$$SD_{16} = \langle x, y | x^8 = y^2 = 1,\ yxy^{-1} = x^3 \rangle.$$

Characteristic 0

SD_{16}
Ordinary characters

g	1	x^4	x^2	x	x^5	y	xy
$\lvert C_G(g)\rvert$	16	16	8	8	8	4	4
χ_1	1	1	1	1	1	1	1
χ_{1a}	1	1	1	1	1	-1	-1
χ_{1b}	1	1	1	-1	-1	1	-1
χ_{1c}	1	1	1	-1	-1	-1	1
χ_{2a}	2	2	-2	0	0	0	0
χ_{2b}	2	-2	0	$i\sqrt{2}$	$-i\sqrt{2}$	0	0
χ_{2c}	2	-2	0	$-i\sqrt{2}$	$i\sqrt{2}$	0	0

Notes for SD_{16}

The comments for D_{16} and Q_{16} also apply here. The quotient by $\langle x^4 \rangle$ is a copy of D_8, and we obtain the top five rows of the character table by inflating (or lifting) the characters from the quotient group. The final two characters are obtained by inducing the characters χ_{ζ_8} and $\chi_{\zeta_8^{-1}}$ from the cyclic subgroup $\langle x \rangle$.

The Non-abelian Group $C_7 \rtimes C_3$ of Order 21

This group has a presentation

$$C_7 \rtimes C_3 = \langle x, y | x^7 = y^3 = 1,\ yxy^{-1} = x^2 \rangle.$$

Characteristic 0

$C_7 \rtimes C_3$
Ordinary characters

g $\lvert C_G(g)\rvert$	1 21	x 7	x^{-1} 7	y 3	y^{-1} 3
χ_1	1	1	1	1	1
χ_{1a}	1	1	1	ζ_3	ζ_3^2
χ_{1b}	1	1	1	ζ_3^2	ζ_3
χ_{3a}	3	$\zeta_7+\zeta_7^2+\zeta_7^4$	$\zeta_7^3+\zeta_7^5+\zeta_7^6$	0	0
χ_{3b}	3	$\zeta_7^3+\zeta_7^5+\zeta_7^6$	$\zeta_7+\zeta_7^2+\zeta_7^4$	0	0

Characteristic 3

$C_7 \rtimes C_3$
Brauer simple $p = 3$

g $\lvert C_G(g)\rvert$	1 21	x 7	x^{-1} 7
ϕ_1	1	1	1
ϕ_{3a}	3	$\zeta_7+\zeta_7^2+\zeta_7^4$	$\zeta_7^3+\zeta_7^5+\zeta_7^6$
ϕ_{3b}	3	$\zeta_7^3+\zeta_7^5+\zeta_7^6$	$\zeta_7+\zeta_7^2+\zeta_7^4$

$C_7 \rtimes C_3$
Brauer projective $p = 3$

g $\lvert C_G(g)\rvert$	1 21	x 7	x^{-1} 7
η_1	3	3	3
η_{3a}	3	$\zeta_7+\zeta_7^2+\zeta_7^4$	$\zeta_7^3+\zeta_7^5+\zeta_7^6$
η_{3b}	3	$\zeta_7^3+\zeta_7^5+\zeta_7^6$	$\zeta_7+\zeta_7^2+\zeta_7^4$

$C_7 \rtimes C_3$
Decomposition matrix $p = 3$

	ϕ_1	ϕ_{3a}	ϕ_{3b}
χ_1	1	0	0
χ_{1a}	1	0	0
χ_{1b}	1	0	0
χ_{3a}	0	1	0
χ_{3b}	0	0	1

$C_7 \rtimes C_3$
Cartan matrix $p = 3$

	η_1	η_{3a}	η_{3b}
ϕ_1	3	0	0
ϕ_{3a}	0	1	0
ϕ_{3b}	0	0	1

Characteristic 7

$C_7 \rtimes C_3$
Brauer simple $p = 7$

g $\lvert C_G(g)\rvert$	1 21	y 3	y^{-1} 3
ϕ_1	1	1	1
ϕ_{1a}	1	ζ_3	ζ_3^2
ϕ_{1b}	1	ζ_3^2	ζ_3

$C_7 \rtimes C_3$
Brauer projective $p = 7$

g $\lvert C_G(g)\rvert$	1 21	y 3	y^{-1} 3
η_1	7	1	1
η_{1a}	7	ζ_3	ζ_3^2
η_{1b}	7	ζ_3^2	ζ_3

$C_7 \rtimes C_3$
Decomposition matrix $p = 7$

	ϕ_1	ϕ_{1a}	ϕ_{1b}
χ_1	1	0	0
χ_{1a}	0	1	0
χ_{1b}	0	0	1
χ_{3a}	1	1	1
χ_{3b}	1	1	1

$C_7 \rtimes C_3$
Cartan matrix $p = 7$

	η_1	η_{1a}	η_{1b}
ϕ_1	3	2	2
ϕ_{1a}	2	3	2
ϕ_{1b}	2	2	3

Notes for $C_7 \rtimes C_3$

The two ordinary characters of degree 3 are induced from the characters χ_{ζ_7} and $\chi_{\zeta_7^3}$ of the cyclic subgroup of order 7. In characteristic 3, they are blocks of defect zero, so remain simple on reduction. In characteristic 7, the indecomposable projectives were constructed in Example 8.3.4, giving the Cartan matrix by a method which did not use characters. In the context of characters, we can simply say that the three 1-dimensional characters are all distinct on the 7-regular classes so give the three simple Brauer characters. This allows us to compute the decomposition matrix and then the Cartan matrix by the theory of Chapters 9 and 10.

The Symmetric Group S_4

Characteristic 0

S_4
Ordinary characters

g $\lvert C_G(g)\rvert$	() 24	(12) 4	(12)(34) 8	(1234) 4	(123) 3
χ_1	1	1	1	1	1
χ_{sign}	1	−1	1	−1	1
χ_2	2	0	2	0	−1
χ_{3a}	3	−1	−1	1	0
χ_{3b}	3	1	−1	−1	0

Characteristic 2

S_4
Brauer simple $p = 2$

g $\lvert C_G(g)\rvert$	() 24	(123) 3
ϕ_1	1	1
ϕ_2	2	−1

S_4
Brauer projective $p = 2$

g $\lvert C_G(g)\rvert$	() 24	(123) 3
η_1	8	2
η_2	8	−1

S_4
Decomposition matrix $p = 2$

	ϕ_1	ϕ_2
χ_1	1	0
χ_{sign}	1	0
χ_2	0	1
χ_{3a}	1	1
χ_{3b}	1	1

S_4
Cartan matrix $p = 2$

	η_1	η_2
ϕ_1	4	2
ϕ_2	2	3

Characteristic 3

S_4
Brauer simple $p = 3$

g $\lvert C_G(g)\rvert$	() 24	(12) 4	(12)(34) 8	(1234) 4
ϕ_1	1	1	1	1
ϕ_{sign}	1	−1	1	−1
ϕ_{3a}	3	−1	−1	1
ϕ_{3b}	3	1	−1	−1

S_4
Brauer projective $p = 3$

g $\lvert C_G(g)\rvert$	() 24	(12) 4	(12)(34) 8	(1234) 4
η_1	3	1	3	1
η_{sign}	3	−1	3	−1
η_{3a}	3	−1	−1	1
η_{3b}	3	1	−1	−1

S_4
Decomposition matrix $p = 3$

	ϕ_1	ϕ_{sign}	ϕ_{3a}	ϕ_{3b}
χ_1	1	0	0	0
χ_{sign}	0	1	0	0
χ_2	1	1	0	0
χ_{3a}	0	0	1	0
χ_{3b}	0	0	0	1

S_4
Cartan matrix $p = 3$

	η_1	η_{sign}	η_{3a}	η_{3b}
ϕ_1	2	1	0	0
ϕ_{sign}	1	2	0	0
ϕ_{3a}	0	0	1	0
ϕ_{3b}	0	0	0	1

Notes for S_4

The ordinary table was constructed in Chapter 3. There are several ways to construct it. We have used an elementary approach, but the reader should be aware that there are extensive combinatorial methods available for the symmetric groups which go beyond the scope of this text. The Brauer tables were constructed in Example 10.1.5.

The Special Linear Group $SL(2,3)$

There is an isomorphism $SL(2,3) \cong Q_8 \rtimes C_3$ which is established in the notes which appear after the character tables. To identify certain elements of this group, we write

$$a = \begin{bmatrix} 0 & 1 \\ -1 & 0 \end{bmatrix} \quad \text{and} \quad y = \begin{bmatrix} 1 & 1 \\ 0 & 1 \end{bmatrix}$$

with entries in $\mathbb{F}_3$. These matrices have orders 4 and 3, respectively.

Characteristic 0

$SL(2,3)$
Ordinary characters

g $\lvert C_G(g)\rvert$	1 24	a^2 24	a 4	y 6	y^2 6	ya^2 6	y^2a^2 6
χ_1	1	1	1	1	1	1	1
χ_{1a}	1	1	1	ζ_3	ζ_3^2	ζ_3	ζ_3^2
χ_{1b}	1	1	1	ζ_3^2	ζ_3	ζ_3^2	ζ_3
χ_{2a}	2	-2	0	-1	-1	1	1
χ_{2b}	2	-2	0	$-\zeta_3$	$-\zeta_3^2$	ζ_3	ζ_3^2
χ_{2c}	2	-2	0	$-\zeta_3^2$	$-\zeta_3$	ζ_3^2	ζ_3
χ_3	3	3	-1	0	0	0	0

Characteristic 2

$SL(2,3)$
Brauer simple $p = 2$

g $\lvert C_G(g)\rvert$	1 24	y 6	y^{-1} 6
ϕ_1	1	1	1
ϕ_{1a}	1	ζ_3	ζ_3^2
ϕ_{1b}	1	ζ_3^2	ζ_3

$SL(2,3)$
Brauer projective $p = 2$

g $\lvert C_G(g)\rvert$	1 24	y 6	y^{-1} 6
η_1	8	2	2
η_{1a}	8	$2\zeta_3$	$2\zeta_3^2$
η_{1b}	8	$2\zeta_3^2$	$2\zeta_3$

$SL(2, 3)$
Decomposition matrix $p = 2$

	ϕ_1	ϕ_{1a}	ϕ_{1b}
χ_1	1	0	0
χ_{1a}	0	1	0
χ_{1b}	0	0	1
χ_{2a}	0	1	1
χ_{2b}	1	0	1
χ_{2c}	1	1	0
χ_3	1	1	1

SL(2, 3)
Cartan matrix $p = 2$

	η_1	η_{1a}	η_{1b}
ϕ_1	4	2	2
ϕ_{1a}	2	4	2
ϕ_{1b}	2	2	4

Characteristic 3

$SL(2, 3)$
Brauer simple $p = 3$

g $\lvert C_G(g)\rvert$	1 24	a^2 24	a 4
ϕ_1	1	1	1
ϕ_2	2	−2	0
ϕ_3	3	3	−1

$SL(2, 3)$
Brauer projective $p = 3$

g $\lvert C_G(g)\rvert$	1 24	a^2 24	a 4
η_1	3	3	3
η_2	6	6	0
η_3	3	3	1

$SL(2, 3)$
Decomposition matrix $p = 3$

	ϕ_1	ϕ_3	ϕ_5
χ_1	1	0	0
χ_{1a}	1	0	0
χ_{1b}	1	0	0
χ_{2a}	0	1	0
χ_{2b}	0	1	0
χ_{2c}	0	1	0
χ_3	0	0	1

SL(2, 3)
Cartan matrix $p = 3$

	η_1	η_2	η_3
ϕ_1	3	0	0
ϕ_2	0	3	0
ϕ_3	0	0	1

Notes for $SL(2, 3)$

We first establish the isomorphism $SL(2,3) \cong Q_8 \rtimes C_3$. The order of $SL(2, 3)$ is 24, by counting ordered bases of $\mathbb{F}_3^2$. We show that there is only one element of order 2 in $SL(2, 3)$, namely, the matrix $-I = \begin{bmatrix} -1 & 0 \\ 0 & -1 \end{bmatrix}$. This is because if $g^2 = 1$ then g has minimal polynomial dividing $X^2 - 1$. If this were the characteristic polynomial then $\det g = -1$, which is not possible, so both eigenvalues of g must be -1. Since g is diagonalizable, g must be the matrix $-I$.

Next, $SL(2,3)$ permutes the four 1-dimensional subspaces of $\mathbb{F}_3^2$ giving a homomorphism to S_4, with kernel $\{\pm I\}$ of order 2. The image is thus a subgroup of S_4 of order 12, so must be A_4. Since A_4 has a normal Sylow 2-subgroup, the preimage of this subgroup in $SL(2,3)$ is a normal subgroup of order 8, which must therefore be a Sylow 2-subgroup of $SL(2,3)$. This subgroup has only one element of order 2, so it is Q_8. A Sylow 3-subgroup is now a complement to this normal subgroup, and the isomorphism $SL(2,3) \cong Q_8 \rtimes C_3$ is established.

We enumerate the conjugacy classes and centralizer orders, and find that all elements of order 4 in Q_8 are conjugate, using conjugacy within Q_8 and the action of the 3-cycle. The elements outside Q_8 commute with no elements of Q_8 apart from the center, so they have centralizers of order 6. This puts them in conjugacy classes of size 4, and we are able to complete the enumeration of conjugacy classes from this information.

Since $SL(2,3)$ has A_4 as a quotient, the character table of A_4 lifts to $SL(2,3)$ giving the characters of degree 1 and 3. Calculating the sum of the squares of the degrees shows that the remaining three characters have degree 2. Since a^2 is central it must act as a scalar on each simple representation, so it acts as ± 2 in the degree 2 representations, and column orthogonal between columns 1 and 2 shows that the entries must be -2 for the characters of degree 2. Column orthogonality now gives the values of these characters as 0 on a. If any remaining entry in the degree 2 characters is nonzero, it gives nonzero values in the same column for the degree two characters, by multiplying by the three degree 1 characters. Thus, any zero entry here gives three zero entries on the degree 2 characters. This is not possible because the product of that column with its complex conjugate must be the centralizer order, which is 6. Hence the remaining entries are all nonzero, and once one of the degree 2 characters is determined the other two are obtained by multiplying by the degree 1 characters. We deduce from column orthogonality that each degree 2 character must have absolute value 1 on y, y^2, ya^2 and y^2a^2. The matrices by which the element y acts has eigenvalues taken from $1, \zeta_3, \zeta_3^2$ and the only possible sums of two of these with absolute value 1 are $-1 = \zeta_3 + \zeta_3^2$, $-\zeta_3 = 1 + \zeta_3^2$ and $-\zeta_3^2 = 1 + \zeta_3$, so these must be the three values of the degree 2 characters on y. The values on y^2 are their conjugates. The three degree 2 characters must arise as one complex conjugate pair and one real character. A similar argument with the possible sums of the eigenvalues of the matrix of ya^2, which must come from $-1, -\zeta_3, -\zeta_3^2$ now determines the remaining character values.

When $p = 2$, it is immediate that the three degree 1 characters reduce to give three distinct simple Brauer characters. When $p = 3$, we have the trivial Brauer character, and there are no more degree 1 characters because the abelianization is C_3. The degree 3 character is a block of defect zero so also gives a simple Brauer character. Finally, the degree 2 character gives a simple Brauer character, because it cannot be written as a sum of degree 1 characters.

The tables for characteristics 2 and 3 illustrate the theory described in Proposition 8.8 and Theorem 8.10 for semidirect products of a p-group and a p'-group. In characteristic 3, they also illustrate the form of the simple representations, as described in Chapter 6, Exercise 25. In characteristic 2 more detailed information about the projectives is obtained in Chapter 8, Exercise 4. However, the calculation of these tables is straightforward and does not require this theory.

The Alternating Group A_5

This simple group is isomorphic to $SL(2, 4)$ and also to $PSL(2, 5)$.

Characteristic 0

A_5
Ordinary characters

g $\lvert C_G(g)\rvert$	() 60	(12)(34) 4	(123) 3	(12345) 5	(13524) 5
χ_1	1	1	1	1	1
χ_{3a}	3	-1	0	$-(\zeta_5^2+\zeta_5^3)$	$-(\zeta_5+\zeta_5^4)$
χ_{3b}	3	-1	0	$-(\zeta_5+\zeta_5^4)$	$-(\zeta_5^2+\zeta_5^3)$
χ_4	4	0	1	-1	-1
χ_5	5	1	-1	0	0

Characteristic 2

$A_5 \cong SL(2, 4)$
Brauer simple $p = 2$

g $\lvert C_G(g)\rvert$	() 60	(123) 3	(12345) 5	(13524) 5
ϕ_1	1	1	1	1
ϕ_{2a}	2	-1	$\zeta_5+\zeta_5^4$	$\zeta_5^2+\zeta_5^3$
ϕ_{2b}	2	-1	$\zeta_5^2+\zeta_5^3$	$\zeta_5+\zeta_5^4$
ϕ_4	4	1	-1	-1

$A_5 \cong SL(2,4)$
Brauer projective $p = 2$

g $\lvert C_G(g)\rvert$	() 60	(123) 3	(12345) 5	(13524) 5
η_1	12	0	2	2
η_{2a}	8	-1	$-(\zeta_5^2+\zeta_5^3)$	$-(\zeta_5+\zeta_5^4)$
η_{2b}	8	-1	$-(\zeta_5+\zeta_5^4)$	$-(\zeta_5^2+\zeta_5^3)$
η_4	4	1	-1	-1

$A_5 \cong SL(2,4)$
Decomposition matrix $p = 2$

	ϕ_1	ϕ_{2a}	ϕ_{2b}	ϕ_4
χ_1	1	0	0	0
χ_{3a}	1	1	0	0
χ_{3b}	1	0	1	0
χ_4	0	0	0	1
χ_5	1	1	1	0

$A_5 \cong SL(2,4)$
Cartan matrix $p = 2$

	η_1	η_{2a}	η_{2b}	η_4
ϕ_1	4	2	2	0
ϕ_{2a}	2	2	1	0
ϕ_{2b}	2	1	2	0
ϕ_4	0	0	0	1

Characteristic 3

A_5
Brauer simple $p = 3$

g $\lvert C_G(g)\rvert$	() 60	(12)(34) 4	(12345) 5	(13524) 5
ϕ_1	1	1	1	1
ϕ_{3a}	3	-1	$-(\zeta_5^2+\zeta_5^3)$	$-(\zeta_5+\zeta_5^4)$
ϕ_{3b}	3	-1	$-(\zeta_5+\zeta_5^4)$	$-(\zeta_5^2+\zeta_5^3)$
ϕ_4	4	0	-1	-1

A_5
Brauer projective $p = 3$

g $\lvert C_G(g)\rvert$	() 60	(12)(34) 4	(12345) 5	(13524) 5
η_1	6	2	1	1
η_{3a}	3	-1	$-(\zeta_5^2+\zeta_5^3)$	$-(\zeta_5+\zeta_5^4)$
η_{3b}	3	-1	$-(\zeta_5+\zeta_5^4)$	$-(\zeta_5^2+\zeta_5^3)$
η_4	9	1	-1	-1

A_5
Decomposition matrix $p = 3$

	ϕ_1	ϕ_{3a}	ϕ_{3b}	ϕ_4
χ_1	1	0	0	0
χ_{3a}	0	1	0	0
χ_{3b}	0	0	1	0
χ_4	0	0	0	1
χ_5	1	0	0	1

A_5
Cartan matrix $p = 3$

	η_1	η_{3a}	η_{3b}	η_4
ϕ_1	2	0	0	1
ϕ_{3a}	0	1	0	0
ϕ_{3b}	0	0	1	0
ϕ_4	1	0	0	2

Characteristic 5

$A_5 \cong PSL(2,5)$
Brauer simple $p = 5$

g	()	(12)(34)	(123)
$\lvert C_G(g)\rvert$	60	4	3
ϕ_1	1	1	1
ϕ_3	3	-1	0
ϕ_5	5	1	-1

$A_5 \cong PSL(2,5)$
Brauer projective $p = 5$

g	()	(12)(34)	(123)
$\lvert C_G(g)\rvert$	60	4	3
η_1	5	1	2
η_3	10	-2	1
η_5	5	1	-1

$A_5 \cong PSL(2,5)$
Decomposition matrix $p = 5$

	ϕ_1	ϕ_3	ϕ_5
χ_1	1	0	0
χ_{3a}	0	1	0
χ_{3b}	0	1	0
χ_4	1	1	0
χ_5	0	0	1

$A_5 \cong PSL(2,5)$
Cartan matrix $p = 5$

	η_1	η_3	η_5
ϕ_1	2	1	0
ϕ_3	1	3	0
ϕ_5	0	0	1

Notes for A_5

To compute the conjugacy classes, compute the centralizer of each element first in S_5, then intersect that centralizer with A_5. The index in A_5 is the number of conjugates of that element. This enables us to see that the class of 5-cycles splits into two in A_5, but the other classes of S_5 do not. One of the 3-dimensional representations can be obtained via the realization of A_5 as the group of rotations of the icosahedron. Note that when computing the trace of a rotation matrix it simplifies the calculation to choose the most convenient basis. The tensor square of this representation contains two copies of itself, determined by taking an inner product. The remaining 3-dimensional summand is simple and new. It can also be obtained as an algebraic conjugate of the first 3-dimensional representation. The permutation representation on five symbols is 2-transitive, so decomposes as the direct sum of the trivial representation and an irreducible 4-dimensional representation, by Lemma 5.5. The remaining 5-dimensional representation can be found by the orthogonality relations. It is also induced from one of the nontrivial 1-dimensional representations of A_4, as can be verified by taking the inner product of this induced representation with the other representations of A_5 so far obtained, using Frobenius reciprocity, to see that the induced representation is new.

The Symmetric Group S_5

Characteristic 0

S_5
Ordinary characters

g $\lvert C_G(g)\rvert$	() 120	(12) 12	(123) 6	(12)(34) 8	(1234) 4	(123)(45) 6	(12345) 5
χ_1	1	1	1	1	1	1	1
χ_{sign}	1	−1	1	1	−1	−1	1
χ_{4a}	4	2	1	0	0	−1	−1
χ_{4b}	4	−2	1	0	0	1	−1
χ_6	6	0	0	−2	0	0	1
χ_{5a}	5	1	−1	1	−1	1	0
χ_{5b}	5	−1	−1	1	1	−1	0

Notes for S_5

The conjugacy classes are determined by cycle type, the derived subgroup is A_5 and there are two 1-dimensional representations. The permutation representation on the five symbols is the direct sum of the trivial representation and a 4-dimensional simple representation χ_{4a} (by Lemma 5.5, but it can be checked using the orthogonality relations). Its tensor with the sign representation gives another 4-dimensional simple. Its exterior square is simple of dimension 6. The symmetric square decomposes with a new summand of dimension 5. The final character is obtained by multiplying by the sign representation.

The General Linear Group $GL(3, 2)$

The group $GL(3, 2) = SL(3, 2)$ is simple and is isomorphic to $PSL(2, 7)$.

Characteristic 0

$GL(3,2)$
Ordinary characters

g	1	2	4	3	$7a$	$7b$
$\|C_G(g)\|$	168	8	4	3	7	7
χ_1	1	1	1	1	1	1
χ_{3a}	3	-1	1	0	α	$\overline{\alpha}$
χ_{3b}	3	-1	1	0	$\overline{\alpha}$	α
χ_6	6	2	0	0	-1	-1
χ_7	7	-1	-1	1	0	0
χ_8	8	0	0	-1	1	1

Here $\alpha = \zeta_7 + \zeta_7^2 + \zeta_7^4$ so that $\overline{\alpha} = \zeta_7^3 + \zeta_7^5 + \zeta_7^6$. In calculating with orthogonality relations it is helpful to know that $\alpha^2 = \overline{\alpha} - 1$ and $\alpha\overline{\alpha} = 2$.

Characteristic 2

$GL(3,2)$
Brauer simple $p = 2$

g	1	3	$7a$	$7b$
$\|C_G(g)\|$	168	3	7	7
ϕ_1	1	1	1	1
ϕ_{3a}	3	0	α	$\overline{\alpha}$
ϕ_{3b}	3	0	$\overline{\alpha}$	α
ϕ_8	8	-1	1	1

$GL(3,2)$
Brauer projective $p = 2$

g	1	3	$7a$	$7b$
$\|C_G(g)\|$	168	3	7	7
η_1	8	2	1	1
η_{3a}	16	1	$\alpha - 1$	$\overline{\alpha} - 1$
η_{3b}	16	1	$\overline{\alpha} - 1$	$\alpha - 1$
η_8	8	-1	1	1

$GL(3,2)$
Decomposition matrix $p = 2$

	ϕ_1	ϕ_{3a}	ϕ_{3b}	ϕ_8
χ_1	1	0	0	0
χ_{3a}	1	0	0	0
χ_{3b}	0	0	1	0
χ_6	0	1	1	0
χ_7	1	1	1	0
χ_8	0	0	0	1

$GL(3,2)$
Cartan matrix $p = 2$

	η_1	η_{3a}	η_{3b}	η_8
ϕ_1	2	1	1	0
ϕ_{3a}	1	3	2	0
ϕ_{3b}	1	2	3	0
ϕ_8	0	0	0	1

Characteristic 7

$GL(3,2) \cong PSL(2,7)$
Brauer simple $p = 7$

g	1	2	4	3
$\lvert C_G(g)\rvert$	168	8	4	3
ϕ_1	1	1	1	1
ϕ_3	3	−1	1	0
ϕ_5	5	1	−1	−1
ϕ_7	7	−1	−1	1

$GL(3,2) \cong PSL(2,7)$
Brauer projective $p = 7$

g	1	2	4	3
$\lvert C_G(g)\rvert$	168	8	4	3
η_1	7	3	1	1
η_3	14	−2	2	−1
η_5	14	2	0	−1
η_7	7	−1	−1	1

$GL(3,2) \cong PSL(2,7)$
Decomposition matrix $p = 7$

	ϕ_1	ϕ_3	ϕ_5	ϕ_7
χ_1	1	0	0	0
χ_{3a}	0	1	0	0
χ_{3b}	0	1	0	0
χ_6	1	0	1	0
χ_7	0	0	0	1
χ_8	0	1	1	0

$GL(3,2) \cong PSL(2,7)$
Cartan matrix $p = 7$

	η_1	η_3	η_5	η_7
ϕ_1	2	0	1	0
ϕ_3	0	3	1	0
ϕ_5	1	1	2	0
ϕ_7	0	0	0	1

Notes for $GL(3,2)$

By counting ordered bases for $\mathbb{F}_2^3$ the order of $GL(3,2)$ is $168 = (2^3-1)(2^3-2)(2^3-2^2)$. It is helpful to identify certain subgroups of $GL(3,2)$. We describe these by indicating the form of the matrices in the subgroups. These matrices must be invertible and, subject to that condition, can have any field element from $\mathbb{F}_2$ (i.e., 0 or 1) where $*$ is positioned. Let

$$B = \begin{bmatrix} 1 & 0 & 0 \\ * & 1 & 0 \\ * & * & 1 \end{bmatrix}, \; P_1 = \begin{bmatrix} * & * & 0 \\ * & * & 0 \\ * & * & 1 \end{bmatrix}, \; U_1 = \begin{bmatrix} 1 & 0 & 0 \\ 0 & 1 & 0 \\ * & * & 1 \end{bmatrix},$$

$$L_1 = \begin{bmatrix} * & * & 0 \\ * & * & 0 \\ 0 & 0 & 1 \end{bmatrix}$$

and

$$P_2 = \begin{bmatrix} 1 & 0 & 0 \\ * & * & * \\ * & * & * \end{bmatrix}, \; U_2 = \begin{bmatrix} 1 & 0 & 0 \\ * & 1 & 0 \\ * & 0 & 1 \end{bmatrix}, \; L_2 = \begin{bmatrix} 1 & 0 & 0 \\ 0 & * & * \\ 0 & * & * \end{bmatrix}.$$

Then

$$
\begin{aligned}
B &\cong D_8, \\
L_1 &\cong L_2 \cong GL(2,2) \cong S_3, \\
U_1 &\cong U_2 \cong C_2 \times C_2, \\
P_1 &= U_1 \rtimes L_1 \cong S_4, \\
P_2 &= U_2 \rtimes L_2 \cong S_4
\end{aligned}
$$

because S_4 also has this semidirect product structure. The subgroup B is a Sylow 2-subgroup of $GL(3,2)$, since it has order 8.

We show that $GL(3,2)$ has a single class of elements of order 2 (involutions) by showing that all involutions in B are conjugate in $GL(3,2)$. Since all Sylow 2-subgroups are conjugate in $GL(3,2)$, this will be sufficient. Conjugacy of involutions within B follows because in P_1 they fall into two conjugacy classes, and in P_2 they fall into two different classes, so that combining this information we see they are all conjugate in $GL(3,2)$. Direct calculation shows that

$$
B = C_{GL(3,2)}\left(\begin{bmatrix} 1 & 0 & 0 \\ 0 & 1 & 0 \\ 1 & 0 & 1 \end{bmatrix} \right)
$$

so that there are $21 = 168/8$ elements of order 2, all conjugate.

The two elements of B of order 4 are conjugate in B, so there is a single class of elements of order 4. The centralizer of such an element also centralizes its square, so is a subgroup of B, and hence has order 4. Thus there are $42 = 168/4$ elements of order 4, all conjugate.

Sylow's Theorem shows that the number of Sylow 3-subgroups must be one of 1, 4, 7 or 28. The first three possibilities would imply that there is an element of order 2 which centralizes a Sylow 3-subgroup, which does not happen since the centralizer of an involution has order 8. Thus there are 28 Sylow 3-subgroups H and $N_{GL(3,2)}(H) \cong S_3$ of order 6, since this group appears as a subgroup of P_1. All elements of order 3 are thus conjugate, and there are $56 = 168/3$ of them.

Sylow's Theorem shows that there are 1 or 8 Sylow 7-subgroups K, and since $GL(3,2)$ does not have a normal 7-cycle, there are 8. Thus $N_{GL(3,2)}(K) \cong K \rtimes H$ of order 21, where H has order 3. The action of H on K is nontrivial because the centralizer of H contains no element of order 7, so if $K = \langle g \rangle$ the action of a generator of H may be taken to send $g \mapsto g^2 \mapsto g^4 \mapsto g$. This means the elements of order 7 fall into two conjugacy classes, represented by g and g^{-1}, each of size 24.

Since $168 = 1 + 21 + 42 + 56 + 24 + 24$, we have accounted for all the elements of $GL(3, 2)$.

We construct the three largest degree characters of $GL(3, 2)$. The induced character from the subgroup $K \rtimes H$ of order 21 is

g $\lvert C_G(g)\rvert$	1 168	2 8	4 4	3 3	$7a$ 7	$7b$ 7
$\chi_1 \uparrow_{K \rtimes H}^{GL(3,2)}$	8	0	0	2	1	1

by the induced character formula. This is because no elements of order 2 or 4 can be conjugated into $K \rtimes H$; an element which conjugates a generator of K into $K \rtimes H$ must normalize K, so lies in $K \rtimes H$ and hence the character is 1 on an element of order 7; the elements of order 3 in $K \rtimes H$ lie in two conjugacy classes under this group, each of size 7. The images of such a class under the action of $GL(3, 2)$ partition the 56 elements of order 3 into eight sets, two of which lie in $K \rtimes H$. From this, we see that the value of this induced character on a 3-element is 2. Now $\chi_1 \uparrow_{K \rtimes H}^{GL(3,2)} - \chi_1$ is verified to be a simple character of degree 7 using the orthogonality relations.

We next compute $\chi_1 \uparrow_{P_1}^{GL(3,2)}$. To do this, it is convenient to observe that $GL(3, 2)$ permutes the seven nonzero elements of $\mathbb{F}_2^3$ transitively, and the stabilizer of $\begin{bmatrix} 0 \\ 0 \\ 1 \end{bmatrix}$ is P_1, so that the induced character is the permutation character on these 7 points. We now take typical elements of orders 2, 4, and 3 such as

$$\begin{bmatrix} 0 & 1 & 0 \\ 1 & 0 & 0 \\ 0 & 0 & 1 \end{bmatrix}, \begin{bmatrix} 1 & 0 & 0 \\ 1 & 1 & 0 \\ 0 & 1 & 1 \end{bmatrix}, \begin{bmatrix} 0 & 1 & 0 \\ 0 & 0 & 1 \\ 1 & 0 & 0 \end{bmatrix}$$

and find the number of fixed points. For elements of order 7, it is clear that they act regularly on the 7 non-zero elements of $\mathbb{F}_2^3$, so there are no fixed points. This shows that the character is

g $\lvert C_G(g)\rvert$	1 168	2 8	4 4	3 3	$7a$ 7	$7b$ 7
$\chi_1 \uparrow_{P_1}^{GL(3,2)}$	7	3	1	1	0	0

and orthogonality relations show that $\chi_1 \uparrow_{P_1}^{GL(3,2)} - \chi_1$ is simple and has degree 6.

To construct the simple character of degree 8, we exploit a slightly more complicated combinatorial structure, namely, the incidence graph of lines and planes in $\mathbb{F}_2^3$. This graph is, in fact, the *building* of $GL(3,2)$. The graph has 14 vertices which are the 7 linear subspaces of $\mathbb{F}_2^3$ of dimension 1 and the 7 linear subspaces of $\mathbb{F}_2^3$ of dimension 2. We place an edge between a 1-dimensional subspace and a 2-dimensional subspace if one is contained in the other, obtaining a bipartite graph. These containment pairs are permuted transitively by $GL(3,2)$ as are the 1-dimensional subspaces (with typical stabilizer P_1) and the 2-dimensional subspaces (with typical stabilizer P_2). The stabilizer of a typical edge is thus $P_1 \cap P_2 = B$, and we deduce that there are $|GL(3,2) : B| = 21$ edges. Regarding this graph as a simplicial complex, the simplicial chain complex over $\mathbb{C}$ has the form

$$0 \to \mathbb{C} \uparrow_B^{GL(3,2)} \to \mathbb{C} \uparrow_{P_1}^{GL(3,2)} \oplus \mathbb{C} \uparrow_{P_2}^{GL(3,2)} \to 0.$$

The simplicial complex is, in fact, connected so that we have an exact sequence

$$0 \to H_1 \to \mathbb{C} \uparrow_B^{GL(3,2)} \to \mathbb{C} \uparrow_{P_1}^{GL(3,2)} \oplus \mathbb{C} \uparrow_{P_2}^{GL(3,2)} \to \mathbb{C} \to 0,$$

where H_1 is a $\mathbb{C}G$-module which is the first homology of the complex. Because $\mathbb{C}G$ is semisimple the sequence is split and we have an alternating sum formula for the character of H_1:

$$\chi_{H_1} = \chi_1 \uparrow_B^{GL(3,2)} - \chi_1 \uparrow_{P_1}^{GL(3,2)} - \chi_1 \uparrow_{P_2}^{GL(3,2)} + \chi_1.$$

We have already computed $\chi_1 \uparrow_{P_1}^{GL(3,2)}$ and $\chi_1 \uparrow_{P_2}^{GL(3,2)}$ is similar with the same answer. The computation of $\chi_1 \uparrow_B^{GL(3,2)}$ is made easier by the fact that B is a 2-group, so that the character is only nonzero on 2-power elements. It is

g	1	2	4	3	$7a$	$7b$
$\|C_G(g)\|$	168	8	4	3	7	7
$\chi_1 \uparrow_B^{GL(3,2)}$	21	5	1	0	0	0

The value on an involution arises because there are five involutions in B, which is the centralizer one of them, and so five coset representatives of B conjugate this involution to an element of B. There are two elements of order 4 in B, and they determine B as the normalizer of the subgroup they generate. Since B is self-normalizing, there is only one coset of B whose representative conjugates an element of order 4 into B. We compute that χ_{H_1} is a simple character using the orthogonality relations, and it has degree 8.

It remains to compute the two 3-dimensional characters of $GL(3,2)$. To do this we can say that the two columns indexed by elements of order 7 must be complex conjugates of each other because the elements are mutually inverse. All the characters constructed so far are real-valued, so the two remaining characters must not be real-valued in order that those two columns should be distinct. This means that the two remaining characters must be complex conjugates of each other. On elements of orders 1, 2, 3, and 4 these characters are real, so must be equal. With this information, we can now determine those characters on these columns using orthogonality relations and the fact that the sum of the squares of the degrees of the characters is 168. Orthogonality gives an equation for the missing complex number α which is also determined in this way.

Dihedral Groups

We let $D_{2n} = \langle x, y \mid x^n = y^2 = 1,\ yxy^{-1} = x^{-1} \rangle$.

Characteristic 0

D_{2n}, n odd
Ordinary characters

g $\lvert C_G(g)\rvert$	1 $2n$	x n	x^2 n	$\cdots$	$x^{\frac{n-1}{2}}$ n	y 2
χ_1	1	1	1	$\cdots$	1	1
χ_{1a}	1	1	1	$\cdots$	1	-1
$\chi_{\zeta_n^s} \uparrow_{\langle x \rangle}^G$ $(1 \le s \le \frac{n-1}{2})$	2	$\zeta_n^s + \overline{\zeta}_n^s$	$\zeta_n^{2s} + \overline{\zeta}_n^{2s}$	$\cdots$	$\zeta_n^{\frac{n-1}{2}s} + \overline{\zeta}_n^{\frac{n-1}{2}s}$	0

D_{2n}, n even
Ordinary characters

g $\lvert C_G(g)\rvert$	1 $2n$	x n	x^2 n	$\cdots$	$x^{\frac{n}{2}}$ $2n$	y 4	xy 4
χ_1	1	1	1	$\cdots$	1	1	1
χ_{1a}	1	1	1	$\cdots$	1	-1	-1
χ_{1b}	1	-1	1	$\cdots$	$(-1)^{\frac{n}{2}}$	1	-1
χ_{1c}	1	-1	1	$\cdots$	$(-1)^{\frac{n}{2}}$	-1	1
$\chi_{\zeta_n^s} \uparrow_{\langle x \rangle}^G$ $(1 \le s \le \frac{n}{2} - 1)$	2	$\zeta_n^s + \overline{\zeta}_n^s$	$\zeta_n^{2s} + \overline{\zeta}_n^{2s}$	$\cdots$	$2(-1)^s$	0	0

Notes for D_{2n}

These tables are constructed in Example 4.3.11 and Chapter 4, Exercise 1. The decomposition and Cartan matrices for D_{30} in characteristic 2 are constructed in Chapter 9, Exercise 7. The representation theory of D_{30} in characteristic 3 is described in Chapter 10, Exercise 3.

Bibliography

[1] J. L. Alperin, Diagrams for modules, *J. Pure Appl. Algebra* 16 (1980), 111–119.

[2] J. L. Alperin, *Local representation theory*, Cambridge Studies in Advanced Mathematics 11, Cambridge University Press, 1986.

[3] D. J. Benson, *Representations and cohomology 1 and 2*, Cambridge Studies in Advanced Mathematics 30 and 31, Cambridge University Press, 1998.

[4] D. J. Benson and J. F. Carlson, Diagrammatic methods for modular representations and cohomology, *Comm. Algebra* 15 (1987), 53–121.

[5] D. J. Benson and J. H. Conway, Diagrams for modular lattices, *J. Pure Appl. Algebra* 37 (1985), 111–116.

[6] N. Blackburn, On a special class of p-groups, *Acta Math.* 100 (1958), 45–92.

[7] S. Brenner, Modular representations of p groups, *J. Algebra* 15 (1970), 89–102.

[8] S. Brenner, *Decomposition properties of some small diagrams of modules*, Symposia Mathematica 13, 127–141, Academic Press, 1974.

[9] W. W. Crawley-Boevey, On tame algebras and bocses, *Proc. London Math. Soc.* 56 (1988), 451–483.

[10] C. W. Curtis and I. Reiner, *Methods of representation theory, vol. I*, John Wiley, New York, 1981.

[11] Yu. A. Drozd, Tame and wild matrix problems: Representations and quadratic forms (in Russian), pp. 39–74, 154, *Akad. Nauk Ukrain. SSR, Inst. Mat., Kiev*, 154 (1979), 3974 translated in *Amer. Math. Soc. Transl.* 128 (1986), 31–55.

[12] K. Erdmann, *Blocks of tame representation type and related algebras*, Lecture Notes in Mathematics 1428, Springer Verlin, 1990.

[13] W. Feit, *The representation theory of finite groups*, North-Holland Mathematical Library 25, North-Holland, Amsterdam, 1982.

[14] L. K. Hua, *Introduction to number theory*, Springer, Berlin, 1982.

[15] M. Prest, Wild representation type and undecidability, *Comm. Algebra* 19 (1991), 919–929.

[16] D. S. Rim, Modules over finite groups, *Ann. of Math.* 69 (1959), 700–712.

[17] C. M. Ringel, The representation type of local algebras, in *Representations of algebras, Ottawa 1974*, Lecture Notes in Mathematics 488, Springer, Berlin, 1975.

[18] C. M. Ringel, The indecomposable representations of the dihedral 2-groups, *Math. Ann.* 214 (1975), 19–34.

[19] R. G. Swan, Projective modules over group rings and maximal orders, *Ann. Math.* 76 (1962), 55–61.

[20] J. G. Thackray, *Modular representations of some finite groups*, Ph.D. thesis, University of Cambridge, 1981.

[21] J. Thévenaz, *G-Algebras and modular representation theory*, Oxford University Press, Oxford, 1995.

Bibliography

Index

Printed in the United States
by Baker & Taylor Publisher Services